New Trends in Fuzzy Set Theory and Related Items

New Trends in Fuzzy Set Theory and Related Items

Special Issue Editors

Esteban Induráin
Humberto Bustince
Javier Fernandez

MDPI • Basel • Beijing • Wuhan • Barcelona • Belgrade

Special Issue Editors
Esteban Induráin
Universidad Pública de Navarra
Spain

Humberto Bustince
Universidad Pública de Navarra
Spain

Javier Fernandez
Universidad Pública de Navarra
Spain

Editorial Office
MDPI
St. Alban-Anlage 66
Basel, Switzerland

This is a reprint of articles from the Special Issue published online in the open access journal *Axioms* (ISSN 2075-1680) from 2017 to 2018 (available at: http://www.mdpi.com/journal/axioms/special_issues/fuzzy_set_theory)

For citation purposes, cite each article independently as indicated on the article page online and as indicated below:

LastName, A.A.; LastName, B.B.; LastName, C.C. Article Title. *Journal Name* **Year**, *Article Number*, Page Range.

ISBN 978-3-03897-123-8 (Pbk)
ISBN 978-3-03897-124-5 (PDF)

Contents

About the Special Issue Editors

Esteban Induráin, Dr., was born in Pamplona, Spain in 1960. He got his degree in Mathematical Sciences from the University of Zaragoza, Spain in 1982 and his doctorate (Ph.D.) in Mathematical Sciences, also from the University of Zaragoza, in 1985. He is currently full professor of Mathematical Analysis at the Public University of Navarre, Pamplona (Spain), since 2000.

Dr. Induráin has published ca. 85 papers in well-ranked journals. His impact factor is h = 14 in "Web of Science" and "Scopus". He has been the main researcher in several projects. Dr. Induráin has co-advised seven Ph.D. theses to-date. One of them was awarded a prize for the best Ph.D. thesis in its ambit and period.

Humberto Bustince, Dr., received his Bs.C. degree on Physics from the Salamanca University, Spain, in 1983 and his Ph.D. degree in Mathematics from the Public University of Navarra, Pamplona, Spain, in 1994. He has been a teacher at the Public University of Navarra since 1991, and he is currently a Full Professor with the Department of Automatics and Computation. He served as subdirector of the Technical School for Industrial Engineering and Telecommunications from 01/01/2003 to 30/10/2008, and he was involved in the implantation of Computer Science courses at the Public University of Navarra. He is currently involved in teaching artificial intelligence for students of computer sciences.

Dr. Bustince has authored more than 120 journal papers (Web of Knowledge) and more than 100 contributions to international conferences. He has also been the co-author of four books on fuzzy theory and extensions of fuzzy sets. His main research interests are interval-valued fuzzy sets (interval type-2 fuzzy sets), Atanassov's intuitionistic fuzzy sets, aggregation functions, implication operators, inclusion measures, image processing, decision making, and approximate reasoning.

Javier Fernandez, Dr., graduated in Mathematics from the University of Zaragoza (1999) and got his doctorate (Ph.D.) in Mathematics from the University of the Basque Country in 2003. He has been a postdoctoral researcher at the CNRS (France) in 2003–2004 and 2005–2007, in the framework of the European Hyperbolic and Kynetic Equations (HYKE) research project. He has been a lecturer in the Public University of Navarre since 2008. He has co-advised two Ph.D. thesis.

Dr. Fernandez is a member of the research Group on Artificial Intelligence and Approximate Reasoning (GIARA) of the Public University of Navarre, whose leader is Dr. Humberto Bustince. His main research interests are fuzzy theory, extensions of fuzzy theory, aggregation functions, computing, differential equations, and unique continuation, as well as image processing.

He has authored more than 30 papers in JCR journals, most of them in Q1. Some of the most relevant ones are on linear orders for intervals.

Preface to "New Trends in Fuzzy Set Theory and Related Items"

Fuzzy set theory is one of the main trend topics in Mathematics and Computer Sciences nowadays. Its wide, multidisciplinary applications, as well as the recent developments in its theoretical bases, provide an excellent set of frameworks for researchers interested in different scientific disciplines.

Owing to these reasons, we have addressed the task of issuing this Special Issue, having in mind the possibility of collecting new contributions devoted to this topic, from many different points of view. In this sense, this book includes works authored by authors working from very different perspectives, but all of them are in the first line of research of their areas.

The articles (twelve thereof) that have been included in this Special Issue cover a large set of the newest developments in this framework of research. Thus, always bearing in mind the notion of a fuzzy set, we can find ideas and new results on orderings, logic, aggregation of functions, functional equations, operators, metric, etc. We have also included some works devoted to applications in, e.g., social choice.

Of course, the field of fuzzy sets theory is so broad nowadays that it is not possible to cover all of them, but we consider that the book provides a rather complete view of the state-of-the-art in the field.

Esteban Induráin, Humberto Bustince, Javier Fernandez
Special Issue Editors

axioms

MDPI

Editorial

Introduction to Special Issue: New Trends in Fuzzy Set Theory and Related Items

Humberto Bustince [1], Javier Fernandez [1] and Esteban Induráin [2,*]

1 Institute of Smart Cities (ISC), Departamento de Estadística, Matemáticas e Informática, Universidad Pública de Navarra, 31006 Pamplona, Spain; bustince@unavarra.es (H.B.); fcojavier.fernandez@unavarra.es (J.F.)
2 Institute for Advanced Materials (InaMat), Departamento de Estadística, Matemáticas e Informática, Universidad Pública de Navarra, 31006 Pamplona, Spain
* Correspondence: steiner@unavarra.es; Tel.: +34-948-169-551

Received: 10 May 2018; Accepted: 1 June 2018; Published: 5 June 2018

Abstract: We focus on the articles recently published in the special issue of Axioms devoted to "New Trends in Fuzzy Set Theory and Related Items".

Keywords: fuzzy set theory and its applications

Since the launch, in 1965, of the key notion of a fuzzy subset by Lotfi A. Zadeh (see [1]), this concept has succeeded in growing interest and development. Nowadays, it is undoubtedly one of the most powerful and appealing branches of mathematics. Its range of applications is wide and multidisciplinary, starting from computer sciences and artificial intelligence, but also touching on a vast set of scientific disciplines (physics, engineering, medicine, economics, social choice, etc.).

Bearing in mind this fact, we accepted the task of organizing an special issue forf the journal, Axioms, to collect and introduce new ideas and trends based on the notion of a fuzzy set. To do so, we suggested possible authors, working in different settings and covering an exhaustive set of possible applications.

Incidentally, as this special issue was starting to become a reality, we received a really sad piece of news: Prof. Lotfi A. Zadeh passed away in September 2017, aged 96. Therefore, this special issue was also, in a way, a homage and tribute to the person who had the clever idea of launching the important concept of a fuzzy set. All of us are scientific heirs of him.

Concerning this special issue, twelve papers have been accepted and published. Five of them come from Spain. Two of them come from Italy. One manuscript has coauthors both from Spain and Venezuela. Other papers come from South Korea, U.S.A., Poland and Pakistan (one from each of these countries).

The topics covered, of course all of them related to fuzzy sets theory and its applications, were quite different from one another. Namely, we received contributions regarding the following items:

1. Scoring and decision-making using linguistic expressions, and suitable orderings (paper by M.J. Campión et al.).
2. Theoretical and practical studies on orderings defined with type-2 fuzzy numbers and their applications in decision-making in models of inference processes where uncertainty, imprecision or vagueness is present (paper by P. Hernández et al.).
3. Aggregation functions and several kinds of operators acting on fuzzy sets, studied from an algebraic point of view (paper by L. Legarreta et al.).
4. Numerical representability of fuzzy binary relations and preferences (paper by P. Bevilacqua et al.).
5. The relationship between indistinguishability operators, introduced with the aim of fuzzifying the crisp notion of equivalence relations, and fuzzy metrics. (paper by J.J. Miñana and O. Valero).

6. Analysis of environmental evaluation practices in a context of multi-criteria decision-making, through managing interacting criteria based on Choquet integrals (paper by T. González-Arteaga et al.).
7. Studies on interval-valued intuitionistic fuzzy sets with applications to different kinds of algebra (paper by Y.B. Jun et al.).
8. Category theory and foundations of Mathematics, where, instead of the basic category of sets, a similar category of fuzzy sets is considered (paper by C. Servin et al.).
9. Studies on ordered fuzzy numbers with a revision of Kosiński's theory (paper by K. Piasecki).
10. Interplay between aggregation functions and aggregation operators, providing a link to social choice theory based on the Arrow's impossibility theorem, using functional equations and presenting some interpretation of the fuzzy framework (paper by J.C. Candeal).
11. Fuzzy graph theory: studying graphs in an intuitionistic fuzzy soft environment to be used then in developing algorithms to deal with some decision-making problems (paper by S. Shahzadi and M. Akram).
12. An study on quantiles in abstract convex structures through a new quantitative framework that generalizes previous ones for the analysis and handling of aggregation operators (paper by M. Cardin).

We want to thank all the authors of these works, which provide a wide view of some of the most recent topics in the field of fuzzy set theory. Also, we acknowledge with thanks the work done by the reviewers who collaborated to make this special issue possible. Our gratitude goes also to the editors of Axioms for the support and help with the preparation of this special issue.

Acknowledgments: This work was partially supported by the research projects MTM2015-63608-P and TIN2016-77356-P (MINECO/ AEI-FEDER, UE).

Conflicts of Interest: The authors declare no conflict of interest.

Reference

1. Zadeh, L.A. Fuzzy Sets. *Inf. Control* **1965**, *8*, 338–353. [CrossRef]

axioms

MDPI

Article

Assigning Numerical Scores to Linguistic Expressions

María Jesús Campión [1], Edurne Falcó [2], José Luis García-Lapresta [3,*] and Esteban Induráin [4]

[1] Inarbe (Institute for Advanced Research in Business and Economics), Departamento de Matemáticas, Universidad Pública de Navarra, 31006 Pamplona, Spain; mjesus.campion@unavarra.es
[2] VIRENA Navarra SL, Beriáin, 31191 Navarra, Spain; edurnefalco@hotmail.com
[3] PRESAD Research Group, BORDA Research Unit, IMUVA, Departamento de Economía Aplicada, Universidad de Valladolid, 47011 Valladolid, Spain
[4] InaMat (Institute for Advanced Materials), Departamento de Matemáticas, Universidad Pública de Navarra, 31006 Pamplona, Spain; steiner@unavarra.es
* Correspondence: lapresta@eco.uva.es; Tel.: +34-983-184-391; Fax: +34-983-423-299

Academic Editor: Radko Mesiar
Received: 7 June 2017; Accepted: 30 June 2017; Published: 6 July 2017

Abstract: In this paper, we study different methods of scoring linguistic expressions defined on a finite set, in the search for a linear order that ranks all those possible expressions. Among them, particular attention is paid to the canonical extension, and its representability through distances in a graph plus some suitable penalization of imprecision. The relationship between this setting and the classical problems of numerical representability of orderings, as well as extension of orderings from a set to a superset is also explored. Finally, aggregation procedures of qualitative rankings and scorings are also analyzed.

Keywords: decision making; imprecision; qualitative scales; linguistic expressions; scores; orderings; aggregation

MSC: 62C86 (Primary); 03E72; 06A06; 91B06; 91B14 (Secondary)

---

1. Introduction

Let U stand for a nonempty set, called *universe*. It is clear that comparing the elements of U through some sort of binary relation $\mathcal{R}$ such that $u \,\mathcal{R}\, v$ is immediately understood as "the element u is not worse than the element v" gives less information than assigning scores to each element of U. A scoring function is a real-valued function $S : U \to \mathbb{R}$. The interpretation is obvious: if the score $S(u)$ of the element $u \in U$ is not smaller than the score $S(v)$ of the element v, then the above conclusion, namely, "the element u is not worse than the element v" is also achieved. However, if an ordering given on U through a binary relation $\mathcal{R}$ is also represented by a scoring function, say $S : U \to \mathbb{R}$, the composition of $f \circ S$ of the map S with no matter what strictly increasing function $f : \mathbb{R} \to \mathbb{R}$ also represents the same ordering $\mathcal{R}$ on U. In other words: $u \,\mathcal{R}\, v \;\Leftrightarrow\; S(u) \geq S(v) \;\Leftrightarrow\; f(S(u)) \geq f(S(v))$ holds true for all $u, v \in U$.

Despite the ordering being the same, the scores have important consequences and collateral meanings. A typical example corresponds to a firm that wants to hire people, and prepares a test or exam consisting of one hundred tasks to be done. Suppose the scores of two candidates u and v have been just $S(u) = 1$ and $S(v) = 2$ (out of 100). Obviously, we would say that v is better than u, but both of them having such poor results, the firm would reject hiring anyone among them. If, instead, the scores were $S(u) = 99$ and $S(v) = 100$, again we would say that v is better than u, but now, quite probably, the firm would be willing to hire both candidates.

The most typical scoring procedures consist of numerical scales. Nevertheless, due to the vagueness and imprecision involved in a wide variety of practical situations in which some sort of evaluation is made, sometimes the use of finite measurement scales based on linguistic terms or expressions seems to be more suitable than just using numerical scales. In this case, if L is a set of linguistic terms (also known as labels), such as $\{$very bad, bad, poor, fair, good, very good, excellent$\}$, a scoring function on U taking values on L is a function $S : U \to L$ that can obviously be interpreted as a qualitative scale of measurement.

From a pure mathematical point of view, whenever L is given a linear order $\succcurlyeq_L$, a scoring procedure on U taking values on L can equivalently be understood through a numerical scoring function $f : U \to \mathbb{R}$, which is obviously a quantitative scale of measurement. Indeed, if $L = \{l_1, l_2, \ldots, l_g\}$ denotes the set of linguistic terms and we understand that $l_h \succcurlyeq_L l_k$ holds true if and only if $h \geq k$, then a scoring function $S : U \to L$ can be immediately interpreted by means of the numerical scoring function $T \circ S : U \to \mathbb{R}$, where $T : L \to \mathbb{R}$ is given by $T(l_h) = h$, $h \in \{1, \ldots, g\}$. In this situation, namely when L is given a linear order, the use of a scoring function taking values on L or an equivalent numerical scoring function is, perhaps, a matter of taste (see García-Lapresta et al. [1]).

However, sometimes, in a higher degree of vagueness or imprecision, and even in the case of using a set $L = \{l_1, l_2, \ldots, l_g\}$ of linguistic terms, it could happen that the agents that assign scores to the elements of a nonempty set U may hesitate when, given an element $u \in U$, the score $S(u)$ needs to be determined. A typical example is an approximation (based on data as climate, composition of soil, amount of rain water per year, etc.) of the quality of the wine produced in a vineyard. An expert could hesitate here, and instead of saying "no doubt, it will be very good", she/he could declare something as "from good to excellent", so using a linguistic expression instead of just a linguistic term.

Our approach concerning the imprecision is based on an adaptation of the *absolute order of magnitude spaces* introduced by Travé-Massuyès and Dague [2] and Travé-Massuyès and Piera [3]; more specifically in the extensions devised by Roselló et al. [4–6] (see also Agell et al. [7] and Falcó et al. [8,9]).

At this stage, several mathematical problems appear, namely

(i) If a set $L = \{l_1, l_2, \ldots, l_g\}$ of linguistic terms has been endowed with a linear order $\succcurlyeq_L$, try to define a linear order on the set of linguistic expressions based on L. Needless to say, this "extended" order should preserve or be compatible with $\succcurlyeq_L$ when acting on the labels of L considered as special cases of linguistic expressions.
(ii) Compare different linear orders (when available) defined on a set of linguistic expressions, trying to explain which is the most suitable one to be used in some practical concrete situation.
(iii) Study problems of aggregation of a finite number of scorings based on linguistic terms and expressions.

Having in mind these questions, the structure of the paper goes as follows. After the Introduction (Section 1) and a compulsory Section 2 of preliminaries, previous results and necessary background, in Section 3, we study sets of linguistic expressions and define suitable orderings on these kinds of sets. In Section 4, we analyze some questions relative to the aggregation of rankings and scorings based on qualitative scales. A final Section 5 of concluding remarks, suggestions for further research and open questions closes the paper.

2. Previous Concepts and Results

Intuitively, a *ranking* on a universe U is some kind of ordering defined on U, usually defined through a binary relation $\mathcal{R}$ on U, such that $u \mathcal{R} v$ means "the element u is at least as good as the element v". A *scoring function* on U is a map $S : U \to \mathbb{R}$. Notice that a scoring function S immediately defines a ranking $\mathcal{R}$ on U, by declaring $u \mathcal{R} v \Leftrightarrow S(u) \geq S(v)$, for all $u, v \in U$. For finite sets, a sort of converse is also true, namely, each ranking can be represented by a scoring function (see e.g., the first three chapters in Bridges and Mehta [10] for further details).

Let us formalize all this through some definitions.

Definition 1. *Let U denote a universe. A preorder* $\succcurlyeq$ *on U is a binary relation on U which is reflexive and transitive.*

An antisymmetric preorder is said to be a partial order. A total preorder $\succcurlyeq$ *on U is a preorder such that* $u \succcurlyeq v$ *or* $v \succcurlyeq u$ *holds true for all* $u, v \in U$. *A total order is also called a linear order.*

If $\succcurlyeq$ is a preorder on U, then the associated *asymmetric* relation is denoted by $\succ$, whereas $\sim$ stands for the associated *equivalence* relation. These relations are respectively defined by $u \succ v \Leftrightarrow ((u \succcurlyeq v)$ and not $(v \succcurlyeq u))$, and $u \sim v \Leftrightarrow ((u \succcurlyeq v)$ and $(v \succcurlyeq u))$, for all $u, v \in U$. Notice that if $\succcurlyeq$ is a linear order, then $u \succ v \Leftrightarrow ((u \succcurlyeq v)$ and $(u \neq v))$.

Definition 2. *Let U denote a universe. Let* $\succcurlyeq$ *stand for a total preorder defined on U. We say that* $\succcurlyeq$ *is representable if there exists a real-valued function* $S : U \to \mathbb{R}$ *such that* $u \succcurlyeq v \Leftrightarrow S(u) \geq S(v)$ *holds true for all* $u, v \in U$.

As commented before, it happens that, provided that U is finite, any total preorder defined on U is representable (see Bridges and Mehta [10]). A function S that represents a total preorder $\succcurlyeq$ defined on U is usually said to be a *utility function* for $\succcurlyeq$.

Remark 1. *Notice that a ranking (e.g., a total preorder) on a universe U is a qualitative scale of measurement, whereas a scoring function is a quantitative scale. Other typical sorts of qualitative scales are those based on labels and linguistic expressions (see next Definition 3). If S is a utility function that represents a total preorder* $\succcurlyeq$ *on a universe U, and* $T : \mathbb{R} \to \mathbb{R}$ *is a strictly increasing function, it is straightforward to see that the composition* $T \circ S : U \to \mathbb{R}$ *is also a utility function that represents* $\succcurlyeq$. *Consequently, there are infinitely many utility functions that represent a given total preorder* $\succcurlyeq$ *on a finite universe U.*

Definition 3. *Let* $L = \{l_1, l_2, \ldots, l_g\}$ *denote a finite nonempty set of linguistic terms (labels), with* $g \geq 3$. *The set of linguistic expressions associated with L is defined as*

$$\mathbb{L} = \{[l_h, l_k] \mid l_h, l_k \in L,\ 1 \leq h \leq k \leq g\},$$

where $[l_h, l_k] = \{l_h, l_{h+1}, \ldots, l_k\}$. *Since the linguistic expression* $[l_h, l_h]$ *is the singleton* $\{l_h\}$, *by convention, it can be replaced by the label* l_h. *In this way,* $L \subset \mathbb{L}$.

Remark 2. *The set of linguistic expressions can be represented by a graph* $G_{\mathbb{L}}$. *In the graph, the lowest layer represents the linguistic terms* $l_h \in L \subset \mathbb{L}$, *the second layer represents the linguistic expressions created by two consecutive labels* $[l_h, l_{h+1}]$, *the third layer represents the linguistic expressions generated by three consecutive labels* $[l_h, l_{h+2}]$, *and so on up to last layer, where we represent the linguistic expression* $[l_1, l_g]$. *As a result, the higher an element of the graph* $G_{\mathbb{L}}$ *is, the more imprecise it becomes.*

The vertices in $G_{\mathbb{L}}$ are the elements of $\mathbb{L}$ and the edges $\mathcal{E} - \mathcal{F}$, where $\mathcal{E} = [l_h, l_k]$ and $\mathcal{F} = [l_h, l_{k+1}]$, or $\mathcal{E} = [l_h, l_k]$ and $\mathcal{F} = [l_{h+1}, l_k]$. Figure 1 shows the graph representation for $g = 5$.

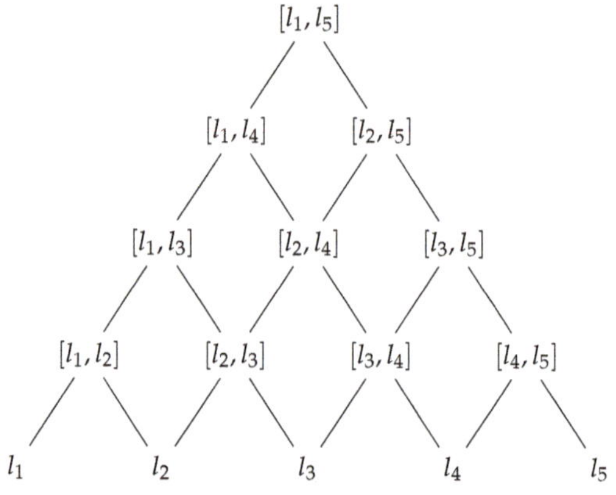

Figure 1. Graph representation of the linguistic expressions for $g = 5$.

3. Defining Orderings on Sets of Linguistic Expressions

Suppose that the set $L = \{l_1, l_2, \ldots, l_g\}$ of linguistic terms is endowed with the linear order $\succcurlyeq$ given by $l_h \succcurlyeq l_k \Leftrightarrow h \geq k$, for all $h, k \in \{1, \ldots, g\}$.

We want now to extend the linear order $\succcurlyeq$ to a new linear order defined on the whole set $\mathbb{L}$ of linguistic expressions on L. This problem has several positive solutions.

Remark 3. *To extend the linear order $\succcurlyeq$ on L to a new linear order defined on $\mathbb{L}$, first, we may consider a partial order $\succcurlyeq_1$ defined on $\mathbb{L}$ as follows: $[l_h, l_k] \succcurlyeq_1 [l_h, l_k]$ (reflexivity), and also $\{l_h\} = [l_h, l_h] \succcurlyeq_1 [l_k, l_k] = \{l_k\} \Leftrightarrow h \geq k$, for all $h, k \in \{1, \ldots, g\}$ (compatibility with $\succcurlyeq$ on L).*

Then, we may apply the well-known Szpilrajn's theorem ([11]) that ensures that any partial order on a given set can be extended to a linear order on that set.

3.1. Constructive Extensions: The Lexicographic Order and the Canonical Order

Instead of using a powerful result as Szpilrajn's theorem as commented in Remark 3, when $L = \{l_1, l_2, \ldots, l_g\}$, we can directly furnish some extensions in a constructive way, as stated in Proposition 1 and Proposition 2, whose straightforward proofs are omitted for the sake of brevity.

Proposition 1. *The binary relation $\succcurlyeq_\ell$ on $\mathbb{L}$ defined as*

$$[l_h, l_k] \succcurlyeq_\ell [l_{h'}, l_{k'}] \Leftrightarrow \begin{cases} h > h' \\ \text{or} \\ h = h' \text{ and } k \geq k' \end{cases}$$

is a linear order that preserves the given order $\succcurlyeq$ on $L \subset \mathbb{L}$. It is called the lexicographic order on $\mathbb{L}$.

Notice that the real valued function $S : \mathbb{L} \to \mathbb{R}$ given by $S([l_h, l_k]) = (g \cdot h) + k$ is a utility representation for $\succcurlyeq_\ell$.

Proposition 2. *([8]) The binary relation $\succcurlyeq_c$ on $\mathbb{L}$ defined as*

$$[l_h, l_k] \succcurlyeq_c [l_{h'}, l_{k'}] \Leftrightarrow \begin{cases} h + k > h' + k' \\ \text{or} \\ h + k = h' + k' \text{ and } k - h \leq k' - h' \end{cases}$$

is a linear order, and it is called the canonical order on $\mathbb{L}$*. It also preserves the given order* $\succcurlyeq$ *on* $L \subset \mathbb{L}$*.*

Example 1. *For* $g = 5$*, the elements of* $\mathbb{L}$ *are ordered as follows with respect to the lexicographic order:*

$$l_5 \succ_\ell [l_4, l_5] \succ_\ell l_4 \succ_\ell [l_3, l_5] \succ_\ell [l_3, l_4] \succ_\ell l_3 \succ_\ell [l_2, l_5] \succ_\ell [l_2, l_4] \succ_\ell$$
$$\succ_\ell [l_2, l_3] \succ_\ell l_2 \succ_\ell [l_1, l_5] \succ_\ell [l_1, l_4] \succ_\ell [l_1, l_3] \succ_\ell [l_1, l_2] \succ_\ell l_1.$$

Moreover, through the canonical order (see also Figure 2), we get:

$$l_5 \succ_c [l_4, l_5] \succ_c l_4 \succ_c [l_3, l_5] \succ_c [l_3, l_4] \succ_c [l_2, l_5] \succ_c l_3 \succ_c [l_2, l_4] \succ_c$$
$$\succ_c [l_1, l_5] \succ_c [l_2, l_3] \succ_c [l_1, l_4] \succ_c l_2 \succ_c [l_1, l_3] \succ_c [l_1, l_2] \succ_c l_1.$$

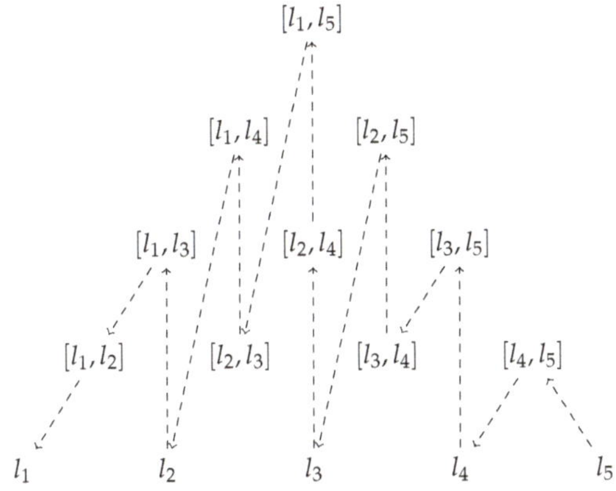

Figure 2. Canonical order in $\mathbb{L}$ for $g = 5$.

3.2. Numerical Representation of the Canonical Order Based on the Geodesic Metric

The canonical order on $\mathbb{L}$ can be represented by utility functions $u : \mathbb{L} \longrightarrow \mathbb{R}$ in such a way that $\mathcal{E} \succ_c \mathcal{F} \Leftrightarrow u(\mathcal{E}) > u(\mathcal{F})$, for all $\mathcal{E}, \mathcal{F} \in \mathbb{L}$.

Among the infinite possible utility representations of the canonical order on $\mathbb{L}$, we are interested in those based on distances. In addition, concerning metrics on a graph, perhaps the most typical one is the so-called *geodesic metric*.

Definition 4. *The geodesic metric* $d_G : \mathbb{L}^2 \to \mathbb{R}$ *is defined as*

$$d_G\big([l_h, l_k], [l_{h'}, l_{k'}]\big) = |h - h'| + |k - k'| \tag{1}$$

for all $\mathcal{E} = [l_h, l_k]$, $\mathcal{F} = [l'_h, l'_k] \in \mathbb{L}$.

Remark 4. *Notice that* $d_G(l_h, l_{h+1}) = 2$ *for every* $h \in \{1, \ldots, g\}$*, so that* $\mathbb{L}$ *becomes uniform, in the sense that adjacent linguistic terms are equidistant (see e.g., Herrera et al. [12] and García-Lapresta and Pérez-Román [13,14] for dealing with non uniform qualitative scales).*

Given $\mathcal{E} = [l_h, l_k] \in \mathbb{L}$, with $\#\mathcal{E}$, we denote the cardinality of $\mathcal{E}$, i.e., the number of labels in the interval $[l_h, l_k]$: $\#\mathcal{E} = k + 1 - h$.

In the following result, we establish that the canonical order on $\mathbb{L}$ introduced in Proposition 2 can be defined through the geodesic distances to the highest possible assessment and the cardinality of the linguistic expressions.

Proposition 3. *For all $\mathcal{E},\mathcal{F} \in \mathbb{L}$, it holds*

$$\mathcal{E} \succ_c \mathcal{F} \Leftrightarrow \begin{cases} d_G(\mathcal{E},l_g) < d_G(\mathcal{F},l_g), \\ \text{or} \\ d_G(\mathcal{E},l_g) = d_G(\mathcal{F},l_g) \text{ and } \#\mathcal{E} < \#\mathcal{F}. \end{cases}$$

Proof. It follows from Proposition 2 and Equation (1). □

Remark 5. *To distinguish between elements $\mathcal{E},\mathcal{F} \in \mathbb{L}$, the second condition that appears in the statement of Proposition 3, namely the one that reads "or $d_G(\mathcal{E},l_g) = d_G(\mathcal{F},l_g)$ and $\#\mathcal{E} < \#\mathcal{F}$" is crucial. In fact, suppose that on $\mathbb{L}$ we use only a scoring function S based on the geodesic metric, and given by $S(\mathcal{E}) = d_G(\mathcal{E},l_g)$. Then, it is straightforward to see that two nodes $\mathcal{E}$, $\mathcal{F}$ that appear in the same vertical in the graph $G_{\mathbb{L}}$ are assigned the same score, which is $S(\mathcal{E}) = d_G(\mathcal{E},l_g) = d_G(\mathcal{F},l_g) = S(\mathcal{F})$. This scoring function S defines an ordering $\succcurlyeq^*$ on $\mathbb{L}$ by declaring $\mathcal{E} \succcurlyeq^* \mathcal{F} \Leftrightarrow S(\mathcal{E}) \geq S(\mathcal{F})$, for all $\mathcal{E},\mathcal{F} \in \mathbb{L}$. However, since $g \geq 3$, it happens that $\succcurlyeq^*$ is a total preorder, but not a linear order. In other words, the geodesic distance is, so-to-say,* miopic *since it cannot detect the difference between two elements in the same vertical in $G_{\mathbb{L}}$. For instance, if $L = \{$very bad, bad, poor, fair, good, very good, excellent$\}$, the linguistic expressions "very good" and "from good to excellent" would be assigned the same score through the geodesic metric.*

3.3. GeodesIc Metric Matching Penalization of Imprecision

Another important feature that carries the second condition in the statement of Proposition 3 and says "or $d_G(\mathcal{E},l_g) = d_G(\mathcal{F},l_g)$ and $\#\mathcal{E} < \#\mathcal{F}$" is warning us about the role played by imprecision. If a linguistic expression $\mathcal{F}$ consists of more labels than another one $\mathcal{E}$, we understand that "$\mathcal{F}$ is more imprecise than $\mathcal{E}$" or, equivalently, "$\mathcal{E}$ is more accurate than $\mathcal{F}$". Taking this fact into account, it seems interesting to search for measurements or scoring methods that, someway, penalize the imprecision when comparing terms in $\mathbb{L}$, as next Proposition 4 does.

Proposition 4. *For every $\alpha \geq 0$, the function $d_\alpha : \mathbb{L}^2 \longrightarrow \mathbb{R}$, defined as*

$$d_\alpha(\mathcal{E},\mathcal{F}) = d_G(\mathcal{E},\mathcal{F}) + \alpha\,|\#\mathcal{E} - \#\mathcal{F}|,$$

i.e.,

$$d_\alpha\left([l_h,l_k],[l_{h'},l_{k'}]\right) = |h-h'| + |k-k'| + \alpha\,|h'-h+k-k'| \tag{2}$$

for all $\mathcal{E} = [l_h,l_k], \mathcal{F} = [l'_{h'},l'_k] \in \mathbb{L}$, is a metric, and it is called the metric associated with α.

Proof. Since d_α is a positive linear combination of the geodesic metric d_G and the pseudometric $d : \mathbb{L}^2 \longrightarrow \mathbb{R}$ defined as $d(\mathcal{E},\mathcal{F}) = |\#\mathcal{E} - \#\mathcal{F}|$, then d_α is a metric. □

Proposition 5. *Given $\alpha \geq 0$, the following statements are equivalent:*

1. $\forall \mathcal{E},\mathcal{F} \in \mathbb{L} \quad (\mathcal{E} \succ_c \mathcal{F} \Leftrightarrow d_\alpha(\mathcal{E},l_g) < d_\alpha(\mathcal{F},l_g))$.
2. *$\alpha \in I_g$, where $I_g = \left(0,\frac{1}{g-2}\right)$, if g is odd, and $I_g = \left(0,\frac{1}{g-1}\right)$, if g is even.*

Proof. 1 ⇒ 2) Consider g is odd. By Proposition 2, we have

$$[l_1,l_g] \succ_c \left[l_{\frac{g-1}{2}},l_{\frac{g+1}{2}}\right].$$

Then, by hypothesis, we have

$$d_\alpha\left([l_1,l_g],l_g\right) < d_\alpha\left(\left[l_{\frac{g-1}{2}},l_{\frac{g+1}{2}}\right],l_g\right). \tag{3}$$

Taking into account Equation (2),

$$d_\alpha\left([l_1,l_g],l_g\right) = |1-g| + |g-g| + \alpha\,|g-1+g-g| = g-1+\alpha\,(g-1)$$

and

$$d_\alpha\left(\left[l_{\frac{g-1}{2}},l_{\frac{g+1}{2}}\right],l_g\right) =$$

$$\left|\frac{g-1}{2}-g\right| + \left|\frac{g+1}{2}-g\right| + \alpha\left|g-\frac{g-1}{2}+\frac{g+1}{2}-g\right| = g+\alpha.$$

Then, from Inequality (3), we obtain $g-1+\alpha(g-1) < g+\alpha$, i.e.,

$$\alpha < \frac{1}{g-2}.$$

In order to prove that $\alpha > 0$, suppose by way of contradiction that $\alpha = 0$. By Proposition 2, we have

$$l_{\frac{g+1}{2}} \succ_c [l_1,l_g].$$

Then, by hypothesis, we have

$$d_\alpha\left(l_{\frac{g+1}{2}},l_g\right) < d_\alpha([l_1,l_g],l_g).$$

Taking into account Equation (2),

$$d_\alpha\left(l_{\frac{g+1}{2}},l_g\right) = g-1 = d_\alpha([l_1,l_g],l_g),$$

which is a contradiction. Consequently, $\alpha \in I_g = \left(0,\frac{1}{g-2}\right)$.

We now consider that g is even. By Proposition 2, we have

$$\left[l_{\frac{g}{2}},l_{\frac{g}{2}+1}\right] \succ_c [l_1,l_g].$$

Then, by hypothesis, we have

$$d_\alpha\left(\left[l_{\frac{g}{2}},l_{\frac{g}{2}+1}\right],l_g\right) < d_\alpha\left([l_1,l_g],l_g\right). \tag{4}$$

Taking into account Equation (2),

$$d_\alpha\left(\left[l_{\frac{g}{2}},l_{\frac{g}{2}+1}\right],l_g\right) = \left|\frac{g}{2}-g\right| + \left|\frac{g}{2}+1-g\right| + \alpha\left|g-\frac{g}{2}+\frac{g}{2}+1-g\right| = g.$$

Then, from Inequality (4), we obtain $g-1+\alpha(g-1) < g$, i.e.,

$$\alpha < \frac{1}{g-1}.$$

In order to prove that $\alpha > 0$, suppose by way of contradiction that $\alpha = 0$. As mentioned above, we have

$$d_\alpha\left(\left[l_{\frac{g}{2}},l_{\frac{g}{2}+1}\right],l_g\right) < d_\alpha([l_1,l_g],l_g).$$

However, taking into account Equation (2),

$$d_\alpha\left(\left[l_{\frac{g}{2}}, l_{\frac{g}{2}+1}\right], l_g\right) = g - 1 = d_\alpha([l_1, l_g], l_g),$$

we obtain a contradiction. Consequently, $\alpha \in I_g = \left(0, \frac{1}{g-1}\right)$.

2 $\Rightarrow$ 1) It is a routine. □

Definition 5. *Given $\alpha \in I_g$, the scoring function $S_\alpha : \mathbb{L} \longrightarrow \mathbb{R}$ is defined as*

$$S_\alpha(\mathcal{E}) = d_\alpha(l_1, l_g) - d_\alpha(\mathcal{E}, l_g),$$

for every $\mathcal{E} \in \mathbb{L}$.

It is easy to see that, for every $[l_h, l_k] \in \mathbb{L}$, it holds

$$S_\alpha([l_h, l_k]) = h + k - 2 - \alpha(k - h). \tag{5}$$

Proposition 6. *Given $\alpha \in I_g$, for all $\mathcal{E}, \mathcal{F} \in \mathbb{L}$, it holds*

$$\mathcal{E} \succ_c \mathcal{F} \Leftrightarrow S_\alpha(\mathcal{E}) > S_\alpha(\mathcal{F}).$$

Proof. By Proposition 5. □

Example 2. *Taking into account the condition appearing in Proposition 5 for $g = 5$, i.e., $0 < \alpha < \frac{1}{3}$, and Equation (5), we have*

$$\begin{aligned}
&S_\alpha(l_5) = 8 > S_\alpha([l_4, l_5]) = 7 - \alpha > S_\alpha(l_4) = 6 > S_\alpha([l_3, l_5]) = 6 - 2\alpha > \\
&S_\alpha([l_3, l_4]) = 5 - \alpha > S_\alpha([l_2, l_5]) = 5 - 3\alpha > S_\alpha(l_3) = 4 > \\
&S_\alpha([l_2, l_4]) = 4 - 2\alpha > S_\alpha([l_1, l_5]) = 4 - 4\alpha > S_\alpha([l_2, l_3]) = 3 - \alpha > \\
&S_\alpha([l_1, l_4]) = 3 - 3\alpha > S_\alpha(l_2) = 2 > S_\alpha([l_1, l_3]) = 2 - 2\alpha > \\
&S_\alpha([l_1, l_2]) = 1 - \alpha > S_\alpha(l_1) = 0.
\end{aligned}$$

3.4. Extensions Based on Different Criteria

As we have already seen, both the lexicographic and the canonical linear orders on $\mathbb{L}$ extend the given order on L. Although it is always possible to extend a linear order from a given finite set U to its power set, a typical question that, as a matter of fact, gave rise to a battery of classical papers from the 1970s (see e.g., Gärdenfors [15], Kannai and Peleg [16], Barberà and Pattanaik [17], Fishburn [18], Bossert [19] and Bossert et al. [20]) is whether or not it is possible to perform an extension that follows a list of criteria imposed a priori.

Sometimes, the extension is not possible because the criteria used are, so-to-say, contradictory. However, perhaps surprisingly, there are other situations in which the extension is not possible because of a combinatorial explosion, which, due to the much bigger cardinality $2^{\#U}$ of the power set of U, does not leave room to accommodate all terms of the power set, in an extended linear order, accomplishing all the criteria. Perhaps the most famous result in this direction is the so-called theorem of impossibility of extension due to Kannai and Peleg [16] (for a detailed account, see Barberà et al. [21]).

Consider the following example.

Example 3. *Let $U = \{u_1, u_2, \ldots, u_p\}$ be a finite universe. Assume that U is endowed with a linear order $\succcurlyeq$ such that $u_i \succcurlyeq u_j \Leftrightarrow i \geq j$. Consider the following two criteria for an extension, say $\succcurlyeq_{ext}$ of $\succcurlyeq$ to the power set of U.*

(i) A criterion [M] based on the idea of a *mean value*, so that if $u_i \succ u_j$ in U, then $\{u_i\} \succ_{\text{ext}} \{u_i, u_j\} \succ_{\text{ext}} \{u_j\}$ holds true.
(ii) A criterion [C] based on the idea of *cardinality*, in the sense "the bigger, the better", so that given $u_i, u_j \in U$ such that $u_i \neq u_j$, then $\{u_i, u_j\} \succ_{\text{ext}} \{u_i\}$ and also $\{u_i, u_j\} \succ_{\text{ext}} \{u_j\}$ hold true.

In this example, it is clear that these two criteria are contradictory. Unless the set U is a singleton, no extension $\succcurlyeq_{\text{ext}}$ can at the same time satisfy [M] and [C].

By the way, suppose that we use the criterion [C], jointly with a rule [B] that pays attention to the best elements of the corresponding subsets, as follows: when two subsets $A, B \subset U$ have the same cardinality, pay attention to the best elements $\max A, \max B$ as regards $\succcurlyeq$. If $\max A = \max B = u_i$, then pay attention to $\max(A \setminus \{u_i\})$ and $\max(B \setminus \{u_i\})$, and so on. We proceed this way until we declare $A \succ B$ or $B \succ A$.

Starting with the set of labels $L = \{l_1, l_2, \ldots, l_g\}$, we extend the usual ordering not to the whole power set of L, but to the set of linguistic expressions $\mathbb{L}$, and accomplishing the criteria [C] and [B], and it is well understood that [C] is accomplished in this context if, for every $[l_h, l_k] \in \mathbb{L}$ such that $k < g$ it holds $[l_h, l_{k+1}] \succ_{\text{ext}} [l_h, l_k]$, and for every $[l_h, l_k] \in \mathbb{L}$ such that $h > 1$, it holds that $[l_{h-1}, l_k] \succ_{\text{ext}} [l_h, l_k]$.

Calling $\succcurlyeq_{\text{cm}}$ to the extended linear order on $\mathbb{L}$, for $g = 5$, we get:

$$[l_1, l_5] \succ_{\text{cm}} [l_2, l_5] \succ_{\text{cm}} [l_1, l_4] \succ_{\text{cm}} [l_3, l_5] \succ_{\text{cm}} [l_2, l_4] \succ_{\text{cm}}$$
$$[l_1, l_3] \succ_{\text{cm}} [l_4, l_5] \succ_{\text{cm}} [l_3, l_4] \succ_{\text{cm}} [l_2, l_3] \succ_{\text{cm}} [l_1, l_2] \succ_{\text{cm}}$$
$$l_5 \succ_{\text{cm}} l_4 \succ_{\text{cm}} l_3 \succ_{\text{cm}} l_2 \succ_{\text{cm}} l_1.$$

The linear order $\succcurlyeq_{\text{cm}}$ is said to be the *cardinality-maximality extension* of $\succcurlyeq$.

Remark 6. *The canonical order $\succcurlyeq_c$ on $\mathbb{L}$ only satisfies the criterion [M], well understood that [M] is accomplished in this context if, for every $l_h \in L$ such that $h < g$, it holds that $\{l_{h+1}\} \succ_{ext} [l_h, l_{h+1}] \succ_{ext} \{l_h\}$.*

The lexicographic order $\succcurlyeq_\ell$ fails to satisfy both [M] and [C].

The cardinality-maximality extension $\succcurlyeq_{cm}$ satisfies [C], but not [M].

Taking into account the last ideas in Example 3, in which an extended order was defined on $\mathbb{L}$ but not necessarily on the whole power set of L, we may realize that, at this stage, an interesting question appears. Suppose that we are given a set of criteria for extension of a linear order from a set U to its power set. Assume also that, as it is the case of Kannai–Peleg's theorem ([16]), such set of criteria provokes an impossibility result. Even if this happens, when we try to extend the ordering $\succcurlyeq$ defined on the set $L = \{l_1, l_2, \ldots, l_g\}$ of labels, to its corresponding set $\mathbb{L}$ of linguistic expressions, it may still happen that an extension that accomplishes the criteria is possible. The reason is that $\mathbb{L}$ is much smaller than the whole power set of L. Sometimes, due to this restriction of domain, the extension could still be obtained. Similar questions, namely impossibility theorems in which the intrinsic impossibility actually disappears when a suitable restriction of domain is done, have been analyzed in depth in the contexts of Social Choice (see e.g., Gaertner [22]).

Let us explore this fact in more detail. Given a finite set $U = \{u_1, u_2, \ldots, u_p\}$ endowed with a linear order $\succcurlyeq$ such that $u_i \succcurlyeq u_j \Leftrightarrow i \geq j$, suppose that we want to define an extension $\succcurlyeq_{\text{ext}}$ on the power set of U. We say that $\succcurlyeq_{\text{ext}}$ satisfies:

(i) The *Gärdenfors principle* [G] (see Gärdenfors [15]) if, for all $A \subseteq U$ and $u_i \in U$, it holds that $u_i \succ \max(A) \Rightarrow A \cup \{u_i\} \succ_{\text{ext}} A$ and $\min(A) \succ u_i \Rightarrow A \succ_{\text{ext}} A \cup \{u_i\}$.
(ii) The *weak monotonicity principle* [W] if, for all $A, B \subseteq U$ and $u_i \notin A \cup B$, it holds that $A \succ_{\text{ext}} B \Rightarrow A \cup \{u_i\} \succcurlyeq_{\text{ext}} B \cup \{u_i\}$.

A version of the key Kannai–Peleg impossibility theorem (see Kannai and Peleg [16], Fishburn [18], Barberà et al. [17] and Bossert [19]) reads as follows.

Theorem 1. *Let $U = \{u_1, u_2, \ldots, u_p\}$ be a finite set with at least six elements $(p \geq 6)$, endowed with the linear order $\succcurlyeq$ defined as $u_i \succcurlyeq u_j \Leftrightarrow i \geq j$. Then, there is no extension $\succcurlyeq_{ext}$ that satisfies both [G] and [W] and is a total preorder on the power set of U.*

However, after a suitable restriction of domain, the impossibility stated in the Kannai–Peleg theorem could actually disappear, if built in our context of linguistic terms and expressions, as the next Remark 7 clearly shows.

Remark 7. *When we consider the set of labels $L = \{l_1, l_2, \ldots, l_6\}$ endowed with the usual linear order $\succcurlyeq$, we may observe that the canonical order $\succcurlyeq_c$ on $\mathbb{L}$ satisfies both [G] and [W], well understood that here in this context where a restriction of domain plays a crucial role, [G] and [W] should read as follows:*

(i) [G] is accomplished in this context if, for every $[l_h, l_k] \in \mathbb{L}$, the following conditions hold:

$$[l_h, l_{k+1}] \succ_{\text{ext}} [l_h, l_k], \text{ whenever } k < g,$$
$$[l_h, l_k] \succ_{\text{ext}} [l_{h-1}, l_k], \text{ whenever } h > 1.$$

(ii) [W] is satisfied in this context if for all $[l_h, l_k], [l_h, l_{k'}] \in \mathbb{L}$, the following conditions hold:

$$[l_h, l_k] \succ_{\text{ext}} [l_h, l_{k'}] \Rightarrow [l_{h-1}, l_k] \succcurlyeq_{\text{ext}} [l_{h-1}, l_k], \text{ whenever } h > 1,$$
$$[l_h, l_k] \succ_{\text{ext}} [l_{h'}, l_k] \Rightarrow [l_h, l_{k+1}] \succcurlyeq_{\text{ext}} [l_{h'}, l_{k+1}], \text{ whenever } k < g.$$

The extension $\succcurlyeq_{\text{ext}}$ is a total preorder order because it is indeed a linear order. Therefore, Kannai–Peleg impossibility theorem is no longer valid when we extend linear orders from L to $\mathbb{L}$, that, obviously, is much smaller than the power set of L.

Another classical question in this approach consists in, starting from a finite set U endowed with a linear order, in order to characterize orderings on the power set of U (but not necessarily extensions of the given order on U) that satisfy a set of criteria imposed a priori.

Let us see an example of this situation.

Example 4. *Suppose that a total preorder $\succcurlyeq$, defined directly on the power set of a finite set U, satisfies the following conditions:*

(i) $\{u_i\} \sim \{u_j\}$ and also $\{u_i, u_j\} \succ \{u_i\}$ hold true for all $u_i, u_j \in U$ such that $u_i \neq u_j$.
(ii) For all $A, B \subset U$ and $u_i \notin A \cup B$, it holds that $A \succcurlyeq B \Rightarrow A \cup \{u_i\} \succcurlyeq B \cup \{u_i\}$.

Then, it can be straightforwardly proved that $\succcurlyeq$ is, exactly, the so-called *cardinal ordering* on the power set of U, namely $A \succcurlyeq B \Leftrightarrow \#A \geq \#B$ holds true for all $A, B \subseteq U$ (see Pattanaik and Xu [23] for a proof of this last fact).

Having these ideas in mind, and coming back to our context of labels and linguistic expressions, we may see that the lexicographic extension $\succcurlyeq_\ell$ on $\mathbb{L}$ is characterized as follows.

Proposition 7. *Consider the set of labels $L = \{l_1, l_2, \ldots, l_g\}$, endowed with the usual linear order $\succcurlyeq$. An extension $\succcurlyeq_{ext}$ to the set $\mathbb{L}$ of linguistic expressions satisfies the following conditions:*

(i) For all $h, k \in \{1, \ldots, g\}$, it holds that $\{l_h\} \succcurlyeq_{ext} \{l_k\} \Leftrightarrow h \geq k$.
(ii) For all $[l_h, l_k]$, $[l_{h'}, l_{k'}] \in \mathbb{L}$, it holds that $h > h' \Rightarrow [l_h, l_k] \succ_{ext} [l_{h'}, l_{k'}]$.
(iii) For all $[l_h, l_k]$, $[l_h, l_{k'}] \in \mathbb{L}$, it holds that $k > k' \Rightarrow [l_h, l_k] \succ_{ext} [l_h, l_{k'}]$,

if and only if $\succcurlyeq_{ext} = \succcurlyeq_\ell$.

Proof. It is in immediate consequence of the mere definition of the lexicographic extension $\succcurlyeq_\ell$ on $\mathbb{L}$. □

The cardinality-maximality extension $\succcurlyeq_{\mathrm{cm}}$ can be characterized as follows.

Proposition 8. *Consider the set of labels* $L = \{l_1, l_2, \ldots, l_g\}$, *endowed with the usual linear order* $\succcurlyeq$. *An extension* $\succcurlyeq_{ext}$ *to the set* $\mathbb{L}$ *of linguistic expressions satisfies the following conditions:*

(i) For all $h, k \in \{1, \ldots, g\}$, *it holds that* $\{l_h\} \succcurlyeq_{ext} \{l_k\} \Leftrightarrow h \geq k$.
(ii) For all $\mathcal{E} = [l_h, l_k]$, $\mathcal{F} = [l_{h'}, l_{k'}] \in \mathbb{L}$, *it holds that* $\#\mathcal{E} > \#\mathcal{F} \Rightarrow \mathcal{E} \succ_{ext} \mathcal{F}$.
(iii) For all $\mathcal{E} = [l_h, l_k]$, $\mathcal{F} = [l_{h'}, l_{k'}] \in \mathbb{L}$ *with* $\#\mathcal{E} = \#\mathcal{F}$, *it holds* $k > k' \Rightarrow \mathcal{E} \succ_{ext} \mathcal{F}$,

if and only if $\succcurlyeq_{ext} = \succcurlyeq_{cm}$.

Proof. It is in immediate consequence of the mere definition of the cardinality-maximality extension $\succcurlyeq_{\mathrm{cm}}$ on $\mathbb{L}$. □

We may also point out that these characterizations can reflect practical situations. In fact, they could reflect particular methods of assigning scores, perhaps discriminating between expressions that have the same score, as the next Examples 5 and 6 show. Notice that sometimes people use more than one score: first, there is one method of scoring, but if this leads to a tie, then a second method of scoring is used, and so on, using more methods of scorings if necessary, until the tie is broken.

Example 5. *Consider the set of labels* $L = \{l_1, l_2, \ldots, l_g\}$, *two scoring functions* $S, T : \mathbb{L} \to \mathbb{R}$ *and the following total preorder on* $\mathbb{L}$

$$[l_h, l_k] \succcurlyeq_{ext} [l_{h'}, l_{k'}] \Leftrightarrow \begin{cases} S([l_h, l_k]) > S([l_{h'}, l_{k'}]) \\ \text{or} \\ S([l_h, l_k]) = S([l_{h'}, l_{k'}]) \text{ and } T([l_h, l_k]) \geq T([l_{h'}, l_{k'}]). \end{cases} \tag{6}$$

(i) If $S([l_h, l_k]) = h + k$ and $T([l_h, l_k]) = h - k$, then $\succcurlyeq_{\mathrm{ext}}$ is the canonical extension $\succcurlyeq_c$ on $\mathbb{L}$.
(ii) If $S([l_h, l_k]) = h$ and $T([l_h, l_k]) = k$, then $\succcurlyeq_{\mathrm{ext}}$ is the lexicographic extension $\succcurlyeq_\ell$ on $\mathbb{L}$.
(iii) If $S([l_h, l_k]) = k - h$ and $T([l_h, l_k]) = k$, then $\succcurlyeq_{\mathrm{ext}}$ is the cardinality-maximality extension $\succcurlyeq_{\mathrm{cm}}$ on $\mathbb{L}$.

Example 6. *An expert is forecasting the quality of the forthcoming new harvest in a piece of land. To do so, she/he uses labels as {very bad, bad, poor, fair, good, very good, excellent}. Moreover she/he can hesitate and show imprecision, so that the forecast could be an expression as, for instance, "from fair to very good". From the point of view of the land keepers, a forecast is more positive than another one if the "average" is more favorable. Among two forecasts with identical average, it is important to avoid imprecision so that the most accurate is declared the more positive. The extension* $\succcurlyeq_{ext}$ *defined this way is the lexicographic one,* $\succcurlyeq_\ell$, *on* $\mathbb{L}$.

4. Aggregation of Qualitative Assessments and Scorings

4.1. Some Facts about the Aggregation of Assessments and Scorings

Suppose now that a finite number of experts, agents, decision-makers or individuals show their assessments on a finite set, called universe of alternatives, $X = \{x_1, x_2, \ldots, x_n\}$. To do so, each individual could establish a *qualitative ranking* on X, based on linguistic terms $L = \{l_1, l_2, \ldots, l_g\}$, or perhaps they show some imprecision and use the set $\mathbb{L}$ of linguistic expressions that come from L. Once the opinions of all the individuals have been established, a typical problem consists in aggregating them into a social ranking that somewhat reflects a decision that could be shared by all the individuals, as a society or collective, following some rules of "common sense" on which all the agents could agree. To put a trivial example, if each agent has ranked the elements of X in an identical manner, it seems compulsory that the social ranking should keep this, making no changes on the label or linguistic expression assigned to each one of the alternatives, and consequently showing respect for *unanimity*. It could also happen that each agent has already defined a sort of *numerical scoring* on the alternatives, so that the problem of aggregation translates in defining some sort of *average* of the individual numerical scores.

Notice that any scoring function $S : X \to \mathbb{R}$ immediately defines a total preorder $\succcurlyeq$ on X by just declaring $x_i \succcurlyeq x_j \Leftrightarrow S(x_i) \geq S(x_j)$, for all $x_i, x_j \in X$. An important fact that comes from the aggregation problem is that, even in the case of any individual scoring function defining a linear order on X—that is, provided that ties never occur at the individual level—the final social aggregated scoring function could fail to be a linear order. Ties could appear after aggregation, so that total preorder is all the best that we may expect a priori as the ordering coming from a *social* scoring function.

Moreover, at this stage, it is relevant to say that even imposing mild restrictions of so-to-say "common sense" to the rules that aggregate individual preferences into a social one, the existence of a social rule in those conditions could become impossible, as proved in the famous *Arrow's impossibility theorem* in Social Choice (see e.g., Arrow [24], Kelly [25], Campbell and Kelly [26] and Campión et al. [27] for further information on this topic).

In the present Section 4, we will not pay attention to these last kinds of questions. Instead, we will introduce some aggregation procedures for qualitative assessments and/or scorings over a finite set, disregarding the fact of the aggregation procedure satisfying the typical conditions (e.g., independence of irrelevant alternatives, non-dictatorship, etc.) arising in the Arrovian model or similar settings.

Let us introduce now some necessary concepts to deal with an aggregation procedure.

Definition 6. *Let $X = \{x_1, x_2, \ldots, x_n\}$ be a finite set of alternatives. Suppose that a finite set of agents $A = \{1, 2, \ldots, m\}$ declare their assessments on the elements of X by means of linguistic expressions. A* profile *V is a matrix (v_i^a) consisting of m rows and n columns (Notice that here the superindex a will correspond to a* row *(that depends on agents), whereas the subindex i corresponds to a* column *(that depends on alternatives)) of linguistic expressions, where the element $v_i^a \in \mathbb{L}$ represents the linguistic assessment given by the agent $a \in A$ to the alternative $x_i \in X$. Then,*

$$V = \begin{pmatrix} v_1^1 & \cdots & v_i^1 & \cdots & v_n^1 \\ \cdots & \cdots & \cdots & \cdots & \cdots \\ v_1^a & \cdots & v_i^a & \cdots & v_n^a \\ \cdots & \cdots & \cdots & \cdots & \cdots \\ v_1^m & \cdots & v_i^m & \cdots & v_n^m \end{pmatrix} = (v_i^a).$$

Therefore, each entry v_i^a is a linguistic expression $[l_h, l_k] \in \mathbb{L}$, which corresponds to the assessment that the a-th agent or expert has done on the i-th element x_i of the set X of alternatives. Both l_h and l_k will depend on the position in which v_i^a is located in the matrix V, namely in the a-th row and i-th column.

Taking into account the interval I_g appearing in Proposition 5 and the scoring function $S_\alpha : \mathbb{L} \longrightarrow \mathbb{R}$ introduced in Definition 5, we now define the scoring function $T_\alpha : X \to \mathbb{R}$ as

$$T_\alpha(x_i) = \frac{1}{m} \sum_{a=1}^{m} S_\alpha\left(v_i^a\right),$$

for each profile $V = \left(v_i^a\right)$.

We now establish the procedure for ranking the alternatives.

Proposition 9. *Given $\alpha \in I_g$ and a profile $V = \left(v_i^a\right)$, the binary relation $\succcurlyeq_\alpha$ on X defined as*

$$x_i \succcurlyeq_\alpha x_j \Leftrightarrow T_\alpha(x_i) \geq T_\alpha(x_j)$$

is a total preorder on X.

Proof. Just notice that any scoring function $T : X \to \mathbb{R}$ immediately defines a total preorder $\succcurlyeq$ on X by declaring $x_i \succcurlyeq x_j \Leftrightarrow T(x_i) \geq T(x_j)$. □

Consider now that agents should assess each alternative from different criteria, so that $C = \{c_1, c_2, \ldots, c_r\}$ represents the set of criteria available to the agents. Then, we have r profiles $V^1 = (v_i^{a,1})$, $V^2 = (v_i^{a,2}), \ldots, V^r = (v_i^{a,r})$, where $v_i^{a,q} \in \mathbb{L}$ is the assessment given by the agent $a \in A$ to the i-th alternative $x_i \in X$ with respect to the q-th criterion $c_q \in C$.

Since each criterion may have different importance in the decision, we will also consider a *weighting vector* $\boldsymbol{w} = (w_1, w_2, \ldots, w_r) \in [0,1]^r$, satisfying $w_1 + w_2 + \cdots + w_r = 1$.

Given $\alpha \in I_g$, the profiles $V^1 = (v_i^{a,1})$, $V^2 = (v_i^{a,2})$, ..., $V^r = (v_i^{a,r})$ and a weighting vector $\boldsymbol{w} = (w_1, w_2, \ldots, w_r)$ satisfying $w_1 + w_2 + \cdots + w_r = 1$, we now define the scoring function $T_{\alpha,\boldsymbol{w}} : X \to \mathbb{R}$ as

$$T_{\alpha,\boldsymbol{w}}(x_i) = \frac{1}{r} \sum_{q=1}^{r} w_q \left(\frac{1}{m} \sum_{a=1}^{m} S_\alpha\left(v_i^{a,q}\right) \right).$$

Proposition 10. *Given $\alpha \in I_g$, the profiles $V^1 = (v_i^{a,1})$, $V^2 = (v_i^{a,2}), \ldots, V^r = (v_i^{a,r})$ and a weighting vector $\boldsymbol{w} = (w_1, w_2, \ldots, w_r)$ satisfying $w_1 + w_2 + \cdots + w_r = 1$, the binary relation $\succcurlyeq_{\alpha,\boldsymbol{w}}$ on X defined as*

$$x_i \succcurlyeq_{\alpha,\boldsymbol{w}} x_j \Leftrightarrow T_{\alpha,\boldsymbol{w}}(x_i) \geq T_{\alpha,\boldsymbol{w}}(x_j)$$

is a total preorder on the set X of alternatives.

Proof. The proof is immediate and leans on the same fact as the proof of Proposition 9. □

Remark 8. *Similar constructions could also be done using, instead of the arithmetic mean, other kinds of aggregation functions as, e.g., medians, trimmed means and so on.*

We note that García-Lapresta and González del Pozo [28] propose a multi-criteria decision-making procedure in the same framework, but using a purely ordinal approach.

4.2. A Field Experiment

As described with more detail in García-Lapresta et al. [1], the proposed use of qualitative rankings and scorings based on linguistic expressions, their numerical representation by means of the geodesic distance plus a penalization of imprecision, as well as the subsequent method of aggregation just described in Proposition 10, have actually been tested in a field experiment carried out in *Trigo* restaurant in Valladolid, Spain (30 November 2013), under appropriate conditions of temperature, light and service.

A total of six experts trained in the sensory analysis of wines and wild mushrooms were recruited through the *Gastronomy and Food Academy of Castile and Leon*. When the test was being carried out, there was no communication between the experts and the samples were given to them without any identification.

The six experts assessed the products included in Table 1 through the five linguistic terms of Table 2 (or the corresponding linguistic expressions, when they hesitated) under three criteria: appearance (A), smell (S) and taste (T).

Five of the six experts sometimes hesitated about their opinions, and they assigned linguistic expressions with two labels. This happened in 19 of the 108 ratings, i.e., 17.59 % of the cases. The ratings provided by the experts appear in Table 3.

Table 1. Alternatives.

x_1	White wine 'Rueda Zascandil 2012'
x_2	Red wine 'Toro Valdelacasa 2007'
x_3	Raw *Boletus pinophilus*
x_4	Raw *Tricholoma portentosum*
x_5	Cooked *Boletus pinophilus*
x_6	Cooked *Tricholoma portentosum*

Table 2. Linguistic terms.

l_1	I don't like it at all
l_2	I don't like it
l_3	I like it
l_4	I rather like it
l_5	I like it so much

Table 3. Ratings.

		Expert 1	Expert 2	Expert 3	Expert 4	Expert 5	Expert 6
	A	l_3	l_4	l_4	l_5	l_3	l_3
x_1	S	$[l_3, l_4]$	l_3	l_4	l_4	l_4	l_4
	T	l_3	l_3	l_4	l_4	l_3	l_3
	A	l_4	l_4	l_2	l_5	l_4	l_4
x_2	S	$[l_4, l_5]$	l_5	l_2	l_5	l_3	l_5
	T	$[l_4, l_5]$	l_5	l_2	l_5	l_3	$[l_4, l_5]$
	A	l_4	l_4	l_5	l_5	l_3	l_4
x_3	S	$[l_4, l_5]$	l_3	$[l_4, l_5]$	l_5	l_3	l_5
	T	l_5	l_5	$[l_4, l_5]$	l_5	l_4	$[l_4, l_5]$
	A	l_4	l_3	l_4	l_5	l_3	$[l_2, l_3]$
x_4	S	l_3	$[l_2, l_3]$	l_4	l_3	l_2	$[l_2, l_3]$
	T	l_3	l_3	$[l_4, l_5]$	l_4	l_2	l_4
	A	l_5	l_4	l_4	l_4	l_3	l_4
x_5	S	l_5	l_5	l_5	l_4	l_3	$[l_4, l_5]$
	T	$[l_4, l_5]$	l_5	l_5	l_5	l_4	$[l_4, l_5]$
	A	l_5	l_3	$[l_4, l_5]$	l_5	$[l_2, l_3]$	l_4
x_6	S	l_5	l_3	l_5	l_3	$[l_3, l_4]$	l_4
	T	l_4	l_4	l_5	l_3	l_4	$[l_3, l_4]$

We now show the outcomes in five cases.

1. When only considering appearance, i.e., $w = (1,0,0)$, and the parameter $\alpha = 0.3$, we obtain:

$$T_{\alpha,w}(x_1) = 5.333, \quad T_{\alpha,w}(x_2) = 5.666, \quad T_{\alpha,w}(x_3) = 6.333,$$
$$T_{\alpha,w}(x_4) = 5.116, \quad T_{\alpha,w}(x_5) = 6.000, \quad T_{\alpha,w}(x_6) = 5.900.$$

Then, $x_3 \succ_{\alpha,w} x_5 \succ_{\alpha,w} x_6 \succ_{\alpha,w} x_2 \succ_{\alpha,w} x_1 \succ_{\alpha,w} x_4$.

2. When only considering smell, i.e., $w = (0,1,0)$, and the parameter $\alpha = 0.3$, we obtain:

$$T_{\alpha,w}(x_1) = 5.450, \quad T_{\alpha,w}(x_2) = 6.116, \quad T_{\alpha,w}(x_3) = 6.233,$$
$$T_{\alpha,w}(x_4) = 3.566, \quad T_{\alpha,w}(x_5) = 6.783, \quad T_{\alpha,w}(x_6) = 5.783.$$

Then, $x_5 \succ_{\alpha,w} x_3 \succ_{\alpha,w} x_2 \succ_{\alpha,w} x_6 \succ_{\alpha,w} x_1 \succ_{\alpha,w} x_4$.

3. When only considering taste, i.e., $w = (0, 0, 1)$, and the parameter $\alpha = 0.3$, we obtain:

$$T_{\alpha,w}(x_1) = 4.666, \quad T_{\alpha,w}(x_2) = 5.900, \quad T_{\alpha,w}(x_3) = 7.233,$$
$$T_{\alpha,w}(x_4) = 4.783, \quad T_{\alpha,w}(x_5) = 7.233, \quad T_{\alpha,w}(x_6) = 5.783.$$

Then, $x_3 \sim_{\alpha,w} x_5 \succ_{\alpha,w} x_2 \succ_{\alpha,w} x_6 \succ_{\alpha,w} x_4 \succ_{\alpha,w} x_1$.

4. Taking into account the weights (These weights are quite usual in this kind of tasting.), $w_1 = 0.2$ for appearance, $w_2 = 0.3$ for smell and $w_3 = 0.5$ for taste, i.e., $w = (0.2, 0.3, 0.5)$, and the parameter $\alpha = 0.3$, we obtain:

$$T_{\alpha,w}(x_1) = 5.035, \quad T_{\alpha,w}(x_2) = 5.918, \quad T_{\alpha,w}(x_3) = 6.753,$$
$$T_{\alpha,w}(x_4) = 4.485, \quad T_{\alpha,w}(x_5) = 6.852, \quad T_{\alpha,w}(x_6) = 5.807.$$

Then, $x_5 \succ_{\alpha,w} x_3 \succ_{\alpha,w} x_2 \succ_{\alpha,w} x_6 \succ_{\alpha,w} x_1 \succ_{\alpha,w} x_4$.

5. Taking into account the weights $w_1 = 0.4$ for appearance, $w_2 = 0.2$ for smell and $w_3 = 0.4$ for taste, i.e., $w = (0.4, 0.2, 0.4)$, and the parameter $\alpha = 0.3$, we obtain the following outcomes:

$$T_{\alpha,w}(x_1) = 5.090, \quad T_{\alpha,w}(x_2) = 5.850, \quad T_{\alpha,w}(x_3) = 6.673,$$
$$T_{\alpha,w}(x_4) = 4.673, \quad T_{\alpha,w}(x_5) = 6.650, \quad T_{\alpha,w}(x_6) = 5.830.$$

Then, $x_3 \succ_{\alpha,w} x_5 \succ_{\alpha,w} x_2 \succ_{\alpha,w} x_6 \succ_{\alpha,w} x_1 \succ_{\alpha,w} x_4$.

Although we have considered the specific value of $\alpha = 0.3$ in the previous calculations, we have to notice that in all cases the alternatives are ranked in the same way for all the values of $\alpha \in I_5 = (0, 1/3)$. Thus, the parameter α has been irrelevant in this field experiment. The reason is the low number of assessments with more than one linguistic term (17.59 % of the cases). However, if applying the procedure of Proposition 9 to the field experiment presented in Agell et al. [29], where five experts assessed five fruits with respect to their suitability to combine with dark chocolate, and 40% of the assessments were linguistic expressions with two or more linguistic terms, the ranking depending on the value of the parameter α.

We may conclude that this method, based on geodesic distances on linguistic expressions, plus a penalization of imprecision, and using a weighting vector according to different criteria, could be considered as a suitable one in these practical experiments, at least in the field of sensory analysis.

5. Conclusions

The definition of ratings and/or scorings on a finite set has been associated with a wide sort of theoretical questions, namely:

(i) The existence of a numerical representation for a total preorder on a finite set, so translating qualitative scales into quantitative (numerical) ones.
(ii) The selection of a suitable method of scoring or numerical representation of a qualitative scale.
(iii) The search for a compatible extension of a ranking defined on a finite set to its power set—or at least to a suitable superset of the given set—following a list of criteria established a priori.
(iv) The implementation of aggregation procedures on profiles of individual ratings and/or scorings, in the search for a new scoring that reflects a social average.
(v) The comparison between different ways of rating, scoring and aggregating, in the search for the most suitable one in view of some practical circumstances to be used in field experiments, and so on.

As a matter of fact, each of these questions, studied separately, could give raise to a wide sort of results in the literature. Indeed most of them have already been considered and analyzed, at least partially, by many authors.

However, many open problems remain. To provide only a few examples, we could say at this point that:

(i) A systematic study of criteria to extend a linear order from a set to its power set, trying to classify the criteria in, so-to-say, contradictory families, so that any two criteria coming from different families give raise to an impossibility result.
(ii) The systematic development of a standard procedure to determine which rating, scoring and/or aggregation procedure is the most suitable, maybe taking into account some previous criteria.

Needless to say, all of this leaves enough room for further research to be developed in the future.

Acknowledgments: This work has been partially supported by the Spanish *Ministerio de Economía y Competitividad* (projects ECO2015-65031-R, MTM2015-63608-P, ECO2016-77900-P and TIN2016-77356-P), ERDF and the Research Services of the Universidad Pública de Navarra (Spain). Thanks are also given to Remedios Marín and Íñigo Arozarena (Departamento de Tecnología de Alimentos, Universidad Pública de Navarra, Pamplona, Spain) for their valuable suggestions and comments.

Author Contributions: The authors contributed equally to this work.

Conflicts of Interest: The authors declare no conflict of interest.

References

1. García-Lapresta, J.L.; Aldavero, C.; de Castro, S. A linguistic approach to multi-criteria and multi-expert sensory analysis. In Proceedings of the International Conference on Information Processing and Management of Uncertainty in Knowledge-Based Systems (IPMU), Montpellier, France, 15–19 July 2014; Volume 443, pp. 586–595.
2. Travé-Massuyès, L.; Dague, P. (Eds.) *Modèles et Raisonnements Qualitatifs*; Hermes Science: Paris, France, 2003.
3. Travé-Massuyès, L.; Piera, N. The orders of magnitude models as qualitative algebras. In Proceedings of the 11th International Joint Conference on Artificial Intelligence, Detroit, Michigan, 20–25 August 1989; pp. 1261–1266.
4. Roselló, L.; Prats, F.; Agell, N.; Sánchez, M. Measuring consensus in group decisions by means of qualitative reasoning. *Int. J. Approx. Reason.* **2010**, *51*, 441–452.
5. Roselló, L.; Prats, F.; Agell, N.; Sánchez, M. A qualitative reasoning approach to measure consensus. In *Consensual Processes, STUDFUZZ*; Herrera-Viedma, E., García-Lapresta, J.L., Kacprzyk, J., Nurmi, H., Fedrizzi, M., Zadrózny, S., Eds.; Springer: Berlin, Germany, 2001; Volume 267, pp. 235–261.
6. Roselló, L.; Sánchez, M.; Agell, N.; Prats, F.; Mazaira, F.A. Using consensus and distances between generalized multi-attribute linguistic assessments for group decision-making. *Inf. Fusion* **2014**, *17*, 83–92.
7. Agell, N.; Sánchez, M.; Prats, F.; Roselló, L. Ranking multi-attribute alternatives on the basis of linguistic labels in group decisions. *Inf. Sci.* **2012**, *209*, 49–60.
8. Falcó, E.; García-Lapresta, J.L.; Roselló, L. Aggregating imprecise linguistic expressions. In *Human-Centric Decision-Making Models for Social Sciences*; Guo, P., Pedrycz, W., Eds.; Springer: Berlin, Germany, 2014; pp. 97–113.
9. Falcó, E.; García-Lapresta, J.L.; Roselló, L. Allowing agents to be imprecise: A proposal using multiple linguistic terms. *Inf. Sci.* **2014**, *258*, 249–265.
10. Bridges, D.S.; Mehta, G.B. *Representations of Preference Orderings*; Springer: Berlin/Heidelberg, Germany, 1995.
11. Szpilrajn, E. Sur l'extension de l'ordre partiel. *Fundam. Math.* **1930**, *16*, 386–389.
12. Herrera, F.; Herrera-Viedma, E.; Martínez, L. A fuzzy linguistic methodology to deal with unbalanced linguistic term sets. *IEEE Trans. Fuzzy Syst.* **2008**, *16*, 354–370.
13. García-Lapresta, J.L.; Pérez-Román, D. Ordinal proximity measures in the context of unbalanced qualitative scales and some applications to consensus and clustering. *Appl. Soft Comput.* **2015**, *35*, 864–872.
14. García-Lapresta, J.L.; Pérez-Román, D. Aggregating opinions in non-uniform ordered qualitative scales. *Appl. Soft Comput.* 2017, in press, doi:10.1016/j.asoc.2017.05.064.
15. Gärdenfors, P. Manipulation of social choice functions. *J. Econ. Theory* **1976**, *13*, 227–228.
16. Kannai, Y.; Peleg, B. A note on the extension of an order on a set to the power set. *J. Econ. Theory* **1984**, *32*, 172–175.

17. Barberà, S.; Pattanaik, P.K. Extending an order on a set to the power set: some remarks on Kannai and Peleg's approach. *J. Econ. Theory* **1984**, *32*, 185–191.
18. Fishburn, P.C. Comment on the Kannai–Peleg impossibility theorem for extending orders. *J. Econ. Theory* **1984**, *32*, 176–179.
19. Bossert, W. On the extension of preferences over a set to the power set: An axiomatic characterization of a quasi-ordering. *J. Econ. Theory* **1989**, *49*, 84–92.
20. Bossert, W.; Pattanaik, P.; Xu, Y. Ranking opportunity sets: an axiomatic approach. *J. Econ. Theory* **1990**, *63*, 326–345.
21. Barberà, S.; Bossert, W.; Pattanaik, P.K. Ranking sets of objects. In *Handbook of Utility Theory, Volume II: Extensions*; Barberà, S., Hammond, P.J., Seidl, C., Eds.; Kluwer Academic Publishers: Dordrecht, The Netherlands, 2004; pp. 893–977.
22. Gaertner, W. *Domain Conditions in Social Choice*; Cambridge University Press: Cambridge, UK, 2001.
23. Pattanaik, P.K.; Xu, Y. On ranking opportunity sets in terms of freedom of choice. *Rech. Écon. Louvain* **1990**, *56*, 383–390.
24. Arrow, K.J. *Social Choice and Individual Values*, 2nd ed.; Wiley: New York, NY, USA, 1963.
25. Kelly, J.S. *Social Choice Theory*; Springer: New York, NY, USA, 1988.
26. Campbell, D.E.; Kelly, J.S. Impossibility theorems in the Arrovian framework. In *Handbook of Social Choice and Welfare*; Arrow, K.J., Sen, A.K., Suzumura, K., Eds.; North Holland: Amsterdam, The Netherlands, 2002; Volume 1, pp. 35–44.
27. Campión, M.J.; Candeal, J.C.; Catalán, R.G.; de Miguel, J.R.; Induráin, E.; Molina, J.A. Aggregation of preferences in crisp and fuzzy settings: Functional equations leading to possibility results. *Int. J. Uncertain. Fuzziness Knowl. Based Syst.* **2011**, *19*, 89–114.
28. García-Lapresta, J.L.; González del Pozo, R. An ordinal multi-criteria decision-making procedure in the context of uniform qualitative scales. In *Soft Computing Applications for Group Decision-Making and Consensus Modeling*; Collan, M., Kacprzyk, J., Eds.; Studies in Fuzziness and Soft Computing; Springer: Berlin/Heidelberg, Germany, 2017; in press.
29. Agell, N.; Sánchez, G.; Sánchez, M.; Ruiz, F.J. Selecting the best taste: A group decision-making application to chocolates design. In Proceedings of the 2013 IFSA-NAFIPS Joint Congress, Edmonton, AB, Canada, 24–28 June 2013; pp. 939–943.

axioms

MDPI

Article

New Order on Type 2 Fuzzy Numbers

Pablo Hernández [1], Susana Cubillo [2,*], Carmen Torres-Blanc [2] and José A. Guerrero [1]

1 Departamento Matemática y Física, Universidad Nacional Experimental del Táchira (UNET), San Cristóbal, 5001 Táchira, Venezuela; hernandezpab@gmail.com (P.H.); jaguerrero4@gmail.com (J.A.G.)
2 Departamento Matemática Aplicada a las TIC, Universidad Politécnica de Madrid, 28660 Boadilla del Monte, Madrid, Spain; ctorres@fi.upm.es
* Correspondence: scubillo@fi.upm.es; Tel.: +34-913-365-474

Received: 5 June 2017 ; Accepted: 24 July 2017; Published: 28 July 2017

Abstract: Since Lotfi A. Zadeh introduced the concept of fuzzy sets in 1965, many authors have devoted their efforts to the study of these new sets, both from a theoretical and applied point of view. Fuzzy sets were later extended in order to get more adequate and flexible models of inference processes, where uncertainty, imprecision or vagueness is present. Type 2 fuzzy sets comprise one of such extensions. In this paper, we introduce and study an extension of the fuzzy numbers (type 1), the type 2 generalized fuzzy numbers and type 2 fuzzy numbers. Moreover, we also define a partial order on these sets, which extends into these sets the usual order on real numbers, which undoubtedly becomes a new option to be taken into account in the existing total preorders for ranking interval type 2 fuzzy numbers, which are a subset of type 2 generalized fuzzy numbers.

Keywords: type 2 fuzzy sets; fuzzy numbers; partial order

---

1. Introduction

In the framework of fuzzy systems, in order to give a response or to make a decision, fuzzy numbers (FNs) and fuzzy quantities (FQs) need to be compared, i.e., to be ranked. In a sense, fuzzy numbers can be considered as real values with degrees of uncertainty, imprecision or vagueness. They are represented as fuzzy sets on the real line, that is, in general, they are not simply real numbers. Therefore, they cannot be compared by means of the usual total order on the set of real numbers. On the other hand, the pointwise partial order of functions, which is the standard order in the fuzzy sets, does not adequately (see Section 2.1) extend the order of real numbers. However, from the horizontal representation of the extension principle, a partial order has been established (see Equation (4)) in the FNs that extends the order of real numbers. Since Jain [1,2] and Dubois and Prade [3] introduced the concept of fuzzy numbers, many authors (see, for example, [2,4–11]) have given different methods that, although they do not produce total orders, allow one to compare the FNs and to order some of their subsets. Among these approaches are: statistical methods (see [7,10]), geometric methods (see [8]), analytical methods (e.g., in [6], the authors work with bounded variation functions), computational methods (e.g., artificial neural networks are applied in [11]) or the combination of some of these.

Bortolan et al. [12], Cheng et al. [13] and, recently, Wang et al. [14,15] reviewed a variety of existing methods for comparing FNs. Moreover, the properties that every method must satisfy to compare reasonably FNs were discussed in [14,15].

On the other hand, since the introduction of extensions of fuzzy sets, the need to define and analyze fuzzy numbers in these extensions has been pointed out. For example, intuitionist fuzzy numbers (IFNs) and generalized intuitionist fuzzy numbers (GIFNs), which are intuitionistic sets on the real line satisfying certain conditions, were defined and studied in [16–18]. Recently, interval type 2 fuzzy numbers (IT2FNs) have been defined in [19–22], as interval type 2 fuzzy sets (IT2FSs) on the real line fulfilling some properties. Besides, in [19,22], total preorders were given to ranking IT2FNs, but such preorders do not hold the property of antisymmetry, which is a theoretical and practical weakness. Moreover, the total preorder given in [19] does not effectively extend the order of real numbers (see Example 2).

Type 2 fuzzy sets (T2FSs) are an extension of type 1 fuzzy sets (FSs), and IT2FSs are a particular case of T2FSs. Because the membership degrees of T2FSs are fuzzy, they are better able to model uncertainty than FSs [23]. Fortunately, new methods have been introduced for the purpose of achieving a computationally-efficient and viable framework for representing T2FSs, as well as the T2FLS (type 2 fuzzy logic system) inference processes (see, for example, [24–27]). Thanks to these computational simplifications, the first applications of generalized T2FSs and not just interval type 2 fuzzy sets (IT2FSs), which is a subset of T2FSs, are now being reported, such as, for example, [24,26,27].

In this paper, we define and compare type 2 fuzzy numbers (T2FNs) and type 2 generalized fuzzy numbers (T2GFNs), which are type 2 fuzzy sets (T2FSs) on the real line, with certain conditions. These sets are more general than the ones considered in [19,20,22], so that in each application, the expert can choose the linguistic labels that best fit the specifications of the problem. We obtain a partial order in T2GFNs and in T2FNs, extending the order of real numbers, as a first step to obtain in future investigations a total order (or a preorder).

The paper is organized as follows. In Section 2, we recall some definitions, basic concepts, partial orders and properties of the FNs and generalized fuzzy numbers (GFNs) of type 1 (Section 2.1) and the T2FSs (Section 2.2). In Section 2.2, a partial order is introduced in the subset of T2FSs with convex membership degrees and with the same height. In Section 3, we define and study the T2FNs and T2GFNs. A partial order is determined on the T2GFNs extending the order of the real numbers. Furthermore, among other examples, two IT2GFNs are ordered with our partial order, and the result is compared with the one obtained applying the preorder established in [19]. Section 4 is devoted to some conclusions.

2. Preliminaries

2.1. Definitions and Properties of Fuzzy Numbers

Throughout the paper, X stands for a non-empty set that represents the universe of discourse. Additionally, $\leq$ denotes the usual order relation in the lattice of real numbers. The operators $\wedge$ and $\vee$ are, respectively, the minimum and maximum operations on real numbers.

Definition 1. *[5,6,28] A fuzzy set (of type 1, FS) A, is characterized by a membership function f_A,*

$$f_A : X \to [0,1],$$

where $f_A(x)$ is called the membership degree of the element $x \in X$ in the set A.

Let $F(X)$ denote the set of all fuzzy sets on X. For each $A \in F(\mathbb{R})$, the height of A is:

$$w_A = \sup_{\leq} \left\{ f_A(x) : x \in \mathbb{R} \right\},$$

and the support is:

$$supp(A) = supp(f_A) = \left\{ x \in \mathbb{R} : f_A(x) > 0 \right\}.$$

We say that A is strongly normal if $f_A(x) = 1$ for some $x \in \mathbb{R}$ and that it is normal if $w_A = 1$. If A is strongly normal, then it is normal, but the opposite in general is not true. For example, the fuzzy set B on $\mathbb{R}$, with membership function:

$$f_B(x) = \begin{cases} 1+x & \text{if } -1 \le x < 0; \\ 0.5 - x & \text{if } 0 \le x \le 0.5; \\ 0 & \text{otherwise,} \end{cases}$$

is normal, but it is not strongly normal, since there is no $x \in \mathbb{R}$ such that $f_B(x) = 1$ (see Figure 1).

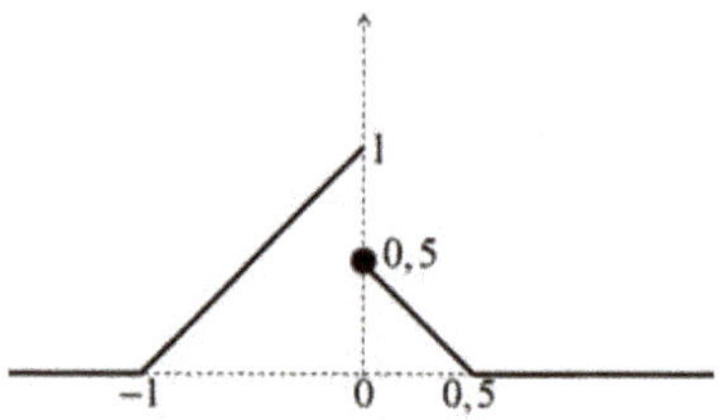

Figure 1. Normal, but non-strongly normal fuzzy set.

The α-cut of an FS A on $\mathbb{R}$, with $0 \le \alpha \le 1$, is $A_\alpha = \{x \in \mathbb{R} : f_A(x) \ge \alpha\}$. We say that A is convex if $f_A(y) \ge (f_A(x) \wedge f_A(z))$, for all $x \le y \le z$ (to simplify the notation, we will also say that f_A is convex). In addition, A is convex if and only if for all α with $0 \le \alpha \le 1$, we have that A_α is a convex subset of the real line, that is an interval or the empty set.

Definition 2. [5,6] *Let X be a subset of $\mathbb{R}$, and let $f : X \to \mathbb{R}$ be a function. We say that f is upper semicontinuous at a point x_0 if, given $\epsilon > 0$, there exists a neighborhood U of x_0 such that $f(x) < f(x_0) + \epsilon$, for all $x \in U$. We say that f is upper semicontinuous if it is upper semicontinuous at any $x \in X$.*

Intuitively, a function f is upper semicontinuous at x_0 if, when approximating to x_0, $f(x)$ approximates to $f(x_0)$ or is below $f(x_0)$. That is, if $\lim_{x \to x_0^+} f(x) \le f(x_0)$ and $\lim_{x \to x_0^-} f(x) \le f(x_0)$. See Figures 2 and 3.

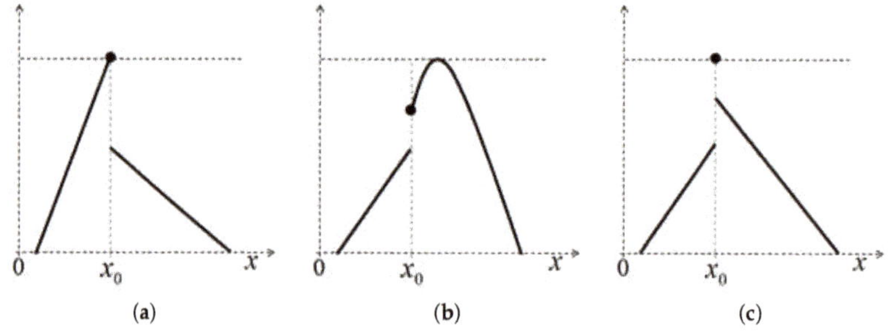

Figure 2. Upper semicontinuous functions: (**a**) Discontinuous from the right; (**b**) Discontinuous from the left; (**c**) Discontinuous from the right and from the left.

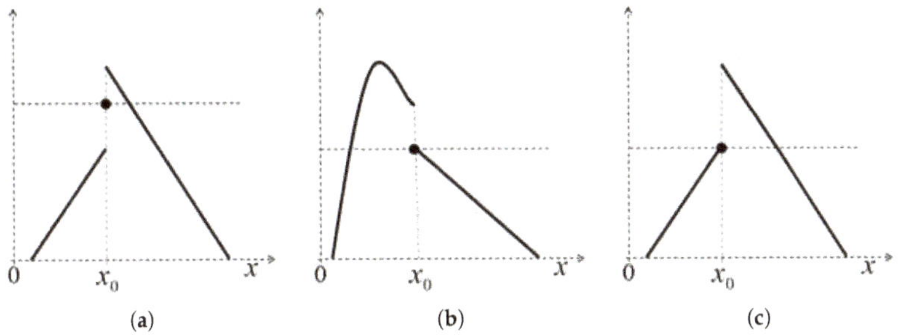

Figure 3. Non-upper semicontinuous functions: (**a**) Discontinuous from the right and from the left; (**b**) Discontinuous from the left; (**c**) Discontinuous from the right.

Definition 3. [5,6] *A generalized fuzzy number (GFN) A is a convex FS on $\mathbb{R}$ whose membership function f_A is upper semicontinuous and with bounded support.*

Any GFN A is determined by a membership function of the form [6]:

$$f_A(x) = \begin{cases} L_A(x) & \text{if } a \leq x \leq b; \\ w_A & \text{if } b \leq x \leq c; \\ R_A(x) & \text{if } c \leq x \leq d; \\ 0 & \text{otherwise,} \end{cases}$$

where $L_A : [a,b] \to [0, w_A]$ is an increasing function and $R_A : [c,d] \to [0, w_A]$ is a decreasing function. Note that the fuzzy set B of the previous example (see Figure 1) is not a GFN since its membership function is not upper semicontinuous. The functions shown in Figure 2 are GFNs, since each function is upper semicontinuous, convex and with bounded support. On the other hand, the functions shown in Figure 3 are not GFNs, because they are not upper semicontinuous functions.

A GFN A is normal if and only if its height w_A equals one. In addition, if a GFN A is normal, then it is also strongly normal, because its membership function, f_A, is upper semicontinuous.

Definition 4. [4–6] *A fuzzy number (FN) A is a GFN that is strongly normal.*

Because a fuzzy number A is strongly normal and convex, then A_α is an interval for all $0 \leq \alpha \leq 1$, and since it is upper semicontinuous and with bounded support, A_α is a closed interval (see [5,6]). The function f_B defined above and shown in Figure 1 is not upper semicontinuous at zero. Consequently, there exists some α, such that B_α is not closed (for example, $B_{0.8} = [-0.2, 0)$). Therefore, B is not a fuzzy number.

A real number a can be extended as a generalized fuzzy number with membership function equal to the characteristic function with height w, $\bar{a}_w : \mathbb{R} \to [0,1]$ where:

$$\bar{a}_w(x) = \begin{cases} w & \text{if } x = a; \\ 0 & \text{if } x \neq a. \end{cases}$$

If $w = 1$, the characteristic function $\bar{a}_1$, denoted $\bar{a}$, will be the extension of the real number a as a fuzzy number (type 1).

Similarly, a closed interval $[a,b]$ can be extended as a generalized fuzzy number with the membership function equal to the characteristic function with height w, $\overline{[a,b]}_w : \mathbb{R} \to [0,1]$, where:

$$\overline{[a,b]}_w(x) = \begin{cases} w & \text{if } x \in [a,b]; \\ 0 & \text{if } x \notin [a,b]. \end{cases}$$

If $w = 1$, the characteristic function $\overline{[a,b]}_1$ will be denoted by $\overline{[a,b]}$.

The usual partial order in the fuzzy sets, which is the partial order defined pointwise, is given by: let $A, B \in F(X)$; we say:

$$A \leq B \Leftrightarrow f_A \leq f_B, \quad \Big(f_A(x) \leq f_B(x), \forall x \in X\Big). \tag{1}$$

As is well known, this partial order is not total, and it is not the most appropriate when comparing fuzzy numbers, since it does not extend the order of real numbers. In fact, suppose we consider two different real numbers a and b, with $a < b$, as fuzzy numbers, by means of their characteristic functions $\bar{a}$ and $\bar{b}$. It is not true that $\bar{a} \leq \bar{b}$, since $\bar{b}(a) = 0 < \bar{a}(a) = 1$. Analogously, $\bar{b} \leq \bar{a}$ is not true either. As a consequence, $\bar{a}$ and $\bar{b}$ are not comparable with this partial order. To summarize, the usual partial order (1) does not extend the usual order $(\leq)$ of the real numbers.

By means of the Zadeh extension principle ([29]), many operations on the real numbers have been extended to the fuzzy numbers. For example, let $\circ$ be a binary and surjective operation on $\mathbb{R}$, then its extension $\bullet$ on the fuzzy numbers is defined by any of the following equivalent expressions ([30–32]):

1. Vertical representation:

$$f_{A \bullet B}(x) = \Big(f_A \bullet f_B\Big)(x) = \sup\Big\{f_A(y) \wedge f_B(z) : y \circ z = x\Big\}. \tag{2}$$

2. Horizontal representation:

$$\Big(A \bullet B\Big)_\alpha = A_\alpha \bullet B_\alpha = \Big\{a \circ b : a \in A_\alpha, b \in B_\alpha\Big\}. \tag{3}$$

Since $\circ$ is surjective, then the function $f_{A \bullet B}$ is defined for all $x \in \mathbb{R}$ and for all fuzzy sets A and B.

From the extensions in horizontal form (3) of the operations $\vee$ and $\wedge$, a new partial order ([9]) is established in the fuzzy numbers:

$$A \preccurlyeq B \Leftrightarrow A_\alpha \leq_I B_\alpha, \; \forall \alpha \in [0,1] \tag{4}$$

where $\leq_I$ is the usual partial order in the closed intervals, that is $[a,b] \leq_I [c,d] \Leftrightarrow a \leq c \text{ y } b \leq d$. As (2) and (3) are equivalent on the fuzzy numbers, the partial order that is obtained from (2), denoted $\sqsubseteq$ (see Section 2.2), on the fuzzy numbers is equivalent to $\preccurlyeq$. Let us remark that the order $\preccurlyeq$ only is defined on fuzzy numbers, while the order $\sqsubseteq$ and the order given in (1) can be applied to any fuzzy sets.

The orders $\preccurlyeq$ and $\sqsubseteq$ are not total, but they are more adequate than the punctual order given in (1) to compare fuzzy numbers, since they extend the usual order of the real numbers.

2.2. *About Type 2 Fuzzy Sets*

From now on, we denote by $\mathbf{M}$ the set of functions from $[0,1]$ to $[0,1]$, i.e., $\mathbf{M} = Map\Big([0,1],[0,1]\Big) = [0,1]^{[0,1]}$.

Definition 5. *[33,34] A type 2 fuzzy set (T2FS) A is characterized by a membership function:*

$$\mu_A : X \to \mathbf{M},$$

where $\mu_A(x)$ is the membership degree of an element $x \in X$ in the set A and a fuzzy set in $[0,1]$. Thus,

$$\mu_A(x) = f_x, \quad \textit{where } f_x : [0,1] \to [0,1].$$

Let $F_2(X) = Map\left(X, \mathbf{M}\right)$ denote the set of all type 2 fuzzy sets on X. It is worthwhile to remember that T2FSs are defined in a different way in [35]; however, both definitions are equivalent.

Example 1. *The following membership function determines a T2FS on* $\mathbb{R}$*:*

$$(\mu(x))(y) = f_x(y) = \begin{cases} -\dfrac{xy}{5} & \text{if } -5 \leq x < 0,\ y \in [0,1]; \\ \dfrac{xy}{7} & \text{if } 0 \leq x \leq 7,\ y \in [0,1]; \\ 0 & \text{if } x \notin [-5,7],\ y \in [0,1]. \end{cases}$$

Definition 6. *[20,36] An interval type 2 fuzzy set (IT2FS) A is a T2FS whose membership degrees $\mu_A(x)$, for all $x \in X$, are characteristic functions (with height $w = 1$) of closed intervals in [0,1].*

Note that these sets are isomorphic to the interval-valued fuzzy sets. Moreover, the IT2FSs given in Definition 6 are a particular case of IT2FSs defined in [37]; however, here, we always use Definition 6.

Walker and Walker show in [36] that the operations on $Map(X, \mathbf{M})$ can be defined naturally from the operations on $\mathbf{M}$ and have the same properties. In fact, given the operation $* : \mathbf{M} \times \mathbf{M} \to \mathbf{M}$, we can define the operation $\star : Map(X, \mathbf{M}) \times Map(X, \mathbf{M}) \to Map(X, \mathbf{M})$ such that, for each pair $f, g \in Map(X, \mathbf{M})$, we have $(f \star g)(x) = f(x) * g(x)$, for all $x \in X$, where $f(x), g(x) \in \mathbf{M}$ (see [36]). This pointwise extension of operations also allows us to define an order in type 2 fuzzy sets from an order on their membership degrees as follows: given two T2FSs A and B with membership functions $\mu_A, \mu_B \in Map(X, \mathbf{M})$, we have $A \preccurlyeq B$ if $\mu_A(x) \preccurlyeq \mu_B(x)$ for all $x \in X$, where $\preccurlyeq$ is an order in $\mathbf{M}$. However, when $X = \mathbb{R}$, this way of defining an order in $Map(\mathbb{R}, \mathbf{M})$ does not extend the order of $\mathbb{R}$, as happened with the usual order $\leq$ of the type 1 fuzzy sets. This is our motivation to define a new partial order in $Map(\mathbb{R}, \mathbf{M})$ (see Definition 15) that extends the order of real numbers (see Example 2 and Remark 6).

The following definition introduces algebraic operations on $\mathbf{M}$ by means of the vertical form (2) of Zadeh's principle of extension [28,29,33,34].

Definition 7. *[36,38,39] In $\mathbf{M}$, the operations $\sqcup$ (generalized maximum), $\sqcap$ (generalized minimum), $\neg$ (complementation) and the elements $\bar{0}$ and $\bar{1}$ are defined as follows:*

$$\begin{aligned}
(f \sqcup g)(x) &= \sup\left\{f(y) \wedge g(z) : y \vee z = x\right\}; \\
(f \sqcap g)(x) &= \sup\left\{f(y) \wedge g(z) : y \wedge z = x\right\}; \\
\neg f(x) &= \sup\left\{f(y) : 1 - y = x\right\} = f(1-x); \\
\bar{0}(x) &= \begin{cases} 1 & \text{if } x = 0; \\ 0 & \text{if } x \neq 0; \end{cases} \\
\bar{1}(x) &= \begin{cases} 1 & \text{if } x = 1; \\ 0 & \text{if } x \neq 1. \end{cases}
\end{aligned}$$

The algebra $\mathbb{M} = \left(\mathbf{M}, \sqcup, \sqcap, \neg, \bar{0}, \bar{1}\right)$ does not have a lattice structure, since the absorption law ([36,39]) is not fulfilled. However, both operations $\sqcup$ and $\sqcap$ define a partial order on $\mathbf{M}$.

Definition 8. *[34,36] In $\mathbf{M}$, the following two partial orders are defined:*

$$f \sqsubseteq g \ \text{ if } f \sqcap g = f \quad \text{and} \quad f \preceq g \ \text{ if } f \sqcup g = g.$$

Generally, these two partial orders do not coincide [34,36]. The map $\bar{1}$ is the greatest element of partial order $\sqsubseteq$, and the map $\bar{0}$ is the least element of partial order $\preceq$. It is also verified that the constant function $g = 0$ $\left(g(x) = 0, \text{ for all } x \in [0,1]\right)$ is the least element of $\sqsubseteq$ and the greatest element of $\preceq$.

In order to facilitate operations on **M**, the following definition is given.

Definition 9. *[36,38,39] For $f \in \mathbf{M}$, the functions f^L, $f^R \in \mathbf{M}$ are defined as:*

$$f^L(x) = \sup_{\leq} \left\{ f(y) : y \leq x \right\} \quad \textit{and} \quad f^R(x) = \sup_{\leq} \left\{ f(y) : x \leq y \right\}.$$

f^L and f^R are increasing and decreasing, respectively, and $(f^R)^L = (f^L)^R$ is a constant function, equal to the supremum of f.

We denote by **C** the set of all convex functions of **M**. In [36], it is shown that $f \in \mathbf{C}$ if and only if $f = f^L \wedge f^R$, and $\mathbf{C}_w$ will represent the set of all convex functions with given height w.

The set of all normal and convex functions of **M** is denoted by **L**. Obviously, $\mathbf{L} = \mathbf{C}_1$. The algebra $\mathbb{L} = \left(\mathbf{L}, \sqcup, \sqcap, \neg, \bar{0}, \bar{1}\right)$ is a subalgebra of $\mathbb{M}$. In **L**, the partial orders $\sqsubseteq$ and $\preceq$ are equivalent, and $\mathbb{L}$ is a bounded and complete lattice ($\bar{0}$ and $\bar{1}$ are the minimum and the maximum, respectively); see [34,36,39,40]. It should be noted that the characteristic functions of closed intervals in $[0,1]$, which are just the membership degrees of the IT2FSs, belong to **L**.

The following characterization of the partial order on **L** is very helpful to establish new results.

Theorem 1. *[39,40] Let $f, g \in \mathbf{L}$. Then, $f \sqsubseteq g$ if and only if:*

$$g^L \leq f^L \quad \textit{and} \quad f^R \leq g^R.$$

In [36,39], the authors did not consider $\mathbf{C}_w$ for any $0 < w < 1$; however, the study on **L** can be extended to $\mathbf{C}_w$, where the partial orders $\sqsubseteq$ and $\preceq$ are also equivalent. Thus, $\left(\mathbf{C}_w, \sqcup, \sqcap, \neg, \bar{0}_w, \bar{1}_w\right)$ is a bounded and complete lattice, where:

$$\bar{0}_w(x) = \begin{cases} w & \text{if } x = 0; \\ 0 & \text{if } x \neq 0; \end{cases} \qquad \text{and} \qquad \bar{1}_w(x) = \begin{cases} w & \text{if } x = 1; \\ 0 & \text{if } x \neq 1, \end{cases}$$

are the minimum and the maximum, respectively. In addition, $\sqsubseteq$ in $\mathbf{C}_w$ satisfies Theorem 1. From here on, the order $\sqsubseteq$ is denoted by $\sqsubseteq_w$ when working in $\mathbf{C}_w$. Note, however, that we denote it by $\sqsubseteq$ to simplify the notation when there is no chance of confusion.

3. Type 2 Fuzzy Numbers

Haven et al. introduce in [20] an extension of the fuzzy numbers, the interval type 2 fuzzy numbers (IT2FNs).

Definition 10. *[20,22] An interval type 2 fuzzy number (IT2FN) is an IT2FS on $\mathbb{R}$, with membership function μ given by $\mu(x) = \overline{[a(x), b(x)]}$, where the functions $a, b : \mathbb{R} \to [0,1]$ are (type 1) fuzzy numbers.*

The function a is the lower membership function, and b is the upper membership function.

It is worth mentioning that this definition is also given in [19] with $a, b : \mathbb{R} \to [0,1]$, where b is a (type 1) fuzzy number, and a is a (type 1) generalized fuzzy number (see Definition 3), i.e., a can be non-normal. Then, we call to these numbers given in [19] the interval type 2 generalized fuzzy numbers (IT2GFNs).

Note that the set of IT2FNs is a subset of IT2GFNs.

Next, we introduce the definitions of type 2 generalized fuzzy number (T2GFN) and type 2 fuzzy number (T2FN). These sets are new extensions of fuzzy numbers and are also extensions of interval type 2 fuzzy numbers (see Proposition 1).

Definition 11. *Given a T2FS on $\mathbb{R}$, with membership function $\mu : \mathbb{R} \to \mathbf{C}_w$, the support of μ is defined as:*

$$Supp(\mu) = \left\{x \in \mathbb{R} : \ \mu(x) \neq \bar{0}_w\right\}.$$

In addition, μ has bounded support if there exists $a, b \in \mathbb{R}$ such that $\mu(x) = \bar{0}_w$, for all $x \notin [a, b]$.

Remark 1. *Given two closed intervals $[a, b]$ and $[c, d]$, then $\overline{[a,b]}_w \sqsubseteq_w \overline{[c,d]}_w$, for all w, if and only if $[a, b] \leq_I [c, d]$.*

Definition 12. *A type 2 generalized fuzzy number (T2GFN) in $(\mathbf{C}_w, \sqsubseteq_w)$ is a T2FS on $\mathbb{R}$, such that its membership function $\mu : (\mathbb{R}, \leq) \to (\mathbf{C}_w, \sqsubseteq_w)$ has bounded support, and moreover, there exists a $z_\mu \in \mathbb{R}$ such that μ is increasing in $(-\infty, z_\mu]$ and decreasing in $[z_\mu, \infty)$.*

Example 3 shows two type 2 generalized fuzzy numbers.

Remark 2. *Given a T2GFN in $\mathbf{C}_w$, with membership function μ, we have that:*

$$\mu(x) \sqsubseteq_w \mu(z_\mu) \ \text{for} \ x \in (-\infty, z_\mu] \quad \text{and} \quad \mu(x) \sqsubseteq_w \mu(z_\mu) \ \text{for} \ x \in [z_\mu, \infty)$$

and therefore, for all $x \in \mathbb{R}$, $\mu(x) \sqsubseteq_w \mu(z_\mu)$. Consequently, $\mu(z_\mu) = \sup_{\sqsubseteq_w} \left\{\mu(x) \ : \ x \in \mathbb{R}\right\}$.

Besides, the set $Z_\mu = \left\{z \in \mathbb{R} : \ \mu(z) = \sup_{\sqsubseteq_w} \left\{\mu(x) : x \in \mathbb{R}\right\}\right\}$ is a point or an interval (not-empty) where μ is constant.

Remark 3. *A type 2 fuzzy set on $\mathbb{R}$ with membership function,*

$$\mu(x) = \begin{cases} \bar{x}_w & \text{if } x \in [0, 1); \\ \bar{0}_w & \text{otherwise,} \end{cases} \tag{5}$$

is not a T2GFN, since there is no z_μ with the required conditions. Note that μ is increasing in $(-\infty, 1)$, but not in $(-\infty, 1]$, since, for instance, $\frac{1}{2} < 1$, but $\mu(1) = \bar{0}_w \sqsubseteq_w \overline{\left(\frac{1}{2}\right)}_w = \mu(\frac{1}{2})$ (see Figure 4; in this figure, only the image of the support of the functions corresponding to membership degrees $\mu(x)$ has been drawn).

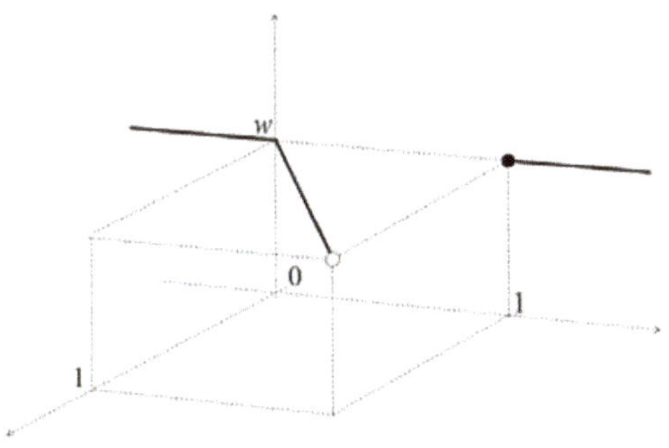

Figure 4. Example of a type 2 fuzzy set (T2FS) that is not a type 2 generalized fuzzy number (T2GFN).

Definition 13. *A type 2 fuzzy number (T2FN) in* $(\mathbf{C}_w, \sqsubseteq_w)$ *is a T2GFN, such that:*

$$\sup_{\sqsubseteq_w}\left\{\mu(x) : x \in \mathbb{R}\right\} = \bar{1}_w.$$

Remark 4. *A type 2 fuzzy set on* $\mathbb{R}$ *with membership function,*

$$\mu(x) = \begin{cases} \overline{(\frac{x}{2})}_w & \text{if } x \in [0,1]; \\ \bar{0}_w & \text{otherwise,} \end{cases} \tag{6}$$

is a type 2 generalized fuzzy number, but not a type 2 fuzzy number, since $\sup_{\sqsubseteq_w}\left\{\mu(x) : x \in \mathbb{R}\right\} = \overline{(\frac{1}{2})}_w$ *(see Figure 5; in this figure, only the image of the support of the functions corresponding to the membership degrees* $\mu(x)$ *has been painted).*

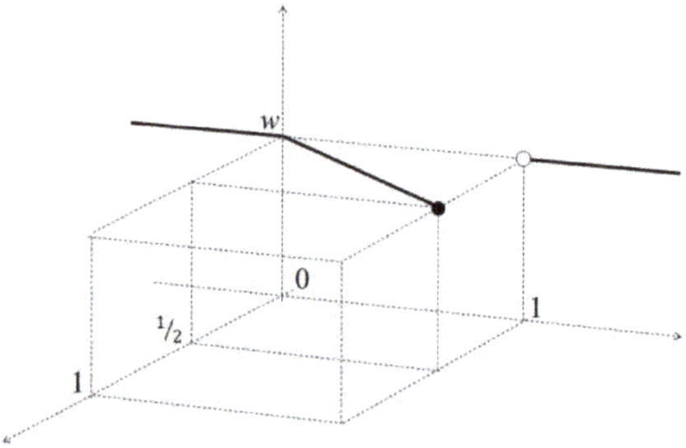

Figure 5. Example of a T2FS that is a T2GFN in $\mathbf{C}_w$.

Proposition 1. *An interval type 2 fuzzy number (IT2FN), defined as in Definition 10, is a T2FN in* $(\mathbf{C}_1, \sqsubseteq_1)$.

Proof. The membership degree $\mu(x)$ of any IT2FN is the characteristic function of an interval. That is, $\mu(x) = \overline{[a(x), b(x)]} \in \mathbf{C}_1 = \mathbf{L}$, where $a(x) \leq b(x)$ for all $x \in \mathbb{R}$. The functions a and b are fuzzy numbers of type 1 (see Definition 4). Since a is strongly normal, we have $a(z) = 1$ for some $z \in \mathbb{R}$. In addition, because $a(z) \leq b(z)$, then $a(z) = b(z) = 1$, which implies that $\mu(z) = \bar{1}_1 = \bar{1}$.

On the other hand, because a and b are fuzzy numbers and $a(z) = b(z) = 1$, we have that a and b are increasing in $(-\infty, z]$ and decreasing in $[z, \infty)$. Thus, $a(x_1) \leq a(x_2)$ and $b(x_1) \leq b(x_2)$ for all $x_1 \leq x_2 \leq z$, which implies that $\overline{[a(x_1), b(x_1)]} \sqsubseteq \overline{[a(x_2), b(x_2)]}$. Consequently, $\mu : ((-\infty, z], \leq) \to (\mathbf{C}_w, \sqsubseteq)$ is increasing.

In a similar way, we can prove that μ is decreasing in $[z, \infty)$.

Therefore, the IT2FNs given in Definition 10 are a particular case of the T2FNs of Definition 13. □

Proposition 2. *An interval type 2 generalized fuzzy number (IT2GFN) is a T2GFN in* $(\mathbf{C}_1, \sqsubseteq_1)$.

Proof. This is similar to the proof made for Proposition 1. □

It should be noted that the set of T2GFNs is a subset of the set of T2FSs on $\mathbb{R}$, whose membership degrees are functions in $\mathbf{C}_w$, so the following partial order, pointwise defined, can be induced: given two T2GFNs A, B, we say that $A \sqsubseteq_w B$ if and only if $\mu_A(x) \sqsubseteq_w \mu_B(x)$ for all $x \in \mathbb{R}$. However, this order is not convenient to compare T2GFNs, since it does not extend the natural order of real numbers, when these are represented as T2GFN, as is shown in Example 2.

Example 2. *Let A and B represent the extension of real numbers 2 and 3 to the T2FSs, respectively, with membership degrees:*

$$\mu_A(x) = \begin{cases} \bar{1} & \text{if } x = 2; \\ \bar{0} & \text{otherwise,} \end{cases} \quad \text{and} \quad \mu_B(x) = \begin{cases} \bar{1} & \text{if } x = 3; \\ \bar{0} & \text{otherwise.} \end{cases}$$

Figure 6a shows the membership degrees $\mu_A(1)$ and $\mu_A(2)$ and only the image of the support of the rest of the membership degrees $\mu_A(x)$. Figure 6b displays the membership degrees $\mu_B(1)$, $\mu_B(2)$ and $\mu_B(3)$ and only the image of the support of the rest of the membership degrees $\mu_B(x)$.

It would be expected that $A \sqsubseteq B$ as $2 \leq 3$, but this does not happen since $\mu_B(2) = \bar{0} \sqsubseteq \bar{1} = \mu_A(2)$ and $\mu_A(3) = \bar{0} \sqsubseteq \bar{1} = \mu_B(3)$, which implies that A and B are not comparable with that partial order.

Furthermore, the preorder given in [19] does not extend the order of $\mathbb{R}$ either. In fact, in the above paper, the authors, before comparing numbers, transform the membership function of any real number, i.e., $\bar{x}$, to the real $\bar{1}$ or $\overline{-1}$, obtaining in this example that $A = B$.

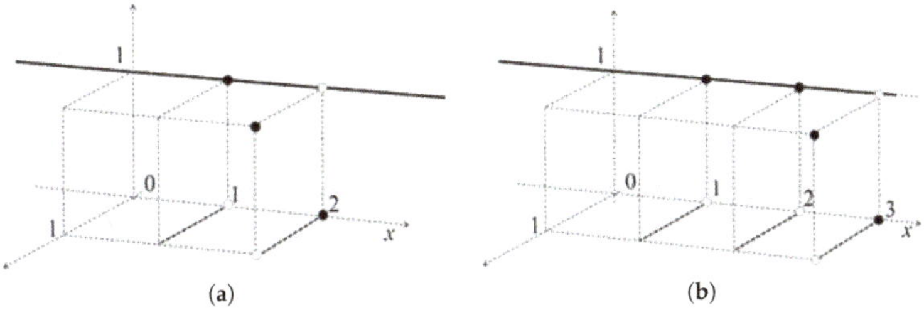

Figure 6. (**a**) Number 2 extended as a T2FN; (**b**) Number 3 extended as a T2FN.

Consequently, a partial order extending the total order of $\mathbb{R}$ needs to be obtained on the T2GFNs. Before, we define two auxiliary functions.

Definition 14. *Given a T2GFN with membership degree $\mu(x) \in (\mathbf{C}_w, \sqsubseteq_w)$, $\forall x \in \mathbb{R}$, we define the functions μ^L, $\mu^R : (\mathbb{R}, \leq) \to (\mathbf{C}_w, \sqsubseteq_w)$ as follows:*

$$\mu^L(x) = \sup_{\sqsubseteq_w} \left\{\mu(y) : y \leq x\right\} = \bigsqcup_{y \leq x} \mu(y),$$

$$\mu^R(x) = \sup_{\sqsubseteq_w} \left\{\mu(y) : y \geq x\right\} = \bigsqcup_{y \geq x} \mu(y).$$

Proposition 3. *Given a T2GFN in $\mathbf{C}_w$ with membership function μ, we have $\mu^L(x)$, $\mu^R(x) \in \mathbf{C}_w$ for all x, and μ^L, μ^R are increasing and decreasing functions, respectively, with respect to the partial order $\sqsubseteq_w$.*

Proof. Since μ is increasing in $(-\infty, z_\mu]$ and decreasing in $[z_\mu, \infty)$ for some $z_\mu \in \mathbb{R}$, then $\mu^L(x) = \mu(x) \in \mathbf{C}_w$, $\forall x \leq z_\mu$ and $\mu^L(x) = \mu(z_\mu) \in \mathbf{C}_w$, for all $x \geq z_\mu$; therefore $\mu^L(x) \in \mathbf{C}_w$, for all $x \in \mathbb{R}$.

Similarly, we have $\mu^R(x) = \mu(x) \in \mathbf{C}_w$ for all $x \geq z_\mu$ and $\mu^R(x) = \mu(z_\mu) \in \mathbf{C}_w$ for all $x \leq z_\mu$; therefore $\mu^R(x) \in \mathbf{C}_w$, for all $x \in \mathbb{R}$. In addition, it follows that μ^L and μ^R are increasing and decreasing functions, respectively, with respect to the partial order $\sqsubseteq_w$. □

Remark 5. *Given a T2GFN in $\mathbf{C}_w$ with membership function μ and $z_\mu \in \mathbb{R}$ a point where μ reaches the supremum, as $\mu(x) \sqsubseteq \sup_{\sqsubseteq} \left\{\mu(x) : x \in \mathbb{R}\right\} = \mu(z_\mu)$, it is clear that:*

$$\sup_{\sqsubseteq}\left\{\mu^L(x):x\in\mathbb{R}\right\} = \sup_{\sqsubseteq}\left\{\mu^R(x):x\in\mathbb{R}\right\}=\sup_{\sqsubseteq}\left\{\mu(x):x\in\mathbb{R}\right\}$$
$$= \mu^L(z_\mu)=\mu^R(z_\mu).$$

Definition 15. *Let A and B be T2GFNs in $\mathbf{C}_w$, with membership function μ_A and μ_B respectively. We define the relation $\tilde{\sqsubseteq}$ as:*

$$A\,\tilde{\sqsubseteq}\,B\ \left(\acute{o}\ \ \mu_A\,\tilde{\sqsubseteq}\,\mu_B\ \right)\ \Leftrightarrow \mu_A^R(x)\ \sqsubseteq_w\ \mu_B^R(x)\quad \text{and}\quad \mu_B^L(x)\ \sqsubseteq_w\ \mu_A^L(x),$$

for all $x\in\mathbb{R}$.

Proposition 4. *Let A and B be T2GFNs in $\mathbf{C}_w$, with membership functions μ_A and μ_B, respectively. If A and B are comparable with $\tilde{\sqsubseteq}$, then:*

$$\sup_{\sqsubseteq_w}\left\{\mu_A(x):x\in\mathbb{R}\right\} = \sup_{\sqsubseteq_w}\left\{\mu_B(x):x\in\mathbb{R}\right\}.$$

Proof. If $A\,\tilde{\sqsubseteq}\,B$ then $\mu_A^R(x)\ \sqsubseteq\ \mu_B^R(x)$ and $\mu_B^L(x)\ \sqsubseteq\ \mu_A^L(x)$ for all $x\in\mathbb{R}$. From the Remark 5, we have:

$$\sup_{\sqsubseteq}\left\{\mu_A(x):x\in\mathbb{R}\right\} = \sup_{\sqsubseteq}\left\{\mu_A^R(x):x\in\mathbb{R}\right\} \sqsubseteq \sup_{\sqsubseteq}\left\{\mu_B^R(x):x\in\mathbb{R}\right\} = \sup_{\sqsubseteq}\left\{\mu_B(x):x\in\mathbb{R}\right\},$$

$$\sup_{\sqsubseteq}\left\{\mu_B(x):x\in\mathbb{R}\right\} = \sup_{\sqsubseteq}\left\{\mu_B^L(x):x\in\mathbb{R}\right\} \sqsubseteq \sup_{\sqsubseteq}\left\{\mu_A^L(x):x\in\mathbb{R}\right\} = \sup_{\sqsubseteq}\left\{\mu_A(x):x\in\mathbb{R}\right\}.$$

Since $\sqsubseteq$ is antisymmetric, we have:

$$\sup_{\sqsubseteq_w}\left\{\mu_A(x):x\in\mathbb{R}\right\} = \sup_{\sqsubseteq_w}\left\{\mu_B(x):x\in\mathbb{R}\right\}.$$

□

Proposition 5. *Let A and B be two T2GFNs in $\mathbf{C}_w$, with membership functions μ_A and μ_B, respectively. If $A\tilde{\sqsubseteq}B$, then $z_{\mu_A}\le z_{\mu_B}$, being $z_{\mu_A}=\inf_{\le}\{x\in\mathbb{R}:\mu_A(x)=\sup_{\sqsubseteq_w}\mu_A\}$ and $z_{\mu_B}=\inf_{\le}\{x\in\mathbb{R}:\mu_B(x)=\sup_{\sqsubseteq_w}\mu_B\}$.*

Proof. If $A\tilde{\sqsubseteq}B$, then $\mu_B^L(x)\ \sqsubseteq_w\ \mu_A^L(x)$. Suppose $z_{\mu_B}<z_{\mu_A}$, then there exists $x\in\mathbb{R}$ such that $z_{\mu_B}<x<z_{\mu_A}$ and $\mu_B^L(x)=\sup_{\sqsubseteq_w}\mu_B=\sup_{\sqsubseteq_w}\mu_A$ by Proposition 4, but $\mu_A^L(x)\sqsubsetneq\sup_{\sqsubseteq_w}\mu_A$. Therefore, $\mu_A^L(x)\sqsubsetneq\mu_B(x)$, which is impossible. □

Proposition 6. *$\tilde{\sqsubseteq}$ is a partial order in the set of T2GFNs in $\mathbf{C}_w$.*

Proof. Let A,B,C be three T2GFNs in $\mathbf{C}_w$, with membership functions μ_A,μ_B and μ_C, respectively.

Reflexivity: As $\sqsubseteq$ is a partial order in $\mathbf{C}_w$, it is reflexive, and thus, $\mu_A^R(x)\ \sqsubseteq\ \mu_A^R(x)$ and $\mu_A^L(x)\sqsubseteq\mu_A^L(x)$, for all x. Therefore, $A\,\tilde{\sqsubseteq}\,A$.

Transitivity: If $C\,\tilde{\sqsubseteq}\,A$ and $A\,\tilde{\sqsubseteq}\,B$, then, for all $x\in\mathbb{R}$, we have that:

$$\mu_C^R(x)\ \sqsubseteq\ \mu_A^R(x),\qquad \mu_A^L(x)\ \sqsubseteq\ \mu_C^L(x),$$
$$\mu_A^R(x)\ \sqsubseteq\ \mu_B^R(x),\qquad \mu_B^L(x)\ \sqsubseteq\ \mu_A^L(x).$$

Since $\sqsubseteq$ is transitive, we have that:

$$\mu_C^R(x) \sqsubseteq \mu_B^R(x) \quad \text{and} \quad \mu_B^L(x) \sqsubseteq \mu_C^L(x),$$

for all x. Therefore, $C \tilde{\sqsubseteq} B$.

Antisymmetry: If $A \tilde{\sqsubseteq} B$, then $\mu_A^R(x) \sqsubseteq \mu_B^R(x)$ and $\mu_B^L(x) \sqsubseteq \mu_A^L(x)$. Additionally, if $B \tilde{\sqsubseteq} A$, then $\mu_B^R(x) \sqsubseteq \mu_A^R(x)$ and $\mu_A^L(x) \sqsubseteq \mu_B^L(x)$.

Since $\sqsubseteq$ is antisymmetric, then:

$$\mu_A^R(x) = \mu_B^R(x) \quad \text{and} \quad \mu_A^L(x) = \mu_B^L(x),$$

for all x.

As A and B are T2GFNs, their membership functions are increasing in $(-\infty, z_a]$ and $(-\infty, z_b]$, respectively, and decreasing in $[z_a, \infty)$ and $[z_b, \infty)$, respectively, for some $z_a, z_b \in \mathbb{R}$. Therefore, $\mu_A^L(x) = \mu_A(x)$, for all $x \leq z_a$; $\mu_B^L(x) = \mu_B(x)$, for all $x \leq z_b$; $\mu_A^R(x) = \mu_A(x)$, for all $x \geq z_a$; and $\mu_B^R(x) = \mu_B(x)$, for all $x \geq z_b$.

Without loss of generality, we suppose $z_a \leq z_b$. If $x \in (-\infty, z_a]$, then we have $\mu_A(x) = \mu_A^L(x) = \mu_B^L(x) = \mu_B(x)$. If $x \in [z_b, \infty)$, then we have $\mu_A(x) = \mu_A^R(x) = \mu_B^R(x) = \mu_B(x)$.

On the other hand, if $z_a < x < z_b$, we obtain $\mu_B(x) \sqsubseteq \mu_B^R(x) = \mu_A^R(x) = \mu_A(x)$ and $\mu_A(x) \sqsubseteq \mu_A^L(x) = \mu_B^L(x) = \mu_B(x)$. Since $\sqsubseteq$ is antisymmetric, we have that $\mu_A(x) = \mu_B(x)$ for all x. Therefore, $A = B$. $\square$

Remark 6. *The partial order of Definition 15 extends the order of $\mathbb{R}$. In fact, let A and B be the extensions in the T2FNs, of the real numbers a and b, respectively, such that $a \leq b$, with membership functions:*

$$\mu_A(x) = \begin{cases} \bar{1} & \text{if } x = a; \\ \bar{0} & \text{otherwise,} \end{cases} \quad \text{and} \quad \mu_B(x) = \begin{cases} \bar{1} & \text{if } x = b; \\ \bar{0} & \text{otherwise.} \end{cases}$$

In this case, we have that:

$$\mu_A^L(x) = \begin{cases} \bar{1} & \text{if } x \geq a; \\ \bar{0} & \text{otherwise,} \end{cases} \qquad \mu_B^L(x) = \begin{cases} \bar{1} & \text{if } x \geq b; \\ \bar{0} & \text{otherwise,} \end{cases}$$

$$\mu_A^R(x) = \begin{cases} \bar{1} & \text{if } x \leq a; \\ \bar{0} & \text{otherwise,} \end{cases} \qquad \mu_B^R(x) = \begin{cases} \bar{1} & \text{if } x \leq b; \\ \bar{0} & \text{otherwise.} \end{cases}$$

One can prove that $\mu_B^L(x) \sqsubseteq \mu_A^L(x)$ and $\mu_A^R(x) \sqsubseteq \mu_B^R(x)$ for all x. Then, $A \tilde{\sqsubseteq} B$, but $A \not\sqsubseteq B$ and $B \not\sqsubseteq A$ (see Example 2).

Note that the methods provided in [19,22] for ranking IT2FNs are total preorders; however, to the best of our knowledge, there is no previous work where a partial order is established extending the order of the real numbers to T2FNs and not just to IT2FNs or IT2GFNs. Moreover, as has already been pointed out, the preorder in [19] does not extend the order of $\mathbb{R}$; meanwhile, it can be proven that the preorder in [22] does extend it. Example 3 shows the ranking of two T2GFNs that are not IT2FNs.

Example 3. *Consider the two T2GFNs A and B given by the degrees of membership $\mu(x), \varphi(x) \in C_{1/2}$, respectively, defined below. Note that A and B are not IT2FNs (see Definition 10), because $w = 0.5 \neq 1$, and some membership degrees of A are not characteristic functions of a point or an interval.*

Let $\mu(x) : [0,1] \to [0,1]$ be defined by:

- if $x \in \mathbb{R}\backslash(0,7)$, then $\mu(x) = \bar{0}_{1/2}$;
- if $x \in (0,3)$, then $(\mu(x))(y) = \begin{cases} \dfrac{10y}{x} - 1 & \text{if } y \in [\frac{x}{10}, \frac{15x}{100}]; \\ 2 - \frac{10y}{x} & \text{if } y \in [\frac{15x}{100}, \frac{2x}{10}]; \\ 0 & \text{if } y \notin [\frac{x}{10}, \frac{2x}{10}]; \end{cases}$
- if $x \in [3,5]$, then $\mu(x) = \overline{[0.7, 0.9]_{1/2}}$;
- if $x \in (5,7)$, then $\mu(x) = \overline{(2.1 - 0.3x)_{1/2}}$.

In Figure 7, we only show the membership degrees of the set A, $\mu(x)$, when $x \in (0,3)$. In the other case, when $x \notin (0,3)$, the membership degrees are characteristic functions of a point or an interval with height $w = 0.5$. In Figure 8, on the left, we have drawn the images of the supports of the membership degrees of the set A, as well as the projection of these values on the plane xy.

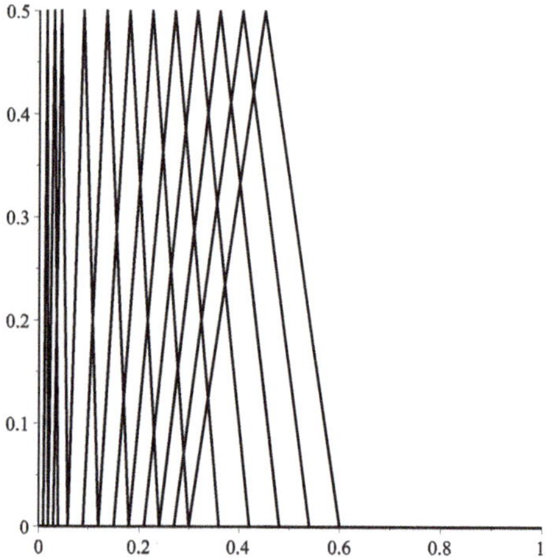

Figure 7. Membership degrees $\mu(x)$, for all $x \in (0,3)$.

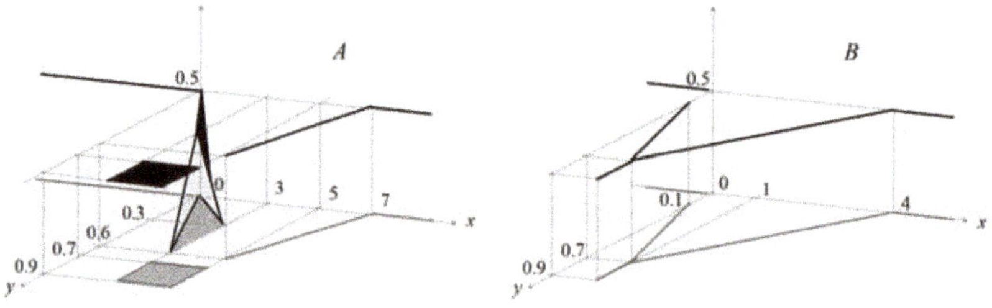

(**a**) (**b**)

Figure 8. Membership functions of T2GFNs (**a**,**b**) given in Example 3.

Additionally, let $\varphi(x) : [0,1] \to [0,1]$ be defined by:

- if $x \in \mathbb{R}\backslash[0,4]$, then $\varphi(x) = \bar{0}_{1/2}$;
- if $x \in [0,1)$, then $\varphi(x) = \overline{(0.6x + 0.1)_{1/2}}$;
- if $x = 1$ then $\varphi(1) = \overline{[0.7, 0.9]_{1/2}}$;
- if $x \in (1,4]$, then $\varphi(x) = \overline{\frac{7}{30}(4 - x)_{1/2}}$.

All membership degrees of the set B are characteristic functions of a point or an interval with height $w = 0.5$. On the right of Figure 8, we have drawn the image of the support of each membership degree of the set B, as well as the projection of these values on the plane xy.

In this example, $w = 1/2$, $Z_\mu = [3,5]$ and $Z_\varphi = \{1\}$; therefore, for all $z_\mu \in [3,5]$ and $z_\varphi = 1$, it is:

$$\sup\left\{\mu(x) : x \in \mathbb{R}\right\} = \mu(z_\mu) = \overline{[0.7, 0.9]}_{1/2} = \varphi(1) = \sup\left\{\varphi(x) : x \in \mathbb{R}\right\},$$

In addition, we have:

$\mu^L : \mathbb{R} \to [0,1]^{[0,1]}$ is such that (see Figure 9):

- if $x < 3$, then $\mu^L(x) = \mu(x)$;
- if $x \geq 3$, then $\mu^L(x) = \mu(3) = \overline{[0.7, 0.9]}_{1/2}$;

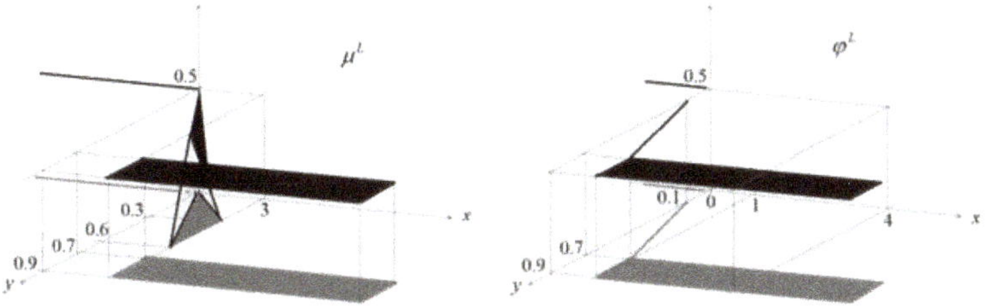

(a) (b)

Figure 9. (**a**) μ^L and (**b**) φ^L according to Example 3.

$\mu^R : \mathbb{R} \to [0,1]^{[0,1]}$ is such that (see Figure 10):

- if $x \leq 5$, then $\mu^R(x) = \mu(5) = \overline{[0.7, 0.9]}_{1/2}$;
- if $x > 5$, then $\mu^R(x) = \mu(x)$.

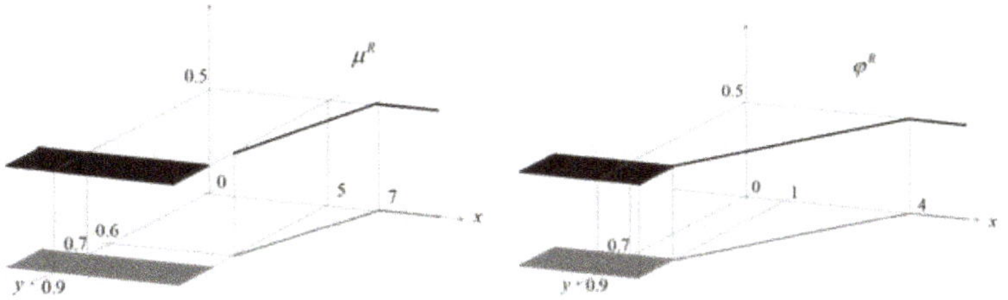

(a) (a)

Figure 10. (**a**) μ^R and (**b**) φ^R according to Example 3.

$\varphi^L : \mathbb{R} \to [0,1]^{[0,1]}$ is such that (see Figure 9):

- if $x < 1$, then $\varphi^L(x) = \varphi(x)$;
- if $x \geq 1$, then $\varphi^L(x) = \varphi(1) = \overline{[0.7, 0.9]}_{1/2}$;

$\varphi^R : \mathbb{R} \to [0,1]^{[0,1]}$ is such that (see Figure 10):

- if $x \leq 1$, then $\varphi(1) = \overline{[0.7, 0.9]}_{1/2}$;
- if $x > 1$, then $\varphi^R(x) = \varphi(x)$.

Comparing $\mu^L(x)$ with $\varphi^L(x)$ and $\mu^R(x)$ with $\varphi^R(x)$ respect to the partial order $\sqsubseteq_{1/2}$, we can prove $\mu^L(x) \sqsubseteq_{1/2} \varphi^L(x)$ and $\varphi^R(x) \sqsubseteq_{1/2} \mu^R(x)$ for all $x \in \mathbb{R}$. Therefore,

$$B \mathrel{\tilde{\tilde{\sqsubseteq}}} A.$$

For example, from Figure 11, it follows that $(\mu^L(2))^R(y) \leq (\varphi^L(2))^R(y)$ and $(\varphi^L(2))^L(y) \leq (\mu^L(2))^L(y)$, for all $y \in [0,1]$, and then, $\mu^L(2) \sqsubseteq_{1/2} \varphi^L(2)$.

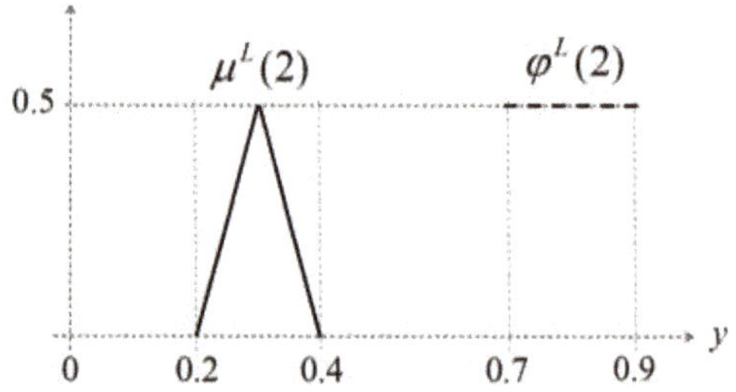

Figure 11. $\mu^L(2)$ and $\varphi^L(2)$ according to Example 3.

Remark 7. *As mentioned before, in [19,22], total preorders for ranking IT2FNs were provided. One of the disadvantages of these preorders is that they do not satisfy the property of antisymmetry; therefore, two elements can be considered as equal, when they are not, which undoubtedly affects the decision making. On the other hand, from a theoretical point of view, operating or working with preorders adversely affects the consolidation of a consistent and coherent algebraic structure. A solid algebraic structure allows, among other things, the correct determination and application of operators (e.g., negations, t-norms, aggregation operators, among others). Besides, these preorders given in [19,22] reduce each IT2FN to a real value and finally compare these real values with the order of the real numbers, which implies some loss of the information (representation of the uncertainty, impression and vagueness) contained in the original form of each IT2FN. On the other hand, the partial order guarantees the algebraic properties of posets. Furthermore, with the partial order* $\tilde{\tilde{\sqsubseteq}}$*, the loss of information is minor compared to those preorders. One of the disadvantages of this partial order is that it is not total; therefore, there are elements that are non-comparable with it. Considering the above, the ideal is to have a total order, extending the order of the real numbers in the T2FNs; however, a partial order is a good starting point to obtain some total order. Anyway, it would be suitable, for ranking IT2FNs or IT2GFNs, firstly to apply the order* $\tilde{\tilde{\sqsubseteq}}$*, and if they are not comparable, then to apply the above-mentioned preorders.*

It should be noted that in [19], it was established that any method to ranking IT2FNs must be consistent with the human intuition, in the sense that the more to the right an IT2FN is, the greater it will be. That is, the more to the right the centroid of an IT2FN is, the greater it will be. This property is satisfied by the order $\tilde{\tilde{\sqsubseteq}}$*.*

Example 4 shows two IT2GFNs previously ranked in [19] with the preorder CPS (centroid point and spread). We rank them with $\tilde{\tilde{\sqsubseteq}}$, obtaining the same decision as the one achieved with the CPS method. However, before, in the next Proposition 7, we give the characterization of $\tilde{\tilde{\sqsubseteq}}$, when ranking IT2GFNs exclusively.

Proposition 7. *Let A, B be two IT2GFNs, with membership degrees* $\mu(x) = \overline{[a_1(x), b_1(x)]}$ *and* $\phi(x) = \overline{[a_2(x), b_2(x)]}$*, respectively,* $\forall x \in \mathbb{R}$*, then:*

$$A \tilde{\tilde{\sqsubseteq}} B \Leftrightarrow a_2^L(x) \leq a_1^L(x),\ a_1^R(x) \leq a_2^R(x),\ b_2^L(x) \leq b_1^L(x),\ b_1^R(x) \leq b_2^R(x),\ \forall x \in \mathbb{R} (\Leftrightarrow a_1 \sqsubseteq_w a_2, b_1 \sqsubseteq b_2).$$

Proof. This is directly according to Definition 15, Remark 1, Proposition 5 and the usual partial order in the closed intervals, $\leq_I$. □

Example 4. *Let A, B be two IT2GFNs, with membership degrees $A(x) = \overline{[a_1(x), b_1(x)]}$ and $B(x) = \overline{[a_2(x), b_2(x)]}$, $\forall x$, respectively, whose supports are shown in Figure 12. In [19], Section 4, Case 3, these IT2GFNs, A, B, were ranked with the CPS method, and it was obtained that B is greater than A. On the other hand, we apply the order $\widetilde{\sqsubseteq}$, according to Figures 13 and 14 and Proposition 7, and we obtain the same above result, i.e, $A \widetilde{\sqsubseteq} B$ ($A \neq B$). In Figures 13 and 14, note that $a_2^L(x) \leq a_1^L(x)$, $a_1^R(x) \leq a_2^R(x)$, $b_2^L(x) \leq b_1^L(x)$, $b_1^R(x) \leq b_2^R(x)$, $\forall x$; therefore, $A \widetilde{\sqsubseteq} B$ ($A \neq B$).*

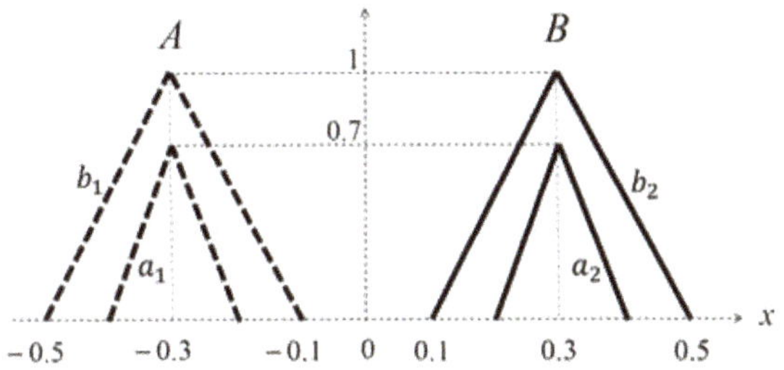

Figure 12. Support of two IT2GFNs A and B, with the functions a_1, b_1, a_2, b_2.

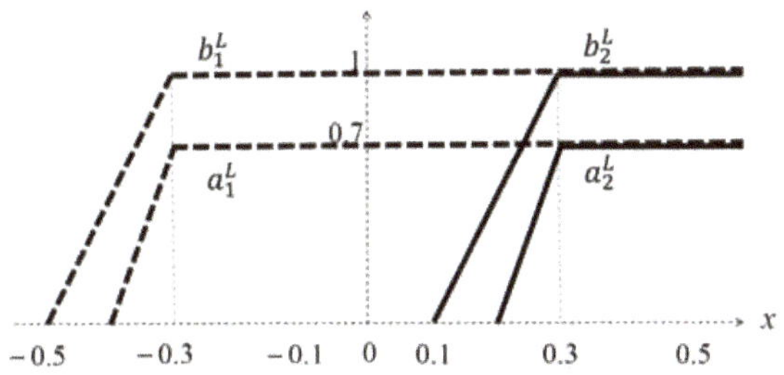

Figure 13. Functions $a_1^L, b_1^L, a_2^L, b_2^L$.

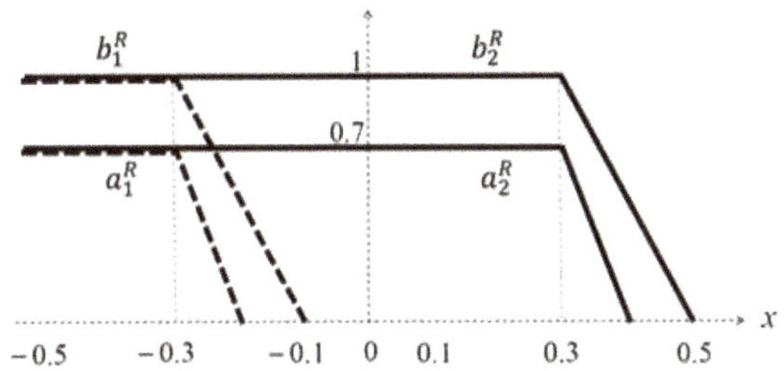

Figure 14. Functions $a_1^R, b_1^R, a_2^R, b_2^R$.

Now, comparing the IT2FNs A and B in Figure 15 with the preorder given in [22] ($\prec_{P_D}$), it results $A \prec_{P_D} B$ and $B \not\prec_{P_D} A$ and also $A \widetilde{\sqsubsetneq} B$. Moreover, it can be proven that for any IT2FNs A and B, if $A \widetilde{\sqsubseteq} B$, then $A \prec_{P_D} B$.

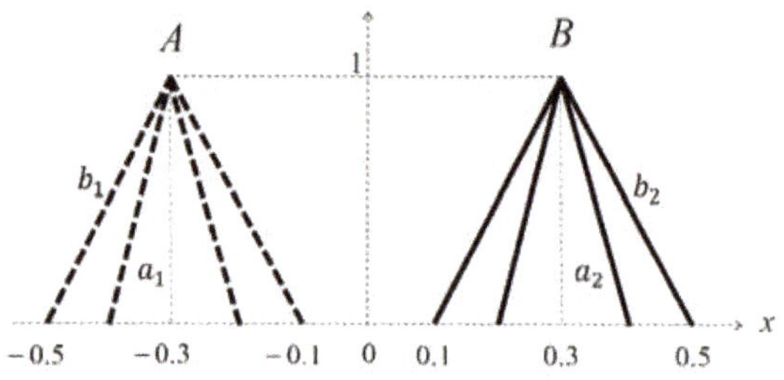

Figure 15. Support of two IT2FNs A and B.

Finally, Table 1 shows the properties fulfilled by each ranking method mentioned in this paper. Remember that:

$\tilde{\sqsubseteq}$ denotes the relation defined in Definition 15;
$\sqsubseteq_w$ denotes pointwise extension of the partial order $\sqsubseteq_w$;
$\leq_I$ denotes pointwise extension of the partial order of the intervals $\leq_I$;
$\prec_{P_{CPS}}$ denotes the relation given in [19];
$\prec_{P_D}$ denotes the relation given in [22].

Table 1. Comparative results for several ranking methods, "Y" = Yes, "N" = Not.

Method	Properties				Sets Can Be Compared			Extends Order of $\mathbb{R}$
	Reflex.	Antisymmetry	Trans.	Total	IT2Ns	IT2GFNs	T2GFNs-T2FNs	
$\tilde{\sqsubseteq}$	Y	Y	Y	N	Y	Y	Y	Y
$\sqsubseteq_w$	Y	Y	Y	N	Y ($\sqsubseteq = \leq_I$)	Y ($\sqsubseteq = \leq_I$)	Y	N
$\leq_I$	Y	Y	Y	N	Y ($\sqsubseteq = \leq_I$)	Y ($\sqsubseteq = \leq_I$)	N	N
$\prec_{P_D}$	Y	N	Y	Y	Y	N	N	Y
$\prec_{CPS}$	Y	N	Y	Y	Y	Y	N	N

4. Conclusions

In this work, we have defined, within the type 2 fuzzy sets (T2FSs), the fuzzy numbers (T2FNs) and the generalized fuzzy numbers (T2GFNs). It has been shown that the order on these sets naturally induced from the order on their membership degrees (pointwise extension) is not adequate, since it does not extend the usual order of real numbers. That is why a new partial order has been proposed on T2GFNs. Previously, taking into account this goal, we have generalized the partial order on $\mathbf{L}$ (set of convex and normal functions from $[0,1]$ to $[0,1]$) to the set of convex functions with the same height. Although, there are methods to ranking interval type 2 fuzzy numbers, the partial order obtained becomes a new option to be applied to ranking such type 2 fuzzy numbers and not only IT2FNs or IT2GFNs.

Some topics remain open for future research. Among them to study some properties of the proposed order and to obtain different ways to compare T2GFNs, allowing us to rank as many type 2 fuzzy sets as possible.

Acknowledgments: This paper was partially supported by UPM (Spain) and UNET (Venezuela).

Author Contributions: This work was carried out in collaboration between Pablo Hernández, Susana Cubillo, Carmen Torres-Blanc and José A. Guerrero. P. Hernández conceived and proposed the first ideas, results and examples of the paper. S. Cubillo, C. Torres-Blanc and A.J. Guerrero completed and improved the results and examples. All authors have read and approved the final manuscript.

Conflicts of Interest: The authors declare no conflict of interest.

References

1. Jain, R. A procedure for multi-aspect decision making using fuzzy sets. *Int. J. Syst. Sci.* **1978**, *8*, 1–7.
2. Jain, R. Decision-making in the presence of fuzzy variables. *IEEE Trans. Syst. Man Cybern. SMC 6* **1976**, *6*, 698–703.
3. Dubois, D.; Prade, H. Operations on fuzzy numbers. *Int. J. Syst. Sci.* **1978**, *9*, 613–626.
4. Abbasbandy, S.; Hajjari, T. A new approach for ranking of trapezoidal fuzzy numbers. *Comput. Math. Appl.* **2009**, *57*, 413–419.
5. Anzilli, L.; Facchinetti, G. Ambiguity of fuzzy quantities and a new poposal for their ranking. *Przegal Elektrotechniczny (Electr. Rev. B)* **2012**, *10*, 280–283.
6. Anzilli, L.; Facchinetti, G. The total variation of bounded variation functions to evaluate and rank fuzzy quantities. *Int. J. Intell. Syst.* **2013**, *28*, 927–956.
7. Dubois, D.; Prade, H. The mean value of a fuzzy number. *Fuzzy Sets Syst.* **1987**, *24*, 279–300.
8. Fortemps, P.; Roubens, M. Ranking and defuzzification methods based on area compensation. *Fuzzy Sets Syst.* **1996**, *82*, 319–330.
9. Kerre, E.; Mareš, M.; Mesiar, R. Generate fuzzy quantities an their orderings. In *Information, Uncertainty and Fusion*; Bouchon-Meunier, B., Yager, R., Zadeh, L., Eds.; Springer Science and Business Media: Dordrecht, The Netherlands, 2012; pp. 119–130.
10. Lee, E.; Li, R. Comparison of fuzzy numbers based on the probability measure of fuzzy events. *Comput. Math. Appl.* **1988**, *15*, 887–896.
11. Requena, I.; Delgado, M.; Verdegay, J. Automatic ranking of fuzzy numbers with the criterion of decision-maker learnt by an artificial neural network. *Fuzzy Sets Syst.* **1994**, *64*, 1–19.
12. Bortolan, G.; Degani, R. A review of some methods for ranking fuzzy numbers. *Fuzzy Sets Syst.* **1985**, *15*, 1–19.
13. Chen, S.J.; Hwang, C.L. Fuzzy ranking methods. In *Fuzzy Multiple Attribute Decision Making*; Spinger: Berlin, Germany, 1992; pp. 101–288.
14. Wang, X.; Kerre, E. Reasonable properties for the ordering of fuzzy quantities (I). *Fuzzy Sets Syst.* **2001**, *118*, 375–385.
15. Wang, X.; Kerre, E. Reasonable properties for the ordering of fuzzy quantities (II). *Fuzzy Sets Syst.* **2001**, *118*, 387–405.
16. Liu, H.-C. Type 2 generalized intuitionistic fuzzy choquet integral operator for multi-criteria decision making. In Proceedings of the IEEE International Symposium on Parallel and Distributed Processing with Applications (ISPA), Taipei, Taiwan, 6–9 September 2010; pp. 605–611.
17. Liu, H.-C. Liu's generalized intuitionistic fuzzy sets. *J. Educ. Meas. Statist.* **2010**, *18*, 1–14.
18. Mondal, T.; Samanta, S. Generalized intuitionistic fuzzy sets. *J. Fuzzy Math.* **2002**, *10*, 839–861.
19. Abu, A.; Naim, K.; Gegov, A. Ranking of interval type 2 fuzzy numbers based on centroid point and spread. In Proceedings of the 2015 7th International Joint Conference on Computational Intelligence (IJCCI), Lisbon, Portugal, 12–14 November 2015; pp. 131–140.
20. Haven, T.; Anderson, D.; Keller, J. A fuzzy choquet integral with an interval type 2 fuzzy number-valued integrand. In Proceedings of the 2010 IEEE International Conference on Fuzzy Systems, Barcelona, Spain, 18–23 July 2010; pp. 1–8.
21. Haven, T.; Anderson, D.; Wagner, C. Data-informed fuzzy measures for fuzzy integration of intervals and fuzzy numbers. *IEEE Trans. Fuzzy Syst.* **2015**, *23*, 1861–1875.
22. Hesamian, G. Measuring similarity and ordering based on interval type 2 fuzzy numbers. *IEEE Trans. Fuzzy Syst.* **2016**, doi:10.1109/TFUZZ.2016.2578342.
23. Linda, O.; Manic, M. Monotone Centroid Flow Algorithm for Type Reduction of General Type-2 Fuzzy Sets. *IEEE Trans. Fuzzy Syst.* **2012**, *20*, 805–819.
24. Linda, O.; Manic, M. General type 2 fuzzy C-means algorithm for uncertain fuzzy clustering. *IEEE Trans. Fuzzy Syst.* **2012**, *20*, 883–897.
25. Liu, F. An efficient centroid type reduction strategy for general type 2 fuzzy logic system. *Inf. Sci.* **2018**, *178*, 2224–2236.
26. Niewiadomski, A. On finity, countability, cardinalities, and cylindric extensions of type 2 fuzzy sets in linguistic summarization of databases. *IEEE Trans. Fuzzy Syst.* **2010**, *18*, 532–545.

27. Wagner, C.; Hagras, H. Toward general type 2 fuzzy logic systems based on zSlices. *IEEE Trans. Fuzzy Syst.* **2010**, *18*, 637–660.
28. Zadeh, L. Fuzzy sets. *Inf. Control* **1965**, *20*, 301–312.
29. Zadeh, L. The concept of a linguistic variable and its application to approximate reasoning-I. *Inf. Sci.* **1975**, *8*, 199–249.
30. Klir, G. The role of constrained fuzzy arithmetic in engineering. In *Uncertainty Analysis in Engineering and Sciences: Fuzzy Logic, Statistics, and Neural Network Approach*; Ayyub, B., Gupta, M., Eds.; Kluwer: Dordrecht, The Netherlands, 1997; pp. 1–19.
31. Klir, G.; Pan, Y. Constrained fuzzy arithmetic. Basic questions and some answers. *Soft Comput.* **1998**, *2*, 100–108.
32. Navara, M. Computation with fuzzy quantities. In Proceedings of the 7th conference of the European Society for Fuzzy Logic and Technology (EUSFLAT-2011) and "les Rencontres Francophones sur la Logique Floue et ses Applications" (LFA-2011), Aix-les-Bains, France, 18–22 July 2011; pp. 209–214.
33. Mizumoto, M.; Tanaka, K. Fuzzy sets of type 2 under algebraic product and algebraic sum. *Fuzzy Sets Syst.* **1981**, *5*, 277–290.
34. Mizumoto, M.; Tanaka, K. Some properties of fuzzy sets of type 2. *Inf. Control* **1976**, *31*, 312–340.
35. Mendel, J.; Jhon, R. Type-2 fuzzy sets made Simple. *IEEE Trans. Fuzzy Syst.* **2002**, *10*, 117–127.
36. Walker, C.; Walker, E. The algebra of fuzzy truth values. *Fuzzy Sets Syst.* **2005**, *149*, 309–347.
37. Bustince, H.; Fernandez, J.; Hagras, H.; Herrera, F.; Pagola, M.; Barrenechea, E. Interval type 2 fuzzy sets are generalization of interval-valued fuzzy sets: Toward a wider view on their relationship. *IEEE Trans. Fuzzy Syst.* **2015**, *23*, 1876–1882.
38. Gera, Z.; Dombi, J. Type-2 implications on non-interactive fuzzy truth values. *Fuzzy Sets Syst.* **2008**, *159*, 3014–3032.
39. Harding, J.; Walker, C.; Walker, E. Lattices of convex normal functions. *Fuzzy Sets Syst.* **2008**, *159*, 1061–1071.
40. Harding, J.; Walker, C.; Walker, E. Convex normal functions revisited. *Fuzzy Sets Syst.* **2010**, *161*, 1343–1349.

axioms

MDPI

Article

Orness For Idempotent Aggregation Functions

Leire Legarreta [1], Inmaculada Lizasoain [2,*] and Iraide Mardones-Pérez [1]

[1] Departamento de Matemáticas, Universidad del País Vasco-Euskal Herriko Unibertsitatea, Apdo. 644, 48080 Bilbao, Spain; leire.legarreta@ehu.eus (L.L.); iraide.mardones@ehu.eus (I.M.-P.)
[2] Departamento de Matemáticas, Institute for Advanced Materials INAMAT, Universidad Pública de Navarra, Campus de Arrosadía, 31006 Pamplona, Spain
* Correspondence: ilizasoain@unavarra.es

Received: 23 August 2017; Accepted: 17 September 2017; Published: 20 September 2017

Abstract: Aggregation functions are mathematical operators that merge given data in order to obtain a global value that preserves the information given by the data as much as possible. In most practical applications, this value is expected to be between the infimum and the supremum of the given data, which is guaranteed only when the aggregation functions are idempotent. Ordered weighted averaging (OWA) operators are particular cases of this kind of function, with the particularity that the obtained global value depends on neither the source nor the expert that provides each datum, but only on the set of values. They have been classified by means of the orness—a measurement of the proximity of an OWA operator to the OR-operator. In this paper, the concept of orness is extended to the framework of idempotent aggregation functions defined both on the real unit interval and on a complete lattice with a local finiteness condition.

Keywords: aggregation functions; lattice operators; idempotence; orness

1. Introduction

Aggregation functions [1] are a family of operators that allow us to fuse data—either quantitative or qualitative—in order to obtain a global value that captures the information of the given data as faithfully as possible.

These functions have been widely used in decision making [2,3], where it is necessary to merge the different opinions of several experts, or in image processing tasks [4], where the values of different pixels must be fused in order to obtain a single one.

OWA (Ordered weighted averaging) operators, defined by Yager in [5] for real values, are a kind of aggregation function that include the most widely used ones, such as the arithmetic mean or the order statistics. An OWA operator is simply a weighted average of data which have been previously ordered in a descending way. The main property of OWA operators is the symmetry, which makes them independent of the order in which the data are provided.

Real OWA operators have been classified by Yager in [6] by means of the orness—a measurement of their proximity to the OR-operator. In this way, the orness of the OR-operator is 1, while the orness of the AND-operator is 0 and the orness of the arithmetic mean is equal to $1/2$.

In [7], OWA operators have been extended from $[0,1]$ to a complete lattice $(L, \leq_L)$ endowed with both a t-norm T and a t-conorm S. The concept of orness has also been generalized to this kind of operator in two ways. On the one hand, a qualitative orness has been defined for OWA operators when the lattice L is finite [4]. On the other hand, when the lattice satisfies a weaker locally-finiteness condition, each OWA operator may be assigned a quantitative orness [8].

The present paper analyzes the possibility of extending both concepts of orness to the wider set of all the aggregation functions lying between the AND and the OR-operators, for which the measurement of their proximity to the OR-operator has sense. These aggregation functions—not necessarily symmetric—are precisely those that are idempotent (see Remark 1). Such maps will be the object of our study when they are defined both on the real case and on some lattices that satisfy the necessary conditions of finiteness.

The extension of the concept of orness to all of the idempotent aggregation functions makes it possible to classify two wide classes of operators: the class consisting of all the Choquet integrals defined on the real unit interval on the one hand, and all the discrete Sugeno integrals defined on an appropriate lattice on the other hand.

The rest of the paper is organized as follows. Section 2 presents the preliminary concepts that will be used further. In Section 3, Yager's concept of orness is extended to idempotent aggregation functions defined on the real unit interval. Section 4 generalizes the concept of qualitative orness to idempotent aggregation functions defined on finite lattices. The extension of the concept of quantitative orness to this kind of function is analyzed in Section 5 where they are defined on lattices with a local finiteness condition. The paper finishes with some conclusions and references.

2. Preliminaries

In this paper, $(L, \leq_L)$ denotes a complete lattice (i.e., a partially ordered set in which all subsets have both a supremum and an infimum). The least element of L is denoted by 0_L and the greatest one by 1_L. When L is finite, we will refer to $(L, \leq_L)$ as a finite bounded lattice. In some cases, we will require the lattice $(L, \leq_L)$ to be *distributive*, which means that any of the following two equivalent properties holds:

$$a \wedge (b \vee c) = (a \wedge b) \vee (a \wedge c) \text{ for any } a,\ b,\ c \in L$$
$$a \vee (b \wedge c) = (a \vee b) \wedge (a \vee c) \text{ for any } a,\ b,\ c \in L$$

A map $T : L \times L \to L$ (resp. $S : L \times L \to L$) is said to be a *t-norm* (resp. *t-conorm*) on $(L, \leq_L)$ if it is commutative, associative, increasing in each component, and has a neutral element 1_L (resp. 0_L). For any $n > 2$, $S(a_1, \ldots, a_n)$ will be used to denote $S(S(\ldots(S(S(a_1, a_2), a_3), \ldots), a_{n-1}), a_n)$.

Throughout this paper, $(L, \leq_L, T, S)$ will denote a complete lattice endowed with a t-norm and a t-conorm.

Recall that an n-ary aggregation function is a map $M : L^n \to L$ such that:

(i) $M(a_1, \ldots, a_n) \leq_L M(a'_1, \ldots, a'_n)$ whenever $a_i \leq_L a'_i$ for $1 \leq i \leq n$.
(ii) $M(0_L, \ldots, 0_L) = 0_L$ and $M(1_L, \ldots, 1_L) = 1_L$.

The function M is said to be *idempotent* if $M(a, \ldots, a) = a$ for every $a \in L$. Besides, M is said to be *symmetric* if for every permutation σ of the set $\{1, \ldots, n\}$, the equality $M(a_1, \ldots, a_n) = M(a_{\sigma(1)}, \ldots, a_{\sigma(n)})$ holds for every $(a_1, \ldots, a_n) \in L^n$.

For example, the AND-operator, the OR-operator and, for $L = [0, 1]$ (or $L \subset \mathbb{R}$) the arithmetic mean are symmetric idempotent aggregation functions.

Remark 1. *Note that by using property (i) of the definition of aggregation function, we have the following equivalence:*

$$M \text{ is idempotent} \iff \inf\{a_1, \ldots, a_n\} \leq_L M(a_1, \ldots, a_n) \leq_L \sup\{a_1, \ldots, a_n\}, \quad \text{for any } (a_1, \ldots, a_n) \in L^n,$$

i.e., the aggregation function M is idempotent iff it lies in between the AND and the OR-operators. Indeed, if M is idempotent, $\inf\{a_1, \ldots, a_n\} = M(\inf\{a_1, \ldots, a_n\}, \ldots, \inf\{a_1, \ldots, a_n\}) \leq M(a_1, \ldots, a_n) \leq M(\sup\{a_1, \ldots, a_n\}, \ldots, \sup\{a_1, \ldots, a_n\}) = \sup\{a_1, \ldots, a_n\}$. *Conversely, for any* $a \in L$, $a = \inf\{a, \ldots, a\} \leq M(a, \ldots, a) \leq \sup\{a, \ldots, a\} = a$.

A particular case of idempotent symmetric aggregation functions are OWA operators, which were defined by Yager for values in the real unit interval $[0,1]$ as follows:

Definition 1 (Yager [5]). *For each weighting vector* $\alpha = (\alpha_1, \dots, \alpha_n) \in [0,1]^n$, *with* $\alpha_1 + \cdots + \alpha_n = 1$, *the map* $F_\alpha : [0,1]^n \to [0,1]$ *given by*

$$F_\alpha(a_1, \dots, a_n) = \alpha_1 a_{\sigma(1)} + \cdots + \alpha_n a_{\sigma(n)}$$

is called an ordered weighted averaging operator *or* OWA operator, *where the n-tuple* $(a_{\sigma(1)}, \dots, a_{\sigma(n)})$ *is a rearrangement of the set* $(a_1, \dots, a_n)$ *of inputs such that* $a_{\sigma(1)} \geq \cdots \geq a_{\sigma(n)}$.

Note that both the AND-operator and the OR-operator are some examples of OWA operators, provided respectively by the weighting vector $(0, \dots, 0, 1)$ and by the weighting vector $(1, 0, \dots, 0)$. The arithmetic mean is also an OWA operator, provided by the weighting vector $(1/n, \dots, 1/n)$.

OWA operators were generalized to complete lattices in [7] in the following way.

Definition 2. *Let* $(L, \leq_L, T, S)$ *be a complete lattice endowed with a t-norm and a t-conorm. A vector* $\alpha = (\alpha_1, \dots, \alpha_n) \in L^n$ *is said to be a* distributive weighting vector *if conditions* $S(\alpha_1, \dots, \alpha_n) = 1_L$ *and*

$$T(a, S(\alpha_1, \dots, \alpha_n)) = S(T(a, \alpha_1), \dots, T(a, \alpha_n)) \text{ for any } a \in L$$

are satisfied.

Definition 3 ([7]). *Let* $(L, \leq_L, T, S)$ *be a complete lattice endowed with a t-norm and a t-conorm. For each distributive weighting vector* $\alpha = (\alpha_1, \dots, \alpha_n) \in L^n$, *the n-ary* OWA operator $F_\alpha : L^n \to L$ *is defined by*

$$F_\alpha(a_1, \dots, a_n) = S(T(\alpha_1, b_1), \dots, T(\alpha_n, b_n)),$$

where for each $(a_1, \dots, a_n) \in L^n$, *the elements* $b_n \leq_L \cdots \leq_L b_1$ *are calculated by means of the n-ary k-th statistics* $\{A_k : L^n \to L \mid 1 \leq k \leq n\}$ *given below:*

- $b_1 = A_1(a_1, \dots, a_n) = a_1 \vee \cdots \vee a_n \in L.$
- $b_2 = A_2(a_1, \dots, a_n) = \Big((a_1 \wedge a_2) \vee \cdots \vee (a_1 \wedge a_n)\Big) \vee \Big((a_2 \wedge a_3) \vee \cdots \vee (a_2 \wedge a_n)\Big) \vee \cdots \vee \Big(a_{n-1} \wedge a_n\Big) \in L.$

  ⋮
- $b_k = A_k(a_1, \dots, a_n) = \bigvee\{a_{j_1} \wedge \cdots \wedge a_{j_k} \mid j_1 < \cdots < j_k \in \{1, \dots, n\}\} \in L.$

  ⋮
- $b_n = A_n(a_1, \dots, a_n) = a_1 \wedge \cdots \wedge a_n \in L.$

It is worth mentioning that when $(L, \leq_L)$ is a distributive lattice and the t-norm and t-conorm considered on L are, respectively, the meet and the join (i.e., $T = \wedge$ and $S = \vee$), OWA operators are particular cases of a wider class of idempotent aggregation functions: the discrete Sugeno integrals.

Definition 4 ([9]). *Let* $(L, \leq_L)$ *be a complete lattice,* $P_n = \{1, \dots, n\}$ *and let* $\mathcal{P}_n$ *be the set of all the subsets of* P_n.

(i) *An* L-fuzzy measure *on* P_n *is a map* $\mu : \mathcal{P}_n \to L$ *with* $\mu(\emptyset) = 0_L$, $\mu(P_n) = 1_L$ *and* $\mu(E) \leq \mu(F)$ *whenever* $E \subseteq F \subseteq P_n$.
In particular, if $L = [0,1]$, *it is called a* fuzzy measure *[10,11].*

(ii) *The discrete Sugeno integral* $S_\mu : L^n \to L$ *is given by*

$$S_\mu(a_1, \ldots, a_n) = \bigvee \left\{ \bigwedge \{a_i \mid i \in X\} \wedge \mu(X) : X \in \mathcal{P}_n \right\}.$$

In fact, when L is a distributive complete lattice, a discrete Sugeno integral is an OWA operator if and only if it is symmetric (see [7]).

3. Orness for Idempotent Aggregation Functions Defined on the Real Unit Interval

The great deal of weighting vectors that can define an OWA operator on the real interval $[0,1]$ makes difficult the election for each particular situation.

With the purpose of classifying OWA operators defined on the real unit interval, in [6] Yager introduced an orness measure for each OWA operator F_α. This measure depends only on the weighting vector $\alpha = (\alpha_1, \ldots, \alpha_n)$, in the following way:

$$\text{orness}(F_\alpha) = \frac{1}{n-1} \sum_{i=1}^{n} (n-i)\alpha_i. \tag{1}$$

It is easy to check that the orness of each OWA operator is a real value situated between 0, corresponding to the AND-operator given by the infimum, and 1, corresponding to the OR-operator given by the supremum. In general, the orness is a measure of the proximity of each OWA operator to the OR-operator. For instance, the orness of the arithmetic mean, provided by the weighting vector $(1/n, \ldots, 1/n)$, is equal to $1/2$.

In order to extend this concept to any idempotent aggregation function, note that for any OWA operator $F_\alpha : [0,1]^n \to [0,1]$, we can write

$$\text{orness}(F_\alpha) = F_\alpha \left(1, \frac{n-2}{n-1}, \ldots, \frac{1}{n-1}, 0\right),$$

which leads to the following generalization.

Definition 5. *Let* $F : [0,1]^n \to [0,1]$ *be an idempotent aggregation function. Define*

$$\text{orness}(F) = F \left(1, \frac{n-2}{n-1}, \ldots, \frac{1}{n-1}, 0\right).$$

Remark 2. *(i) Since F always takes values in* $[0,1]$*, it is clear that* $0 \leq \text{orness}(F) \leq 1$*. Moreover, the orness of the AND-operator is equal to 0, whereas the orness of the OR-operator is 1.*

(ii) It is immediate to see that $\text{orness}(F) \leq \text{orness}(G)$ *if* $F \leq G$ *(with the pointwise order). However, as Example 1 below will show, the converse is not true.*

(iii) Definition 5 could have been given for any aggregation function $F : [0,1]^n \to [0,1]$*, but it does not make any sense to measure the proximity of F to the OR-operator when there does not exist any order relation between them.*

Moreover, if $F(a_1, \ldots, a_n) \leq \inf(a_1, \ldots, a_n)$ *for all* $(a_1, \ldots, a_n)$*, we would have* $\text{orness}(F) = 0$*. Similarly, if* $F(a_1, \ldots, a_n) \geq \sup(a_1, \ldots, a_n)$ *for all* $(a_1, \ldots, a_n)$*, then* $\text{orness}(F)$ *would be equal to 1.*

For the aforementioned reasons, we have extended the definition of orness only to the class of idempotent aggregation functions, which are exactly those lying between the AND and the OR-operators.

The previous extension of the concept of orness makes it possible to classify a wider kind of idempotent aggregation functions which include OWA operators: the so called discrete Choquet integrals. Their definition is as follows:

Definition 6 ([12]). *Let $m : \mathcal{P}_n \to [0,1]$ be a fuzzy measure. The discrete Choquet integral with respect to m is the map $C_m : [0,1]^n \to [0,1]$ defined by*

$$C_m(x_1,\ldots,x_n) = \sum_{i=1}^{n} x_{\sigma(i)} \left(m(\{\sigma(i),\ldots,\sigma(n)\}) - m(\{\sigma(i+1),\ldots,\sigma(n)\})\right)$$

where $\sigma : \{1,\ldots,n\} \to \{1,\ldots,n\}$ is a permutation such that $x_{\sigma(1)} \leq \cdots \leq x_{\sigma(n)}$ and $\{\sigma(n+1),\sigma(n)\} = \emptyset$ by convention.

Remark 3. *As was shown in [13], a discrete Choquet integral is an OWA operator if and only if it is symmetric. The latter is equivalent to the fuzzy measure being symmetric, i.e., $m(E) = m(F)$ for any $E, F \subseteq P_n$ such that $card(E) = card(F)$.*

Example 1. *Let $n = 4$ and consider the fuzzy measures $m, \nu : \mathcal{P}_4 \to [0,1]$ given by*

$$\begin{aligned}
&m(\{1\}) = m(\{4\}) = 0 = \nu(\{1\}) = \nu(\{2\}) \\
&m(\{2\}) = m(\{3\}) = \tfrac{1}{4} = \nu(\{3\}) = \nu(\{4\}); \\
&m(\{i,j\}) = \tfrac{1}{3} = \nu(\{i,j\}) \quad 1 \leq i < j \leq 4; \\
&m(\{i,j,k\}) = \tfrac{1}{2} = \nu(\{i,j,k\}) \quad 1 \leq i < j < k \leq 4; \\
&m(P_4) = 1 = \nu(P_4).
\end{aligned}$$

The orness of each of the corresponding non-symmetric discrete Choquet integrals $C_m, C_\nu : [0,1]^4 \to [0,1]$ is given by

$$\begin{aligned}
orness(C_m) &= C_m\left(1, \tfrac{2}{3}, \tfrac{1}{3}, 0\right) \\
&= \tfrac{1}{3}\left(\tfrac{1}{2} - \tfrac{1}{3}\right) + \tfrac{2}{3}\left(\tfrac{1}{3} - 0\right) = \tfrac{5}{18} \\
&= C_\nu\left(1, \tfrac{2}{3}, \tfrac{1}{3}, 0\right) \\
&= orness(C_\nu)
\end{aligned}$$

so, in particular, $orness(C_\nu) \leq orness(C_m)$. However,

$$\begin{aligned}
C_m\left(0,0,\tfrac{1}{2},1\right) &= \tfrac{1}{2}\left(\tfrac{1}{3} - 0\right) = \tfrac{1}{6} \\
C_\nu\left(0,0,\tfrac{1}{2},1\right) &= \tfrac{1}{2}\left(\tfrac{1}{3} - \tfrac{1}{4}\right) + 1\left(\tfrac{1}{4} - 0\right) = \tfrac{7}{24}
\end{aligned}$$

and therefore, $C_\nu \not\leq C_m$. The latter shows that the converse of Remark 2(ii) is not true.

4. Qualitative Orness for Idempotent Aggregation Functions Defined on Finite Lattices

In this section, $(L, \leq_L)$ will be a bounded finite lattice.

In order to classify OWA operators when they are defined on this kind of lattice, a qualitative orness was introduced in [4] generalizing the one given by Yager for the real case. In this section, this concept is extended to any idempotent aggregation function defined on $(L, \leq_L)$. Note that, by [4], Remark 3, the definition given in [4] can be reformulated so that it still has sense in this wider context.

Definition 7. *Let $L = \{a_1, \ldots, a_l\}$ be a finite bounded lattice and let $1_L = b_1 \geq_L \cdots \geq_L b_l = 0_L$ be the descending chain introduced in Definition 3, involving all the elements of the lattice. For any idempotent n-ary aggregation function $F : L^n \to L$, a qualitative* orness measure *is calculated by means of a descending chain $d_1 \geq_L \cdots \geq_L d_n$ that consists of some* equidistant elements *in the lattice, which are performed following some steps:*

(i) *Call $s = l(n-1)$.*

(ii) *Consider the descending chain* $c_1 \geq_L \cdots \geq_L c_s$ *defined by*

$$c_1 = \cdots = c_{n-1} = b_1;\ c_n = \cdots = c_{2(n-1)} = b_2; \ldots; c_{(l-1)(n-1)+1} = \cdots = c_{l(n-1)} = b_l.$$

Note that $c_1 = b_1 = 1_L$ *and* $c_{l(n-1)} = b_l = 0_L$.

(iii) *Build a descending subchain of* $\{c_1, \ldots, c_s\}$, $d_1 \geq_L \cdots \geq_L d_n$ *by means of*

$$d_1 = 1_L,\ d_2 = c_l,\ d_3 = c_{2l}, \ldots, d_n = c_{(n-1)l} = 0_L;$$

i.e., $d_1 = 1_L$, *and for each* $j \in \{1, \ldots, n-1\}$, $d_{j+1} = c_{jl} = b_k$ *with* $k = 1 + \left\lfloor \frac{jl-1}{n-1} \right\rfloor$, *where the symbol* $\left\lfloor \frac{a}{b} \right\rfloor$ *denotes the integer part of* $\frac{a}{b}$.

(iv) *Call* $orness(F) = F(d_1, \ldots, d_n)$.

As in [4], we have the following observations.

Remark 4. *(i) Note that the same chain* $d_1 \geq_L \cdots \geq_L d_n$ *is obtained if we let s be any common multiple of* $\{l, n-1\}$. *In that case, if* $s = le$ *with* $e \in \mathbb{N}$, *the chain* $c_1 \geq_L \cdots \geq_L c_s$ *must be* $c_1 = \cdots = c_e = b_1;\ c_{e+1} = \cdots = c_{2e} = b_2; \ldots; c_{(l-1)e+1} = \cdots = c_{le} = b_l$ *and, if* $s = (n-1)h$ *with* $h \in \mathbb{N}$, *then the chain* $d_1 \geq_L \cdots \geq_L d_n$ *must be*

$$d_1 = 1_L,\ d_2 = c_h,\ d_3 = c_{2h}, \ldots, d_n = c_{(n-1)h} = 0_L.$$

Therefore, the chain $d_1 \geq_L \cdots \geq_L d_n$ *obtained in this way is the same as that obtained in Definition 7. Indeed, for each* $j \in \{1, \ldots, n-1\}$, $d_{j+1} = c_{jh} = b_k$ *with* $k = 1 + \left\lfloor \frac{jh-1}{e} \right\rfloor$.

(ii) When the aggregation function is an OWA operator, the previous definition agrees with that given in [4] for OWA operators. Hence, in particular, the orness corresponding to the AND-operator is 0, and the one corresponding to the OR-operator is 1.

(iii) Definition 7 could have been given for any aggregation function (not necessarily idempotent), but the same reasons explained in Remark 2(iii) lead us to consider this restriction.

Example 2. *Let* $L = \{0_L, a, b, c, 1_L\}$ *be a partially ordered set with the relation given by* $0_L \leq a \leq b \leq 1_L$ *and* $0_L \leq c \leq 1_L$. *Note that* $(L, \leq_L)$ *is a non-distributive lattice with* $l = 5$. *Consider the discrete Sugeno integral* $S_\mu : L^3 \to L$ *defined by means of the L-fuzzy measure* $\mu : \mathcal{P}_3 \to L$ *with*

$$\mu(\{1\}) = \mu(\{3\}) = 0_L;\ \mu(\{2\}) = a;$$
$$\mu(\{i, j\}) = a,\ 1 \leq i < j \leq 3;\ \mu(\{1,2,3\}) = 1_L,$$

$$S_\mu(a_1, a_2, a_3) = (a_2 \wedge a) \vee (a_1 \wedge a_2 \wedge a) \vee (a_1 \wedge a_3 \wedge a) \vee (a_2 \wedge a_3 \wedge a) \vee (a_1 \wedge a_2 \wedge a_3).$$

Note that since S_μ *is not symmetric, it is not an OWA operator. Nevertheless, we can compute its qualitative orness with the new definition we have just given. In order to calculate it, we follow the next steps:*

Find $b_1 = 1_L, b_2 = 1_L, b_3 = a, b_4 = 0_L, b_5 = 0_L$.
Call $s = l(n-1) = 5 \cdot 2 = 10$.
Consider the descending chain $c_1 \geq_L \cdots \geq_L c_{10}$ *defined by*
$c_1 = c_2 = b_1;\ c_3 = c_4 = b_2;\ c_5 = c_6 = b_3;\ c_7 = c_8 = b_4;\ c_9 = c_{10} = b_5$.
Build the descending subchain $d_1 \geq_L d_2 \geq_L d_3$, *by means of*
$d_1 = 1_L;\ d_2 = c_5 = a;\ d_3 = c_{10} = 0_L$.

Therefore, $orness(S_\mu) = S_\mu(1_L, a, 0_L) = a$.

In [14] it is shown that the class of all the n-ary aggregation functions defined on a complete lattice $(L, \leq_L)$—denoted by $\mathcal{A}_n(L)$—is also a complete lattice with the pointwise order relation; i.e., with the relation $\leq$ defined for any $F, G \in \mathcal{A}_n(L)$ as $F \leq G$ if and only if $F(a_1, \dots, a_n) \leq G(a_1, \dots, a_n)$ for every $(a_1, \dots, a_n) \in L^n$.

The smallest and the greatest aggregation functions in $\mathcal{A}_n(L)$ are, respectively (see [15]):

$$A_\perp(a_1, \dots, a_n) = \begin{cases} 1, & \text{if } a_i = 1 \text{ for all } 1 \leq i \leq n \\ 0, & \text{otherwise} \end{cases}$$

$$A_\top(a_1, \dots, a_n) = \begin{cases} 0, & \text{if } a_i = 0 \text{ for all } 1 \leq i \leq n \\ 1, & \text{otherwise} \end{cases}$$

We use the symbols $\wedge$, $\vee$ to denote both the meet and the join on L and the following operations on $\mathcal{A}_n(L)$:

$$(F \wedge G)(a_1, \dots, a_n) = F(a_1, \dots, a_n) \wedge G(a_1, \dots, a_n)$$
$$(F \vee G)(a_1, \dots, a_n) = F(a_1, \dots, a_n) \vee G(a_1, \dots, a_n),$$

for any $(a_1, \dots, a_n) \in L^n$.

Note [14] that if $(L, \leq_L)$ is a distributive lattice, then $\left(\mathcal{A}_n(L), \leq\right)$ is a distributive lattice as well.

Proposition 1. *Let $(L, \leq_L)$ be a bounded finite lattice and consider $F, G : L^n \to L$, idempotent aggregation functions.*

(i) *If $F \leq G$, then $orness(F) \leq orness(G)$.*
(ii) *Both $F \wedge G$ and $F \vee G$ are idempotent.*
(iii) *It holds that*

$$orness\left(F \wedge G\right) = orness(F) \wedge orness(G)$$
$$orness\left(F \vee G\right) = orness(F) \vee orness(G)$$

Proof. Consider the chain $d_1 \geq_L \cdots \geq_L d_n$ obtained in Definition 7.

(i) Since $F \leq G$, then $\text{orness}(F) = F(d_1, \dots, d_n) \leq_L G(d_1, \dots, d_n) = \text{orness}(G)$.
(ii) For any $a \in L$, $(F \wedge G)(a, \dots, a) = F(a, \dots, a) \wedge G(a, \dots, a) = a \wedge a = a$ by using the idempotency of F and G. In the same way, the idempotency of $F \vee G$ can be proved.
(iii) By Definition 7, $\text{orness}(F \wedge G) = (F \wedge G)(d_1, \dots, d_n) = F(d_1, \dots, d_n) \wedge G(d_1, \dots, d_n) = \text{orness}(F) \wedge \text{orness}(G)$. Analogously, $\text{orness}(F \vee G) = (F \vee G)(d_1, \dots, d_n) = F(d_1, \dots, d_n) \vee G(d_1, \dots, d_n) = \text{orness}(F) \vee \text{orness}(G)$.

□

Remark 5. *As for the orness messure of Section 3, the converse of Proposition 1 (i) is not true. Indeed, let $L = \{0_L, a, b, c, 1_L\}$ and S_μ be, respectively, the lattice and the discrete Sugeno integral defined in Example 2.*

Consider now the non-symmetric discrete Sugeno integral $S_\nu : L^3 \to L$ defined by means of the L-fuzzy measure $\nu : \mathcal{P}_3 \to L$ with

$$\nu(\{i\}) = 0_L, 1 \leq i \leq 3;\ \nu(\{1, 2\}) = a,\ \nu(\{1, 3\}) = \nu(\{2, 3\}) = c;\ \nu(\{1,2,3\}) = 1_L,$$

$$S_\nu(a_1, a_2, a_3) = (a_1 \wedge a_2 \wedge a) \vee (a_2 \wedge a_3 \wedge c) \vee (a_1 \wedge a_3 \wedge c) \vee (a_1 \wedge a_2 \wedge a_3).$$

Therefore, $orness(S_\nu) = S_\nu(1_L, a, 0_L) = a = orness(S_\mu)$.
However, $S_\mu(c, a, c) = a \not\leq c = S_\nu(c, a, c)$.

5. Quantitative Orness for Idempotent Aggregation Functions

In this section, $(L, \leq_L)$ will be a complete lattice satisfying the following additional condition of local finiteness, which is called *maximal finite chain* condition or simply *condition (MFC)*:

(MFC) *For any* $a, b \in L$ *with* $a \leq_L b$*, there exists some maximal chain with a finite length, named* l*, between* a *and* b*,*

$$a = a_0 <_L a_1 <_L \cdots <_L a_l = b.$$

The maximality means that for any $0 \leq i \leq l-1$ there is no $c \in L$ with $a_i <_L c <_L a_{i+1}$. Obviously, if $a = b$, the maximal chain has length $l = 0$. The length of the shortest maximal chain between a and b—which will be denoted by $d(a, b)$—will be called *distance from a to b*.

Remark 6. *Note that any finite lattice satisfies condition (MFC).*

A quantitative orness was defined in [8] in the following way for any OWA operator defined on $(L, \leq_L, T, S)$.

Definition 8 ([8]). *Let* $(L, \leq_L, T, S)$ *be a complete lattice, endowed with a t-norm and a t-conorm satisfying condition (MFC). For any distributive weighting vector,* $\alpha = (\alpha_1, \ldots, \alpha_n) \in L^n$*, define the* qualitative quantifier $Q_\alpha : \{0, 1, \ldots, n\} \to L$ *by means of*

$$Q_\alpha(0) = 0_L,$$
$$Q_\alpha(k) = S(\alpha_1, \ldots, \alpha_k) \text{ for } k = 1, \ldots, n.$$

Definition 9 ([8]). *Let* $(L, \leq_L, T, S)$ *be a complete lattice satisfying condition (MFC). For any distributive weighting vector* $\alpha = (\alpha_1, \ldots, \alpha_n) \in L^n$*, let* F_α *be the n-ary OWA operator associated to it. Consider the qualitative quantifier* $Q_\alpha : \{0, 1, \ldots, n\} \to L$ *defined in Definition 8. For each* $k = 1, \ldots, n$*, call* $m(k) = d(Q_\alpha(k-1), Q_\alpha(k))$*. If* $m = m(1) + \cdots + m(n)$*, then define*

$$orness(F_\alpha) = \frac{1}{n-1} \sum_{i=1}^{n} (n-i) \frac{m(i)}{m}.$$

The following remark contains the key points that allow us to generalize this concept to the setting of idempotent aggregation functions.

Remark 7 ([8]). *Let* $\alpha = (\alpha_1, \ldots, \alpha_n) \in L^n$ *be a distributive weighting vector in* $(L, \leq_L, T, S)$*. Then*

(i) Q_α *is a monotonically increasing function.*
(ii) *For any* $k = 1, \ldots, n$, $Q_\alpha(k) = F_\alpha(1_L, \ldots, \overset{(k)}{1_L}, 0_L, \ldots, 0_L)$.
(iii) *Consequently, if* $\beta = (\beta_1, \ldots, \beta_n) \in L^n$ *is a distributive weighting vector in* $(L, \leq_L)$ *with* $F_\alpha = F_\beta$*, then* $Q_\alpha = Q_\beta$.

Definition 10. *Let* $(L, \leq_L)$ *be a complete lattice satisfying condition (MFC). For any idempotent aggregation function* $F : L^n \to L$*, consider the qualitative quantifier* $Q : \{0, 1, \ldots, n\} \to L$ *defined by means of*

$$Q(0) = 0_L,$$
$$Q(k) = F(1_L, \ldots, \overset{(k)}{1_L}, 0_L, \ldots, 0_L) \text{ for } k = 1, \ldots, n.$$

For each $k = 1, \ldots, n$, *call* $m(k) = d(Q(k-1), Q(k))$. *If* $m = m(1) + \cdots + m(n)$, *then define*

$$orness(F) = \frac{1}{n-1} \sum_{i=1}^{n} (n-i) \frac{m(i)}{m}.$$

Remark 8. *(i) Note that for arbitrary idempotent aggregation functions, the definition of orness does not directly depend on the t-norm and the t-conorm considered on the lattice L.*

(ii) When the aggregation function is an OWA operator, Remark 7(ii) shows that the previous definition agrees with the one given in [8] for OWA operators. In particular, as shown in [8], Proposition 3.6, the orness corresponding to the AND-operator is 0 and the one corresponding to the OR-operator is 1.

(iii) Observe that, contrary to the case of OWA operators (see [8]), Proposition 3.6(ii), there might be idempotent aggregation functions different from the AND-operator whose orness is 0. For instance, take $L = \{0_L, c, 1_L\}$ *with* $0_L \leq c \leq 1_L$. *Define* $F : L^3 \to L$ *as* $F(a_1, a_2, a_3) = a_3$. *Then, F is an idempotent aggregation function different from the AND-operator whose orness is 0.*

(iv) Definition 10 could have been given for any aggregation function, not necessarily idempotent, but the same reasons explained in Remark 2(iii) lead us to consider this restriction.

Proposition 2. *Let* $(L, \leq_L)$ *be a complete lattice satisfying condition (MFC). For any idempotent aggregation function* $F : L^n \to L$, *it is satisfied that* $orness(F) \in [0, 1]$.

Proof. Clearly, $\text{orness}(F) \geq 0$. Moreover,

$$\text{orness}(F) = \frac{1}{m(n-1)} \sum_{i=1}^{n} (n-i) m(i) \leq \frac{1}{m(n-1)} \sum_{i=1}^{n} (n-1) m(i) = 1$$

as desired. □

Example 3. *Let* $L = \{0_L, a, b, 1_L\} \cup \{c_i : i \in \mathbb{N}\}$, *with* $c_i \neq c_j$ *for* $i \neq j$, *be the lattice described as follows:* $0_L \leq a \leq b \leq 1_L$ *and* $0_L \leq c_i \leq 1_L$ *for any* $i \in \mathbb{N}$. *Then, L is a non-finite lattice satisfying condition (MFC). Consider the discrete Sugeno integral* $S_\mu : L^3 \to L$ *defined by means of the L-fuzzy measure* $\mu : \mathcal{P}_3 \to L$ *with*

$$\mu(\{1\}) = \mu(\{3\}) = 0_L;\ \mu(\{2\}) = a;$$
$$\mu(\{i, j\}) = a, 1 \leq i < j \leq 3;\ \mu(\{1,2,3\}) = 1_L,$$

$$S_\mu(a_1, a_2, a_3) = (a_2 \wedge a) \vee (a_1 \wedge a_2 \wedge a) \vee (a_1 \wedge a_3 \wedge a) \vee (a_2 \wedge a_3 \wedge a) \vee (a_1 \wedge a_2 \wedge a_3).$$

Even if S_μ *is not an OWA operator (since it is not symmetric), we can compute its quantitative orness with the new definition we have just given. In order to calculate it, we follow these steps:*

Define $Q : \{0, 1, 2, 3\} \to L$ *by means of* $Q(0) = 0_L$, $Q(1) = S_\mu(1_L, 0_L, 0_L) = 0_L$,
$Q(2) = S_\mu(1_L, 1_L, 0_L) = a$, $Q(3) = S_\mu(1_L, 1_L, 1_L) = 1_L$.
Calculate $m(1) = d(Q(0), Q(1)) = 0$; $m(2) = d(Q(1), Q(2)) = 1$;
$m(3) = d(Q(2), Q(3)) = 2$ *and* $m = m(1) + m(2) + m(3) = 3$.
Therefore, $orness(S_\mu) = \frac{1}{2}\left(2 \cdot \frac{m(1)}{3} + 1 \cdot \frac{m(2)}{3}\right) = \frac{1}{6}$.

6. Conclusions

The concept of orness defined by Yager for OWA operators on the real interval $[0, 1]$ can be extended to the more general setting of idempotent aggregation functions, all of them lying between the AND and the OR-operators. This includes the class of all the discrete Choquet integrals, which are not necessarily symmetric. In addition, on finite bounded lattices, the concept of qualitative orness defined for OWA operators can be extended to this wide set of all the idempotent aggregation functions. Finally, for complete lattices satisfying a weaker condition of local finiteness, the concept of quantitative orness defined for OWA operators can be extended to all the idempotent aggregation functions. The latter includes the set of all the discrete Sugeno integrals defined on that kind of lattice, which are not necessarily symmetric.

For future research, the application of these concepts of orness to practical situations may be analyzed.

Acknowledgments: This work has been partially supported by the research projects MTM2015-63608-P of the Spanish Government and IT974-16 of the Basque Government.

Author Contributions: Leire Legarreta, Inmaculada Lizasoain and Iraide Mardones-Pérez contributed equally to this work.

Conflicts of Interest: The authors declare no conflict of interest.

References

1. Beliakov, G.; Pradera, A.; Calvo, T. *Aggregation Functions: A Guide for Practitioners*; Springer: New York, USA, 2008.
2. De Miguel, L.; Bustince, H.; Pekala, B.; Bentkowska, U.; Da Silva, I.; Bedregal, B.; Mesiar, R.; Ochoa, G. Interval-Valued Atanassov Intuitionistic OWA Aggregations Using Admissible Linear Orders and Their Application to Decision Making. *IEEE Trans. Fuzzy Syst.* **2016**, *24*, 1586–1597, doi:10.1109/TFUZZ.2016.2543744.
3. De Miguel, L.; Sesma-Sara, M.; Elkano, M.; Asiain, M.J.; Bustince, H. An algorithm for group decision making using n-dimensional fuzzy sets, admissible orders and OWA operators. *Inf. Fusion* **2017**, *37*, 126–131, doi:10.1016/j.inffus.2017.01.007.
4. Ochoa, G.; Lizasoain, I.; Paternain, D.; Bustince, H.; Pal, N.R. From Quantitative to Qualitative orness for lattice OWA operators. *Int. J. Gen. Syst.* **2017**, doi:10.1080/03081079.2017.1319364.
5. Yager, R.R. On ordered weighting averaging aggregation operators in multicriteria decision-making. *IEEE Trans. Syst. Man Cybern.* **1988**, *18*, 183–190.
6. Yager, R.R. Families of OWA operators. *Fuzzy Sets Syst.* **1993**, *59*, 125–148.
7. Lizasoain, I.; Moreno, C. OWA operators defined on complete lattices. *Fuzzy Sets Syst.* **2013**, *224*, 36–52.
8. Paternain, D.; Ochoa, G.; Lizasoain, I.; Bustince, H.; Mesiar, R. Quantitative orness for lattice OWA operators. *Inf. Fusion* **2016**, *30*, 27–35.
9. Couceiro, M.; Marichal, J.-L. Characterizations of discrete Sugeno integrals as polynomial functions over distributive lattices. *Fuzzy Sets Syst.* **2010**, *161*, 694–707.
10. Paternain, D.; Bustince, H.; Pagola, M.; Sussner, P.; Kolesárová, A.; Mesiar, R. Capacities and overlap indexes with an application in fuzzy rule-based classification systems. *Fuzzy Sets Syst.* **2016**, *305*, 70–94.
11. Torra, V.; Narukawa, Y.; Sugeno, M. (Eds.) *Non-Additive Measures, Theory and Applications (Studies in Fuzziness and Soft Computing Series 310)*; Springer: New York, USA, 2014.
12. Choquet, G. Theory of capacities. *Ann. l'Inst. Fourier* **1954**, *5*, 131–292.
13. Mesiar, R.; Stupnanova, A.; Yager, R. Generalizations of OWA operators. In Proceedings of the 16th World Congress of the International Fuzzy Systems Association (IFSA) and the 9th Conference of the European Society for Fuzzy Logic and Technology (EUSFLAT), Gijón, Spain, 30 June–3 July 2015.
14. Karaçal, F.; Mesiar, R. Aggregation functions on bounded lattices. *Int. J. Gen. Syst.* **2017**, *46*, 37–51.
15. Komorníková, M.; Mesiar, R. Aggregation functions on Bounded Partially Ordered Sets and their classification. *Fuzzy Sets Syst.* **2011**, *175*, 48–56.

axioms

MDPI

Article

Existence of Order-Preserving Functions for Nontotal Fuzzy Preference Relations under Decisiveness

Paolo Bevilacqua [1], **Gianni Bosi** [2,*,†] and **Magalì Zuanon** [3]

[1] Dipartimento di Ingegneria e Architettura, Università di Trieste, 34127 Trieste, Italy; paolo.bevilacqua@dia.units.it
[2] Dipartimento di Scienze Economiche, Aziendali, Matematiche e Statistiche "Bruno de Finetti", Università di Trieste, 34127 Trieste, Italy
[3] Dipartimento di Economia e Management, Università di Brescia, 25122 Brescia, Italy; magali.zuanon@unibs.it
* Correspondence: gianni.bosi@deams.units.it; Tel.: +39-040-558-7115
† Current address: via Università 1, 34123 Trieste, Italy.

Received: 6 October 2017; Accepted: 26 October 2017; Published: 28 October 2017

Abstract: Looking at decisiveness as crucial, we discuss the existence of an order-preserving function for the nontotal crisp preference relation naturally associated to a nontotal fuzzy preference relation. We further present conditions for the existence of an upper semicontinuous order-preserving function for a fuzzy binary relation on a crisp topological space.

Keywords: fuzzy binary relation; fuzzy Suzumura consistency; order-preserving function; maximal element

---

1. Introduction

The literature concerning the (continuous) real representation of a fuzzy binary relation $R : X \times X \mapsto [0,1]$ is concentrated on the cases of fuzzy total preorders (see e.g., Billot [1], Fono and Andjiga [2], Fono and Salles [3], and Agud et al. [4]), and respectively fuzzy interval orders (see e.g., Mishra et Srivastava [5]). For any two points $x, y \in X$, $R(x,y)$ is interpreted as "the degree to which the alternative $x \in X$ is at least as good as the alternative $y \in X$", or equivalently as "the degree to which $x \in X$ is weakly preferred to $y \in X$".

We recall that a fuzzy binary relation is said to be total (or connected, or else complete) if $R(x,y) + R(y,x) \geq 1$ for all $x, y \in X$. Some authors call "complete" a fuzzy binary relation satisfying the stronger property according to which $\max(R(x,y), R(y,x)) = 1$ for all $x, y \in X$ (see e.g., Beg et Ashraf [6]). Such definitions of completeness clearly correspond to the concept of completeness of a crisp binary relation $\precsim$ (a particular case of a fuzzy binary relation R such that $R(x,y) \in \{0,1\}$ for all $x, y \in X$). Clearly, the existence of a utility function for a fuzzy binary relation $R : X \times X \mapsto [0,1]$ (i.e., u is a real-valued function on X such that $R(x,y) \geq R(y,x)$ is equivalent to $u(x) \geq u(y)$ for all $x, y \in X$) is strictly related to different definitions of transitivity of R (like bin-transitivity, min-transitivity, and product transitivity, among the others).

On the other hand, it should be noted that there is a large number of contributions to the representation of nontotal binary relations $\precsim$ (in particular nontotal preorders) in the crisp case (see e.g., Chapter 5 in Bridges amd Mehta [7], Herden and Levin [8], Herden [9], Mehta [10], Bosi et al. [11], Bosi and Mehta [12], and Bosi and Zuanon [13]). This is the case when there are elements $x, y \in X$ such that neither $x \precsim y$ nor $y \precsim x$. In this case a utility function (i.e., a real-valued function u on X such that $x \precsim y$ is equivalent to $u(x) \geq u(y)$ for all $x, y \in X$) cannot exist, and one can only ask for the existence of an order-preserving function (i.e., of a real-valued function u on X such that, for all $x, y \in X$, $x \precsim y$ implies $u(x) \geq u(y)$, and $x \succ y$ implies $u(x) > u(y)$, where $\succ$ stands for the strict part of $\precsim$).

While the literature is mainly concerned with the existence of order-preserving functions for not necessarily total preorders (i.e., for reflexive and transitive binary relation), it should be noted that the existence of an order-preserving function for a binary relation $\precsim$ on a set X does not imply the transitivity of $\precsim$, but only the property called Suzumura consistency (see Suzumura [14]), according to which, for example, $(x \precsim y)$ and $(y \precsim z)$ imply $not(z \succ x)$ for all $x, y, z \in X$. This property is much weaker than transitivity, which—as it is well known—requires that, for all $x, y, z \in X$, $x \precsim z$ whenever $(x \precsim y)$ and $(y \precsim z)$. Suzumura introduced this sort of consistency in order to generalize the Szpilrajn theorem (see Szpilrajn [15]), and, for example, Bosi and Herden [16]), according to which every (pre)order admits an extension by means of a total (pre)order (see also Cato [17]).

In this paper, we present some initial results concerning the existence of an order-preserving function $u : (X, R) \mapsto (\mathbb{R}, \geq)$ for a nontotal fuzzy binary relation R on a set X under decisiveness. This means that we look for the existence of a real-valued function u on X such that for all $(x, y) \in X \times X$ with $R(x, y) + R(y, x) \geq 1$, $R(x, y) \geq R(y, x)$ implies $u(x) \geq u(y)$, and respectively $R(x, y) > R(y, x)$ implies $u(x) > u(y)$. We refer to decisiveness since the presence of a fuzzy binary relation which is nontotal in our sense is related to the existence of pairs $(x, y) \in X \times X$ such that $R(x, y) + R(y, x) < 1$. In other cases, the representation of nontotal (or else partial) (fuzzy) binary relations is performed by allowing partial definition of some utility representation functions (see e.g., Bosi et al. [18]) or of the fuzzy binary relation R itself (see e.g., Khalid and Beg [19]).

It is clear that if R is total in the sense above, then an order-preserving function u is a utility for R, and this clearly implies that every result concerning the existence of an order-preserving function for a fuzzy binary relation generalizes the corresponding result for a fuzzy total binary relation, where "corresponding" essentially means "under the same assumptions of transitivity of R".

We use this kind of argument in order to present conditions for the existence of an upper semicontinuous order-preserving function u for a fuzzy binary relation R on a crisp topological space (X, τ), which guarantees the existence of a maximal element for R whenever τ is a compact topology.

The paper is structured as follows. Section 2 contains the notation and the preliminary results concerning crisp binary relations. Section 3 concerns the existence of an order-preserving function for fuzzy binary relations under decisiveness. In Section 4 we present a condition for the existence of an upper semicontinuous order-preserving function for a fuzzy binary relation on a crisp topological space. The conclusions are contained in Section 5.

2. Notation and Preliminary Results on the Real Representation of Crisp Binary Relations

We start with the presentation of the classical definitions and properties concerning the real representation of (crisp) binary relations, which are needed in this paper.

Definition 1. *A binary relation R on a nonempty crisp set X is a (crisp) subset of the Cartesian product $X \times X$; i.e., a mapping $R : X \times X \mapsto \{0, 1\}$. For any pair $(x, y) \in X \times X$, we write, for the sake of simplicity, xRy instead of $(x, y) \in R$. xRy has to be read as "the alternative $x \in X$ is at least as good as the alternative $y \in X$", or equivalently "$x \in X$ is weakly preferred to $y \in X$".*

For the sake of convenience, and since our agreement on the interpretation of the binary relation R, from now on we shall denote by $\precsim$ a (crisp) binary relation on a set X. The following definitions are found, for example, in Herden and Levin [8].

Definition 2. *The strict relation $\succ$, the indifference relation $\sim$, and the incomparability relation $\bowtie$ associated to a binary relation $\precsim$ on a set X are defined as follows for all $x, y \in X$:*

(i) $x \succ y \Leftrightarrow (x \precsim y)$ *and* $not(y \precsim x)$;
(ii) $x \sim y \Leftrightarrow (x \precsim y)$ *and* $(y \precsim x)$;
(iii) $x \bowtie y \Leftrightarrow not(x \precsim y)$ *and* $not(y \precsim x)$.

The following definition of *transitive closure* is found, for example, in Suzumura [14].

Definition 3. *The transitive closure* $\precsim^T$ *of a binary relation* $\precsim$ *on a set* X *is defined as follows for every* $x, y \in X$:

$$x \precsim^T y \Leftrightarrow \exists n \in \mathbb{N}^+, \exists x_0, x_1, ..., x_n \in X \text{ such that } [x = x_0 \precsim x_1 \precsim ... \precsim x_n = y].$$

Definition 4. *Consider a binary relation* $\precsim \subset X \times X$. *Then* $\precsim$ *is said to be:*

(i) *reflexive if* $x \precsim x$ *for all* $x \in X$;
(ii) *transitive if* $(x \precsim y)$ *and* $(y \precsim z)$ *imply* $x \precsim z$ *for all* $x, y, z \in X$;
(iii) *Suzumura consistent if for all* $x, y \in X$, $x \precsim^T y$ *implies not*$(y \succ x)$;
(iv) *a preorder if* $\precsim$ *is reflexive and transitive;*
(v) *a quasi-preorder if* $\precsim$ *is reflexive and Suzumura consistent;*
(vi) *total if* $(x \precsim y)$ *or* $(y \precsim x)$ *for all* $x, y \in X$.

Please notice that all the above definitions are standard and very well-known in utility theory (see e.g., Herden [9]), despite for the concept of a *quasi-preorder*. Indeed, Suzumura consistency was introduced by Suzumura [14] in order to generalize *Szpilrajn theorem* (see Szpilrajn [15]) by introducing a condition which is more general than the classical transitivity assumption, but it does not appear specifically in connection with the real representation of binary relations (see Example 1 below).

Definition 5. *Given a quasi-preorder* $\precsim$ *on a set* x, *let us define, for every point* $x \in X$, *the following subsets of* X:

$$l_{\precsim}(x) = \{y \in X \mid x \succ y\}, \quad r_{\precsim}(x) = \{y \in X \mid y \succ x\}$$
$$d_{\precsim}(x) = \{z \in X \mid x \precsim z\}, \quad i_{\precsim}(x) = \{z \in X \mid z \precsim x\}.$$

We say that $l_{\precsim}(x)$ ($r_{\precsim}(x)$) is the *strict lower section* (respectively, *strict upper section*) associated to $x \in X$, while $d_{\precsim}(x)$ ($i_{\precsim}(x)$) is the *weak lower section* (respectively, *weak upper section*) associated to $x \in X$ (see e.g., Bridges and Mehta [7]).

Definition 6. *A subset* D *of a quasi-preordered set* $(X, \precsim)$ *is said to be* $\precsim$*-decreasing if* $d_{\precsim}(x) \subset D$ *for all* $x \in D$.

In the sequel, $\mathbb{R}$ stands for the real line, and $\geq$ is the natural (total) order on $\mathbb{R}$.

Definition 7. *Consider a binary relation* $\precsim$ *on a set* X, *and a mapping* $u : (X, \precsim) \mapsto (\mathbb{R}, \geq)$. *Then* u *is said to be:*

(i) $\precsim$*-increasing, if for all pairs* $(x, y) \in X \times X$, $[x \precsim y \Rightarrow u(x) \geq u(y)]$;
(ii) *a weak utility for* $\succ$, *if for all pairs* $(x, y) \in X \times X$, $[x \succ y \Rightarrow u(x) > u(y)]$;
(iii) $\precsim$*-order-preserving, if* u *is both* $\precsim$*-increasing and a weak utility for* $\succ$;
(iv) *a utility for* $\precsim$, *if for all pairs* $(x, y) \in X \times X$, $[x \precsim y \Leftrightarrow u(x) \geq u(y)]$.

Let us recall the concept of a *maximal element* for a (quasi-)preorder (see e.g., Bosi and Zuanon [20]).

Definition 8. *Let* $\precsim$ *be a quasi-preorder on a set* X. *Then an element* $x_0 \in X$ *is said to be a maximal element for* $\precsim$ *if for no* $z \in X$ *it happens that* $z \succ x_0$.

It is immediate to check that a binary relation admitting a utility (representation) is a total preorder. The distinction between the concept of a $\precsim$-order-preserving function and that of a weak utility for the strict part $\succ$ of $\precsim$ proved to be useful in connection with the existence of maximal elements of a preorder (see Bosi and Zuanon [20]).

It is very well known that the existence of an order-preserving function u does not characterize a nontotal (quasi-)preorder $\precsim$, since whenever $u(x) > u(y)$ it may happen that either $(x \succ y)$ or

($x \bowtie y$). Nevertheless, it is enough for optimization purposes; i.e., in order to guarantee the existence of maximal elements (see e.g., the introduction in Alcantud et al. [21]). More generally, it is clear that if u is a weak utility for the strict part $\succ$ of a quasi-preorder $\precsim$ on a set X, and u attains its maximum at a point x_0, then x_0 is a maximal element for $\precsim$.

Let us consider the following very simple proposition, which illustrates the importance of Suzumura consistency in connection with the existence of $\precsim$-order-preserving functions.

Proposition 1. *If there exists a $\precsim$-order-preserving function $u : (X, \precsim) \mapsto (\mathbb{R}, \geq)$ for a (reflexive) binary relation $\precsim$ on a set X, then $\precsim$ is Suzumura consistent (a quasi-preorder).*

Proof. Let $u : (X, \precsim) \mapsto (\mathbb{R}, \geq)$ be a $\precsim$-order-preserving function for a (reflexive) binary relation $\precsim$ on a set X. By contraposition, assume that $\precsim$ is not Suzumura consistent. Then there exist $n \in \mathbb{N}^+$, and elements $x_0, x_1, ..., x_n \in X$ such that ($x = x_0 \precsim x_1 \precsim ... \precsim x_n = y$) and ($y \succ x$), implying that $u(x) = u(x_0) \geq u(x_1) \geq ... \geq u(x_n) = u(y) > u(x)$, a contradiction. □

The following example shows that the existence of a $\precsim$-order-preserving function $u : (X, \precsim) \mapsto (\mathbb{R}, \geq)$ does not imply that $\precsim$ is a preorder (in particular, does not imply that $\precsim$ is transitive) on X.

Example 1. *Consider the binary relation $\precsim$ on the real interval $X = [0, 1]$ defined as follows:*

$$x \precsim y \Leftrightarrow \begin{cases} (x \geq y) \text{ and } (x, y \in [0, \frac{1}{2}]) \\ \text{or} \\ (x \geq y) \text{ and } (x, y \in [\frac{1}{2}, 1]) \\ \text{or} \\ (x \geq \frac{3}{4}) \text{ and } (y \leq \frac{1}{4}) \end{cases}$$

It is clear that $\precsim$ is reflexive. Notice that $x \bowtie y$ for all $x, y \in [0, 1]$ such that $\frac{3}{4} > x \geq \frac{1}{2} > y > \frac{1}{4}$. We have that $\precsim$ is not transitive since, for example, $\frac{3}{5} \precsim \frac{1}{2} \precsim \frac{1}{3}$ but $\frac{3}{5} \bowtie \frac{1}{3}$. On the other hand, it is easily seen that the identity function $id_{[0,1]}$ on $[0, 1]$ is a $\precsim$-order-preserving function, and therefore $\precsim$ is Suzumura consistent (actually a quasi-preorder) on $[0, 1]$ by Proposition 1.

The following concept of an *extension* (or *refinement*) of a binary relation by a total preorder is well known since the seminal work of Szpilrajn [15].

Definition 9. *A total preorder $\precsim^*$ is said to be an extension of a quasi-preorder $\precsim$ on a set X if, for all $x, y \in X$, $[(x \precsim y \Rightarrow x \precsim^* y)$ and $(x \succ y \Rightarrow x \succ^* y)]$.*

The following concept of *separability* is found, for example, in Herden and Levin [8].

Definition 10. *A quasi-preorder $\precsim$ on a set X is said to be separable if there exists a countable subset D of X such that, for all pairs $(x, y) \in X \times X$ such that $x \succ y$, there exists $d \in D$ such that $x \precsim d \precsim y$.*

The following proposition provides an immediate characterization of the existence of a $\precsim$-order-preserving function in terms of the existence of a *representable* (i.e., admitting a utility function $u : (X, \precsim) \mapsto (\mathbb{R}, \geq)$) total preorder $\precsim^*$ providing an extension of $\precsim$. Please notice that in the sequel, the symbol $\precsim$ will stand for a quasi-preorder on a (nonempty) set X.

Proposition 2. *The following conditions are equivalent on a quasi-preorder $\precsim$ on a set X:*

(i) *There exists a $\precsim$-order-preserving function $u : (X, \precsim) \mapsto (\mathbb{R}, \geq)$;*
(ii) *There exists a representable total preorder $\precsim^*$ which is an extension of $\precsim$;*
(iii) *There exists a total preorder $\precsim^*$ which is a separable extension of $\precsim$.*

Proof. The equivalence of conditions (ii) and (iii) is well known (see e.g., Theorem 1.4.8 in Bridges and Mehta [7]). Therefore, we only have to show the equivalence of conditions (i) and (ii). In order to show that (i) implies (ii), just consider that if $u : (X, \precsim) \mapsto (\mathbb{R}, \geq)$ is a $\precsim$-order-preserving function, the representable total preorder $\precsim^*$ on X defined as follows for all $x, y \in X$: $[x \precsim^* y \Leftrightarrow u(x) \geq u(y)]$ is an extension of $\precsim$. In order to show that (ii) implies (i), let $\precsim^*$ be a representable extension of $\precsim$, and let u be a utility function for $\precsim^*$. Then, it is immediate to check that u is in particular a $\precsim$-order-preserving function. This consideration completes the proof. □

Conditions for the existence of an *upper semicontinuous* and respectively *continuous* order-preserving function u for a preorder $\precsim$ on a set X endowed with a (crisp) *topology* τ are found, for example, in Bosi and Mehta [12].

Corollary 1. *Let $\precsim$ be a reflexive binary relation on a countable set X. Then the following conditions are equivalent:*

(i) *There exists a $\precsim$-order-preserving function $u : (X, \precsim) \mapsto (\mathbb{R}, \geq)$;*
(ii) *$\precsim$ is a quasi-preorder.*

Proof. The implication "(i) ⇒ (ii)" is a consequence of Proposition 1. The implication "(ii) ⇒ (i)" is a consequence of Proposition 2, implication "(iii) ⇒ (i)", and of *Suzumura's extension theorem* [14] (see also Theorem 2 in Cato [17]), which states that a binary relation $\precsim$ is Suzumura consistent if and only if it admits an extension by means of a total preorder $\precsim^*$. On the other hand, such a total preorder $\precsim^*$ is separable since X is a countable set, and therefore it is representable by means of a utility function u, which is clearly a $\precsim$-order-preserving function. □

3. Order-Preserving Functions for Fuzzy Binary Relations under Decisiveness

We first recall the classical concept of a *fuzzy binary relation R*, in order to then introduce the definition of its *strict part under decisiveness* P_R.

Definition 11. *A fuzzy subset h of a nonempty crisp set X is a mapping $h : X \mapsto [0,1]$ from X into the closed bounded real interval $[0,1]$. We denote by $F(X)$ the set of all the fuzzy subsets of X.*

Definition 12. *A fuzzy binary relation R on a nonempty crisp set X is a fuzzy subset of the Cartesian product $X \times X$; i.e. a mapping $X \times X \mapsto [0,1]$. For any pair $(x,y) \in X \times X$, $R(x,y)$ has to be thought of as "the degree to which the alternative $x \in X$ is at least as good as the alternative $y \in X$". We write $R \in F(X,X)$.*

Definition 13. *Consider $R \in F(X,X)$. Then the strict part P_R of R ($P_R \in F(X,X)$) is defined to be, for all $x, y \in X$,*

$$P_R(x,y) := \begin{cases} R(x,y) & \text{if } (R(x,y) > R(y,x)) \text{ and } (R(x,y) + R(y,x) \geq 1) \\ 0 & \text{otherwise} \end{cases}$$

Therefore, for any pair $(x,y) \in X \times X$, $P_R(x,y)$ has to be thought of as "the degree to which the alternative $x \in X$ is strictly preferred to the alternative $y \in X$ with decisiveness".

Remark 1. *Our definition of the strict part P_R of a fuzzy binary relation R takes into account the possibility of dealing with nontotal fuzzy relations R on X (i.e., when there exist $x, y \in X$ such that $R(x,y) + R(y,x) < 1$), in such a way that "decisiveness applies" (i.e., only pairs at which R is decisive are considered as important). Clearly, when we consider total binary relations (preorders) (i.e. when $R(x,y) + R(y,x) \geq 1$ for all $x, y \in X$), the above definition coincides with the usual definition of the strict part of a binary relation, as found for example in Fono and Salles [3].*

Definition 14. *A fuzzy binary relation* $R \in F(X,X)$ *is said to be decisive at a pair* $(x,y) \in X \times X$ *if* $R(x,y) + R(y,x) \geq 1$.

Let us now define the binary relation $\precsim_R$ on X which is naturally associated to a fuzzy binary relation R on X in the case when "decisiveness" applies.

Definition 15. *Let* $R : X \times X \mapsto [0,1]$ *be a fuzzy binary relation on a set* X. *Then the binary relation* $\precsim_R$ *on* X *which is naturally associated to* R *in the case of decisiveness is defined to be, for all* $x, y \in X$,

$$x \precsim_R y \Leftrightarrow (R(x,y) \geq R(y,x)) \text{ and } (R \text{ is decisive at } (x,y)).$$

Definition 16. *Consider* $R \in F(X,X)$. *Then* R *is said to be:*

(i) *reflexive if* $R(x,x) = 1$ *for all* $x \in X$;
(ii) *min-transitive if* $R(x,z) \geq \min\{R(x,y), R(y,z)\}$ *for all* $x,y,z \in X$;
(iii) *bin-transitive if* $(R(x,y) \geq R(y,x))$ *and* $(R(y,z) \geq R(z,y))$ *imply* $R(x,z) \geq R(z,x)$ *for all* $x,y,z \in X$;
(iv) *fuzzy decisively Suzumura consistent if for every positive integer* n, *and for all elements* $x_0, x_1, ..., x_n \in X$ *such that* R *is decisive at* (x_h, x_{h+1}) *for all* $h \in \{0, ...n-1\}$,

$$[R(x_h, x_{h+1}) \geq R(x_{h+1}, x_h) \text{ for all } h \in \{0, ...n-1\}] \Rightarrow P_R(x_n, x_0) = 0;$$

(v) *a fuzzy preorder if* R *is reflexive and min-transitive;*
(vi) *a fuzzy decisive quasi-preorder if* R *is reflexive and fuzzy decisively Suzumura consistent;*
(vii) *total if* R *is decisive at every point* $(x,y) \in X \times X$.

The reader can notice that all the previous definitions are classical and frequently found in the literature (perhaps with slightly different terminologies), except for the *fuzzy decisive Suzumura consistency*. In particular the term "total" is replaced by "*connected*" in Fono and Andjiga [2] and Fono and Salles [3].

Remark 2. *In the particular case when* $R : X \times X \mapsto \{0,1\}$ *is a (crisp) binary relation on the set* X *(i.e.,* $R \subset X \times X$*), axioms (i), (ii) (equivalently (iii)) and (iv) in Definition 16 become axioms (i), (ii), and respectively (iii) in Definition 4.*

Remark 3. *It should be noted that* $\precsim_R$ *is not transitive in general, even if there exists a* $\precsim_R$*-order-preserving function* u *on* X. *Further,* $\precsim_R$ *is reflexive if and only if the fuzzy binary relation* $R \in F(X,X)$ *is reflexive. On the other hand, it is clear that if* $R \in F(X,X)$ *is total, then the associated binary relation* $\precsim_R$ *in Definition 15 is the total preorder* $\precsim_R$ *on* X *defined as follows for all* $x,y \in X$:

$$x \precsim_R y \Leftrightarrow R(x,y) \geq R(y,x).$$

This is precisely the case of the utility representation of a fuzzy total preorder considered in Billot [1], Fono and Andjiga [2], and Fono and Salles [3].

Remark 4. *It is immediate to observe that a reflexive binary relation* $R \in F(X,X)$ *is a fuzzy quasi-preorder if and only if the associated binary relation* $\precsim_R$ *in Definition 15 is a quasi-preorder according to Definition 4, (v).*

Definition 17. *Consider a fuzzy binary relation* $R \in F(X,X)$, *and mapping* $u : (X,R) \mapsto (\mathbb{R}, \geq)$. *Then* u *is said to be:*

(i) *R-increasing, if, for all pairs* $(x,y) \in X \times X$ *at which* R *is decisive,* $[R(x,y) \geq R(y,x) \Rightarrow u(x) \geq u(y)]$;
(ii) *a weak utility for* P_R, *if for all pairs* $(x,y) \in X \times X$, $[P_R(x,y) > 0 \Rightarrow u(x) > u(y)]$;
(iii) *R-order-preserving, if* u *is both R-increasing and a weak utility for* P_R;
(iv) *a utility for* R *if* R *is total and, for all pairs* $(x,y) \in X \times X$, $[R(x,y) \geq R(y,x) \Leftrightarrow u(x) \geq u(y)]$.

It should be noted that the terminology above is perfectly coherent with the usual one recalled in Definition 7 concerning crisp relations (preorders), since in this case $R(x,y) + R(y,x) \geq 1$ is equivalent to $\max(R(x,y), R(y,x)) = 1$.

The following definition of *separability* of a fuzzy binary relation is found in Fono and Salles [3].

Definition 18. *A fuzzy binary relation $R \in F(X,X)$ is said to be separable if there exists a countable subset D of X such that, for all pairs $(x,y) \in X \times X$ at which R is decisive and $R(x,y) > R(y,x)$ there exists $d \in D$ such that R is decisive both at (x,d) and (d,y) and $R(x,d) \geq R(d,x)$, $R(d,y) \geq R(y,d)$.*

As an immediate consequence of the above definitions and Proposition 2, we get the following proposition, whose immediate proof is left to the reader.

Proposition 3. *The following conditions are equivalent on fuzzy binary relation $R \in F(X,X)$ and a real-valued function u on X:*

(i) *$u : (X,R) \mapsto (\mathbb{R}, \geq))$ is an R-order-preserving function;*
(ii) *$u : (X, \precsim_R) \mapsto (\mathbb{R}, \geq))$ is a $\precsim_R$-order-preserving function, where $\precsim_R$ is the binary relation naturally associated to R according to Definition 15;*
(iii) *u is the utility function for some total preorder $\precsim^*$ which is an extension of $\precsim_R$.*

Proposition 4. *Consider a fuzzy binary relation $R \in F(X,X)$. Then there exists an R-order-preserving function $u : (X,R) \mapsto (\mathbb{R}, \geq)$ provided that there exists a fuzzy total preorder $R^* : X \times X \mapsto [0,1]$ satisfying the following conditions:*

(i) *R^* is separable;*
(ii) *The following conditions are verified for all $(x,y) \in X \times X$ at which R is decisive,*
   (a) *$R(x,y) \geq R(y,x)$ implies $R^*(x,y) \geq R^*(y,x)$;*
   (b) *$P_R(x,y) > 0$ implies $P_{R^*}(x,y) > 0$;*
(iii) *For every $x,y,z \in X$, $R^*(x,y) = R^*(y,x) = R^*(y,z) = R^*(z,y) \Rightarrow R^*(x,z) = R^*(z,x)$.*

Proof. Assume that there exists a fuzzy total preorder $R^* : X \times X \mapsto [0,1]$ satisfying the above conditions (i), (ii), and (iii). Then, from Proposition 2 in Fono and Salles [3], since conditions (i) and (iii) hold true, there exists a utility function $u : (X,R^*) \mapsto (\mathbb{R}, \geq)$. Such utility function u is R-order-preserving by condition (ii). This consideration completes the proof. □

The following theorem provides a characterization of the existence of an R-order-preserving function for a fuzzy binary relation $R \in F(X,X)$ in case that the set X is countable.

Theorem 1. *Consider a fuzzy binary relation $R \in F(X,X)$. If R is reflexive and X is countable, then the following conditions are equivalent:*

(i) *There exists an R-order-preserving function $u : (X,R) \mapsto (\mathbb{R}, \geq)$;*
(ii) *R is a fuzzy decisive quasi-preorder.*

Proof. (i) $\Rightarrow$ (ii). If there exists an order-preserving function $u : (X,R) \mapsto (\mathbb{R}, \geq)$, then, by Proposition 3, $u : (X, \precsim_R) \mapsto (\mathbb{R}, \geq))$ is an $\precsim_R$-order-preserving function, where $\precsim_R$ is the binary relation naturally associated to R according to Definition 15. Hence, we have that $\precsim_R$ is a quasi-preorder by Proposition 2, and therefore R is a fuzzy decisive quasi-preorder from the definition of $\precsim_R$.

(ii) $\Rightarrow$ (i). If $R \in F(X,X)$ is a fuzzy decisive quasi-preorder, then $\precsim_R$ is a quasi-preorder, and therefore there exists a $\precsim_R$-order-preserving function by Corollary 1. Clearly, u is also R-order-preserving. This consideration completes the proof. □

Since for a fuzzy total binary relation $R \in F(X,X)$ a function $u : (X,R) \mapsto (\mathbb{R}, \geq)$ is R-order-preserving if and only if it is a utility function for R, we immediately get as a corollary the following result proved by Billot [1] (see also Proposition 2 in Fono and Andjiga [2]).

Corollary 2. *Consider a fuzzy total binary $R \in F(X,X)$. If X is countable, then the following conditions are equivalent:*

(i) *There exists a utility function $u : (X,R) \mapsto (\mathbb{R}, \geq)$;*
(ii) *R is bin-transitive.*

4. Upper Semicontinuous Order-Preserving Functions for Fuzzy Binary Relation on A Topological Space

In order to motivate the contents of this section, let us first introduce the concept of a *maximal element* for a fuzzy binary relation under decisiveness.

Definition 19. *A point $x_0 \in X$ is said to be maximal with respect to fuzzy binary relation $R \in F(X,X)$ if for no $z \in X$ it occurs that $P_R(z, x_0) > 0$ (i.e., $r_{\succsim_R}(x_0) = \emptyset$).*

It is clear that if a binary relation $R \in F(X,X)$ admits a representation by means of an R-order-preserving u, and u attains its maximum at a point $x_0 \in X$, then x_0 is a maximal element for R. In the sequel, the symbol $\succsim_R^T$ stands for the transitive closure of the binary relation $\succsim_R$ associated to some $R \in F(X,X)$.

Definition 20. *Let $R \in F(X,X)$ be a fuzzy binary relation on a set X, endowed with a topology τ. Then R is said to be:*

(i) *T-τ-upper semicontinuous, if $l_{\succsim_R^T}(x)$ is an open subset of X for all $x \in X$:*
(ii) *T-τ-upper semiclosed, if $i_{\succsim_R^T}(x)$ is a closed subset of X for all $x \in X$.*

It should be noted that if R is a crisp binary relation, then the above definitions of T-τ-upper semicontinuity and T-τ-upper semiclosedness slightly generalize the concepts of *upper semicontinuity of type 1* and respectively *upper semicontinuity of type 2* as defined in Herden and Levin [8] in the case of a preorder $\succsim$ on a topological space (X, τ).

We recall that a *(crisp) topology* τ on a set X is a family of subsets of X which contains X and the empty set $\emptyset$ and is closed under arbitrary unions and finite intersections. A set $O \in \tau$ is said to be *open*. We denote by τ_{nat} the *natural (interval)* topology on $\mathbb{R}$. Further, a real-valued function u on a *topological space* (X, τ) is said to be *τ-upper semicontinuous* if the set $u^{-1}(]-\infty, \alpha[) = \{z \in X : u(z) < \alpha\}$ is open for every $\alpha \in \mathbb{R}$.

Proposition 5. *Let R be a fuzzy decisive quasi-preorder on a set X, endowed with a topology τ. Then there exists a τ-upper semicontinuous R-order-preserving function $u : (X, R, \tau) \mapsto (\mathbb{R}, \geq, \tau_{nat})$ provided that the following conditions are verified:*

(i) *There exists a countable subset D of X such that for all pairs $(x,y) \in X \times X$ such that $x \succ_R y$, there exists $d \in D$ such that $x \succsim_R d \succ_R^T y$;*
(ii) *R is either T-τ–upper semicontinuous or T-τ–upper semiclosed.*

Proof. Let R be a fuzzy decisive quasi-preorder on a set X, endowed with a topology τ, and assume that R satisfies the above conditions (i) and (ii). Let $D = \{d_n : n \in \mathbb{N}^+\}$ be a countable subset of X verifying the above condition (i), and define the function $u_n : X \mapsto [0,1]_{\mathbb{R}}$ as follows for every $n \in \mathbb{N}^+$:

$$u_n(x) = \begin{cases} 0 & \text{if } x \in l_{\succsim_R^T}(d_n) \; (x \in X \setminus i_{\succsim_R^T}(d_n)) \\ 1 & \text{otherwise} \end{cases} .$$

Then define $u := \sum_{n=1}^{\infty} 2^{-n} u_n$. Observe that u is $\precsim_R$-increasing since the fact that $l_{\succsim_R^T}(z)$ $(X \setminus i_{\succsim_R^T}(z))$ is $\succsim_R^T$-decreasing for all $z \in X$ implies that u_n is $\succsim_R^T$-increasing for all $n \in \mathbb{N}^+$, and therefore that u is $\succsim_R$-increasing. Further, it is easily seen that the above condition (i) implies that u is a weak utility for $\succ_R$. Indeed, if $x \succ_R y$, then there exists $n \in \mathbb{N}^+$ such that $x \succsim_R d_n \succ_R^T y$, implying that $u_n(y) = 0$ and $u_n(x) = 1$. Since R is T-τ-upper semicontinuous (T-τ-upper semiclosed), it is easy to be seen that u_n is τ-upper semicontinuous for all $n \in \mathbb{N}^+$, implying that u is also τ-upper semicontinuous. This consideration completes the proof. □

Since every upper semicontinuous function u attains its maximum on a *compact topological space* (X, τ) (i.e., when for every family $\mathcal{O} = \{O_\alpha\}_{\alpha \in I}$ of open subsets of X such that $\bigcup_{\alpha \in I} O_\alpha = X$ there exists a finite subfamily $\mathcal{O}' = \{O_1, ..., O_n\}$ of $\mathcal{O}$ such that $\bigcup_{n=1}^{n} O_n = X$), we immediately get the following corollary.

Corollary 3. *Let R be a fuzzy decisive quasi-preorder on a set X, endowed with a topology τ. If R satisfies the assumptions of Proposition 5, and the topology τ is compact, then there exists a maximal element x_0 for R.*

5. Conclusions

In this paper we have presented conditions for the existence of an order-preserving function u for a fuzzy not necessarily total binary relation R on a set X when *decisiveness* applies, in the sense that u is required to "preserve" and "to strictly preserve" the inequalities $R(x,y) \geq R(y,x)$ and respectively $R(x,y) > R(y,x)$ only in the case of a pair $(x,y) \in X \times X$ such that $R(x,y) + R(y,x) \geq 1$ (i.e., when R is *decisive* at (x,y) according to the terminology adopted here). By using this approach, it is possible to generalize the corresponding results concerning the existence of a *utility function* u for a *fuzzy total preorder* R on X; i.e., when $R(x,y) + R(y,x) \geq 1$ for all pairs $(x,y) \in X \times X$.

We have also presented conditions implying the existence of an upper semicontinuous order-preserving function in case that the set X is endowed with a crisp topology.

The possibility of characterizing the existence of an (upper semicontinuous or continuous) order-preserving function for a not necessarily total fuzzy preference relation under decisiveness will hopefully be examined in a future paper.

Author Contributions: The authors contributed equally to this work.

Conflicts of Interest: The authors declare no conflict of interest.

References

1. Billot, A. An existence theorem or fuzzy utility functions: A new elementary proof. *Fuzzy Sets Syst.* **1995**, *74*, 271–276.
2. Fono, L.A.; Andjiga, N.G. Utility function of fuzzy preferences on a countable set under max-∗-transitivity. *Soc. Choice Welf.* **2007**, *28*, 667–683.
3. Fono, L.A.; Salles, M. Continuity of utility functions representing fuzzy preferences. *Soc. Choice Welf.* **2011**, *37*, 669–682.
4. Agud, L.; Catalan, R.G.; Dıaz, S.; Indurain, E.; Montes, S. Numerical representability of fuzzy total preorders. *Int. J. Comput. Intell. Syst.* **2012**, *5*, 996–1009
5. Mishra, S.; Srivastava, R. Representability of fuzzy biorders and fuzzy weak orders. *Int. J. Uncertain. Fuzziness Knowl.-Based Syst.* **2016**, *24*, 9117–9935.
6. Beg, I.; Ashraf, F. Numerical representation of product transitive complete fuzzy orderings. *Math. Comput. Model.* **2011**, *53*, 9617–9623.
7. Bridges, D.S.; Mehta, G.B. *Representations of Preference Orderings*; Springer: Berlin, Germany, 1995.

8. Herden, G., Levin, V.L. Utility representation theorems for Debreu separable preorders. *J. Math. Econ.* **2012**, *48*, 148–154.
9. Herden, G. On the existence of utility functions. *Math. Soc. Sci.* **1989**, *17*, 297–313.
10. Mehta, G.B. Some general theorems on the existence of order-preserving functions. *Math. Soc. Sci.* **1988**, *15*, 135–146.
11. Bosi, G.; Campíon, M.J.; Candeal, J.C.; Induráin, E.; Zuanon, M. Numerical isotonies of preordered semigroups through the concept of a scale. *Math. Pannonica* **2005**, *16*, 65–77.
12. Bosi, G.; Mehta, G.B. Existence of a semicontinuous or continuous utility function: A unified approach and an elementary proof. *J. Math. Econ.* **2002**, *38*, 311–328.
13. Bosi, G.; Zuanon, M. Conditions for the Upper Semicontinuous Representability of Preferences with Nontransitive Indifference. *Theor. Econ. Lett.* **2014**, *4*, 371–377.
14. Suzumura, K. Remarks on the theory of collective choice. *Economica* **1976**, *43*, 381–390.
15. Szpilrajn, E. Sur l'extension de l'ordre partiel. *Fundam. Math.* **1930**, *16*, 386–389.
16. Bosi, G.; Herden, G. On a Strong Continuous Analogue of the Szpilrajn Theorem and its Strengthening by Dushnik and Miller. *Order* **2005**, *22*, 329–342.
17. Cato, S. Szpilrajn, Arrow and Suzumura: Concise proofs of extensions theorems and an extension. *Metroeconomica* **2012**, *63*, 235–249.
18. Bosi, G.; Estevan, A.; Zuanon, M. Partial representations of orderings. *Int. J. Uncertain. Fuzziness Knowl.-Based Syst.* **2017**, in press.
19. Khalid, A.; Beg, I. Incomplete Hesitant Fuzzy Preference Relations in Group Decision Making. *Int. J. Fuzzy Syst.* **2017**, *19*, 637–645.
20. Bosi, G.; Zuanon, M. Maximal elements of quasi upper semicontinuous preorders on compact spaces. *Econ. Theory Bull.* **2017**, *5*, 109–117.
21. Alcantud, J.C.R.; Bosi, G.; Zuanon, M. Richter-Peleg multi-utility representations of preorders. *Theory Decis.* **2016**, *80*, 443–450.

axioms

MDPI

Article

On Indistinguishability Operators, Fuzzy Metrics and Modular Metrics

Juan-José Miñana and **Oscar Valero** *

Department of Mathematics and Computer Science, Universitat de les Illes Balears, Carretera de Valldemossa km. 7.5, 07122 Palma, Spain; jj.minana@uib.es

* Correspondence: o.valero@uib.es; Tel.: +34-971-259817

Received: 20 November 2017; Accepted:12 December 2017; Published: 15 December 2017

Abstract: The notion of indistinguishability operator was introduced by Trillas, E. in 1982, with the aim of fuzzifying the crisp notion of equivalence relation. Such operators allow for measuring the similarity between objects when there is a limitation on the accuracy of the performed measurement or a certain degree of similarity can be only determined between the objects being compared. Since Trillas introduced such kind of operators, many authors have studied their properties and applications. In particular, an intensive research line is focused on the metric behavior of indistinguishability operators. Specifically, the existence of a duality between metrics and indistinguishability operators has been explored. In this direction, a technique to generate metrics from indistinguishability operators, and vice versa, has been developed by several authors in the literature. Nowadays, such a measurement of similarity is provided by the so-called fuzzy metrics when the degree of similarity between objects is measured relative to a parameter. The main purpose of this paper is to extend the notion of indistinguishability operator in such a way that the measurements of similarity are relative to a parameter and, thus, classical indistinguishability operators and fuzzy metrics can be retrieved as a particular case. Moreover, we discuss the relationship between the new operators and metrics. Concretely, we prove the existence of a duality between them and the so-called modular metrics, which provide a dissimilarity measurement between objects relative to a parameter. The new duality relationship allows us, on the one hand, to introduce a technique for generating the new indistinguishability operators from modular metrics and vice versa and, on the other hand, to derive, as a consequence, a technique for generating fuzzy metrics from modular metrics and vice versa. Furthermore, we yield examples that illustrate the new results.

Keywords: indistinguishability operator; fuzzy (pseudo-)metric; modular (pseudo-)metric; continuous Archimedean t-norm; additive generator; pseudo-inverse

---

1. Introduction

Throughout this paper, we will use the following notation. We will denote by $\mathbb{R}$ the set of real numbers, and we will denote by $[a,b]$, $]a,b]$, $[a,b[$ and $]a,b[$, open, semi-open and closet real intervals, respectively, whenever $a,b \in \mathbb{R} \cup \{-\infty,\infty\}$ with $a < b$.

In 1982, Trillas, E. introduced the notion of indistinguishability operator with the purpose of fuzzifying the classical (crisp) notion of equivalence relation (see [1]). Let us recall that an indistinguishability operator, for a t-norm $*$, on a non-empty set X is a fuzzy set $E : X \times X \to [0,1]$, which satisfies, for each $x,y,z \in X$, the following axioms:

(E1) $E(x,x) = 1$; (Reflexivity)
(E2) $E(x,y) = E(y,x)$; (Symmetry)
(E3) $E(x,y) * E(y,z) \leq E(x,z)$. (Transitivity)

If, in addition, E satisfies for all $x,y \in X$ the following condition:

(E1′) $E(x,y) = 1$ implies $x = y$,

then it is said that E separates points.

According to [1] (see also [2]), the numerical value $E(x,y)$ provides the degree up to which x is indistinguishable from y or equivalent to y. Thus, the greater $E(x,y)$, the more similar x and y are. In particular, $E(x,y) = 1$ when $x = y$.

In the light of the preceding definition, the concept of t-norm plays an essential role in the framework of indistinguishability operators. In fact, t-norms are involved in axiom (**E3**) in order to express that x is indistinguishable from z whenever x is indistinguishable from y and z. Throughout this paper, we will assume that the reader is familiar with the basics of triangular norms (see [3] for a deeper treatment of the topic).

Since Trillas introduced the indistinguishability operators, many authors have studied their properties and applications. We refer the reader to [2], and references therein, for an exhaustive treatment of the topic. Among the different properties that such operators enjoy, the metric behavior can be highlighted. In particular, the existence of a duality relationship between metrics and indistinguishability operators in [2–8] has been explored. In this direction, a technique to generate metrics from indistinguishability operators, and vice versa, has been developed by several authors in the literature. Concretely, an indistinguishability operator can be provided from a (pseudo-)metric as follows:

Theorem 1. *Let X be a non-empty set and let $*$ be a t-norm with additive generator $f_* : [0,1] \to [0,\infty]$. If $\diamond$ is a t-norm, then the following assertions are equivalent:*

(1) $* \leq \diamond$ *(i.e., $x * y \leq x \diamond y$ for all $x, y \in [0,1]$).*
(2) *For any indistinguishability operator E on X for $\diamond$, the function $d^{E,f_*} : X \times X \to [0,\infty]$ defined, for each $x, y \in X$, by*
$$d^{E,f_*}(x,y) = f_*(E(x,y)),$$
is a pseudo-metric on X.
(3) *For any indistinguishability operator E on X for $\diamond$ that separates points, the function $d^{E,f_*} : X \times X \to [0,\infty]$ defined, for each $x, y \in X$, by*
$$d^{E,f_*}(x,y) = f_*(E(x,y)),$$
is a metric on X.

Reciprocally, a technique to construct an indistinguishability operator from a (pseudo-)metric can be given as the next result shows.

Theorem 2. *Let X be a non-empty set and let $*$ be a continuous t-norm with additive generator $f_* : [0,1] \to [0,\infty]$. If d is a pseudo-metric on X, then the function $E^{d,f_*} : X \times X \to [0,1]$ defined, for all $x, y \in X$, by*
$$E^{d,f_*}(x,y) = f_*^{(-1)}(d(x,y)),$$
is an indistinguishability operator for $*$, where $f_*^{(-1)}$ denotes the pseudo-inverse of the additive generator f_*. Moreover, the indistinguishability operator E^{d,f_*} separates points if and only if d is a metric on X.

It must be stressed that, in the statement of the preceding results, and, throughout this paper, the considered (pseudo-)metrics can take the value ∞, which are also known as extended (pseudo-)metrics in [9].

Recently, applications of the techniques exposed in Theorems 1 and 2 to the task allocation problem in multi-agent (multi-robot) systems have been given in [10–12]. In particular, in the preceding references, indistinguishability operators have shown to be appropriate to model response functions when response threshold algorithms (in swarm-like methods) are under consideration in order to solve the aforesaid task allocation problem.

Nowadays, in many applications, the degree of similarity is measured relative to a parameter (see, for instance, [13–15]). In this case, the indistinguishability operators are not able to measure such a graded similarity and so a new measurement becomes indispensable instead. The aforesaid measurements are called fuzzy metrics and they were introduced in 1975 by Kramosil, I. and Michalek, J. in [16]. However, currently, the fuzzy metric axioms used in the literature are those given by Grabiec, M. in [17] and by George, A. and Veeramani, P. in [18]. It must be pointed out that the axioms by Grabiec and by George and Veeramani, P. are just a reformulation of those given by Kramosil and Michalek.

Let us recall, on account of [17,18], that a fuzzy metric on a non-empty set X is a pair $(M, *)$ such that $*$ is a continuous t-norm and M is a fuzzy set on $X \times X \times [0, \infty[$, satisfying the following conditions, for all $x, y, z \in X$ and $s, t > 0$:

(KM1) $M(x, y, 0) = 0$;
(KM2) $M(x, y, t) = 1$ for all $t > 0$ if and only if $x = y$;
(KM3) $M(x, y, t) = M(y, x, t)$;
(KM4) $M(x, y, t) * M(y, z, s) \leq M(x, z, t + s)$;
(KM5) The function $M_{x,y} : [0, \infty[\to [0, 1]$ is left-continuous, where $M_{x,y}(t) = M(x, y, t)$.

Similar to the classical case, we will say that $(M, *)$ is a fuzzy pseudo-metric on X provided that axiom **(KM2)** is replaced by the following weaker one:

(KM2') $M(x, x, t) = 1$ for all $t > 0$.

Moreover, given a fuzzy (pseudo-)metric $(M, *)$ on X, we will also say that $(X, M, *)$ is a fuzzy (pseudo-)metric space.

According to [18], the numerical value $M(x, y, t)$ yields the degree of similarity between x and y relative to the value t of the parameter. Of course, it must be clarify that, according to the exposed interpretation, axiom **(KM1)** does not provide any information from a measurement framework because the rest of axioms are enough in order to define a fuzzy measurement. Motivated by this fact, we will assume that a fuzzy metric $(M, *)$ is a fuzzy set M on $X \times X \times]0, \infty[$ that satisfies all the preceding axioms except the axiom **(KM1)**. Of course, the left-continuity of axiom **(KM5)** will be satisfied for the the function $M_{x,y} :]0, \infty[\to [0, 1]$.

The following is a well-known example of fuzzy metric.

Example 1. *Let d be a metric on a non-empty set X. Let M_d be a fuzzy set on $X \times X \times]0, \infty[$ defined, for each $x, y \in X$, by*

$$M_d(x, y, t) = \frac{t}{t + d(x, y)},$$

whenever $t > 0$. On account of [18], $(M_d, \wedge)$ is a fuzzy metric on X, where $\wedge$ denotes the minimum t-norm. The fuzzy metric M_d is called the standard fuzzy metric induced by d.

Following [19], a fuzzy metric $(M, *)$ is said to be stationary provided that the function $M_{x,y} :]0, \infty[\to [0, 1]$ defined by $M_{x,y}(t) = M(x, y, t)$ is constant for each $x, y \in X$.

The next example gives an instance of stationary fuzzy metric.

Example 2. *Let X be a non-empty set X and let $G : X \times X \to]0, \frac{1}{2}[$ be a function such that $G(x, y) = G(y, x)$ for all $x, y \in X$. Consider the fuzzy set M_G on $X \times X \times]0, \infty[$ given by $M_G(x, y, t) = G(x, y)$ for all $t > 0$ and $x, y \in X$ such that $x \neq y$ and $M_G(x, x, t) = 1$ for all $t > 0$. According to [13], $(M_G, *_L)$ is a stationary fuzzy metric, where $*_L$ is the Luckasievicz t-norm.*

Notice that, as in the case of indistinguishability operators, t-norms are crucial in the definition of a fuzzy metric. However, now the unique t-norms under consideration are the continuous ones. Thus, it constitutes a considerable difference between indistinguishability operators and fuzzy metrics. Moreover, another significant difference between these two kinds of fuzzy measurement is that fuzzy metrics include in their definition a parameter. Therefore, none of these types of similarity measurements generalizes the other.

In the light of the preceding fact, it seems natural to try to unify both notions, fuzzy (pseudo-)metrics and indistinguishability operators, under a new one. Thus, the aim of this paper is twofold. On the one hand, we introduce a new type of operator, which we have called modular indistinguishability operator (the name will be justified in Section 3), which provides a degree of similarity or equivalence relative to a parameter and retrieves as a particular case fuzzy (pseudo-)metrics and classical indistinguishability operators. On the other hand, we explore the metric behavior of this new kind of operators. Specifically, we study the duality relationship between modular indistinguishability operators and metrics in the spirit of Theorems 1 and 2. The new results extend the aforementioned results to the new framework and, in addition, allow us to explore also the aforesaid duality relationship when fuzzy (pseudo-)metrics are considered instead of indistinguishability operators.

2. The New Indistinguishability Operators

As we have mentioned before, we are interested in proposing a new type of operator that unify the notion of fuzzy (pseudo-)metric and indistinguishability operator in such a way that a unique theoretical basis can be supplied to develop a wide range of applications. To this end, we introduce the notion of modular indistinguishability operator as follows:

Definition 1. *Let X be a non-empty set and let $*$ be a t-norm, we will say that fuzzy set $F : X \times X \times]0,\infty[\to [0,1]$ is a modular indistinguishability operator for $*$ if for each $x, y, z \in X$ and $t, s > 0$ the following axioms are satisfied:*

(ME1) $F(x, x, t) = 1$;
(ME2) $F(x, y, t) = F(y, x, t)$;
(ME3) $F(x, z, t + s) \geq F(x, y, t) * F(y, z, s)$.

If, in addition, F satisfies for all $x, y \in X$, the following condition:

(ME1') *$F(x, y, t) = 1$ for all $t > 0$ implies $x = y$,*

we will say that F separates points.

Moreover, we will say that F is stationary provided that the function $F_{x,y} :]0,\infty[\to [0,1]$ defined by $F_{x,y}(t) = F(x, y, t)$ is constant for each $x, y \in X$.

Notice that the numerical value $F(x, y, t)$ can understood as the degree up to which x is indistinguishable from y or equivalent to y relative to the value t of the parameter. Moreover, the greater $F(x, y, t)$, the more similar are x and y relative to the value t of the parameter. Clearly, $F(x, y, t) = 1$ for all $t > 0$ when $x = y$.

It is worth mentioning that the classical notion of indistinguishability operator is recovered when the modular indistinguishability operator F is stationary. In addition, it is clear that a modular indistinguishability operator can be considered as a generalization of the concept of fuzzy (pseudo-)metric. However, there are examples of modular indistinguishability operators that are not a fuzzy (pseudo-)metrics such as the next example shows.

Example 3. *Consider a metric d on a non-empty set X. Define the fuzzy set F_d on $X \times X \times]0,\infty[$ as follows:*

$$F_d(x, y, t) = \begin{cases} 0, & \text{if} \quad 0 < t < d(x, y) \text{ and } d(x, y) \neq 0, \\ 1, & \text{if} \quad t \geq d(x, y) \text{ and } d(x, y) \neq 0, \\ 1, & \text{if} \quad d(x, y) = 0. \end{cases}$$

It is easy to check that F_d is a modular indistinguishability operator for the product t-norm $*_P$. Nevertheless, $(F_d, *_P)$ is not a fuzzy (pseudo-)metric because the function $F_{d_{x,y}} :]0,\infty[\to [0,1]$, defined by $F_{d_{x,y}}(t) = F_d(x,y,t)$ is not left-continuous.

The concept of modular indistinguishability operator also generalizes the notion of fuzzy (pseudo-)metric in another outstanding aspect. Observe that, in Definition 1, the continuity on the t-norm is not required. Naturally, the assumption of continuity of the t-norm is useful from a topological viewpoint, since the continuity is necessary in order to define a topology by means of a family of balls in a similar way like in the pseudo-metric case. However, such an assumption could be limiting the range of applications of such fuzzy measurements in those case where (classical) indistinguishability operators works well. In this direction, modular indistinguishability operators present an advantage with respect to fuzzy (pseudo-)metrics because the involved t-norms are not assumed to be continuous.

The following example illustrates the preceding remark providing an instance of modular indistinguishability operator for the Drastic t-norm $*_D$, which is not a modular indistinguishability operator for any continuous t-norm.

Example 4. *Let φ be the function defined on $]0,\infty[$ by $\varphi(t) = \frac{t}{1+t}$. We define the fuzzy set F_D on $[0,1[\times[0,1[\times]0,\infty[$ as follows:*

$$F_D(x,y,t) = \begin{cases} 1, & \text{for each } t > 0, \quad \text{if} \quad x = y, \\ \max\{x,y,\varphi(t)\}, & \text{for each } t > 0, \quad \text{if} \quad x \neq y. \end{cases}$$

First of all, note that, for each $x,y \in [0,1[$ and $t > 0$, we have that $F_D(x,y,t) \in [0,1[$, since $x,y,\varphi(t) \in [0,1[$. Thus, F_D is a fuzzy set on $[0,1[\times[0,1[\times]0,\infty[$.

Now, we will see that F_D is a modular indistinguishability operator on $[0,1[$ for $*_D$. To this end, let us recall that $*_D$ is defined by

$$a *_D b = \begin{cases} 0, & \text{if } a,b \in [0,1[, \\ \min\{a,b\}, & \text{elsewhere.} \end{cases}$$

*It is clear that F_D satisfies axioms **(ME1)** and **(ME2)**. Next, we show that F_D satisfies **(ME3)**, i.e.,*

$$F_D(x,z,t+s) \geq F_D(x,y,t) *_D F_D(y,z,s)$$

for all $x,y,z \in [0,1[$ and $t,s > 0$.

Notice that we can assume that $x \neq z$. Otherwise, the preceding inequality holds trivially. Next, we distinguish two cases:

1. *Case 1. $x \neq y$ and $y \neq z$. Then, $F_D(x,y,t) = \max\{x,y,\varphi(t)\} < 1$ and $F_D(y,z,s) = \max\{y,z,\varphi(s)\} < 1$, since $x,y,z \in [0,1[$ and $\varphi(t) < 1$ for each $t > 0$. Thus, $F_D(x,y,t) *_D F_D(y,z,s) = 0$ attending to the definition of $*_D$. It follows that $F_D(x,z,t+s) \geq F_D(x,y,t) *_D F_D(y,z,s)$.*
2. *Case 2. $x = y$ or $y = z$ (suppose, without loss of generality, that $x = y$). Then, $F_D(x,y,t) = 1$ and so*

$$F_D(x,z,t+s) = F_D(y,z,t+s) = \max\{y,z,\varphi(t+s)\} \geq \max\{y,z,\varphi(s)\} = F_D(y,z,s),$$

since φ is an increasing function. Thus, $F_D(x,z,t+s) \geq F_D(y,z,s) = F_D(x,y,t) *_D F_D(y,z,s)$.

Furthermore, the modular indistinguishability operator F_D separates points. Indeed, let $x,y \in [0,1[$ and $t > 0$. Since $x,y,\varphi(t) \in [0,1[$ for each $t > 0$, we have that, if $x \neq y$, then $F_D(x,y,t) = \max\{x,y,\varphi(t)\} < 1$. Thus, $F_D(x,y,t) = 1$ implies $x = y$.

*Finally, we will prove that F_D is not a modular indistinguishability operator for any continuous t-norm. To this end, we will show that axiom **(ME3)** is not fulfilled for any t-norm continuous at $(1,1)$.*

Let $*$ be a continuous t-norm at $(1,1)$. Then, for each $\epsilon \in]0,1[$, we can find $\delta \in]0,1[$ such that $\delta * \delta > 1-\epsilon$.

Now, consider $x = 0$, $z = \frac{1}{2}$ and $t = s = 1$. Then,

$$F_D(x,z,t+s) = \max\left\{0, \frac{1}{2}, \frac{2}{3}\right\} = \frac{2}{3}.$$

Taking $\epsilon = \frac{1}{3}$, we can find $\delta \in]0,1[$ such that $\delta * \delta > \frac{2}{3}$. Note that, in this case, $\delta > \frac{2}{3}$. Therefore, if we take $y = \delta$, we have that

$$F_D(x,y,t) * F_D(y,z,s) = \max\left\{0, y, \frac{1}{2}\right\} * \max\left\{y, \frac{1}{2}, \frac{1}{2}\right\} = y * y > \frac{2}{3} = F_D(x,z,t+s).$$

*Thus, (**ME3**) is not satisfied.*

We end the section with a reflection on axiom (**KM1**). When such an axiom is considered in the definition of fuzzy (pseudo-)metric (i.e., the fuzzy (pseudo-)metric is considered as a fuzzy set on $X \times X \times [0,\infty[$ instead on $X \times X \times]0,\infty[$), one could wonder whether modular indistinguishability operators would be able to extend the notion of fuzzy (pseudo-)metric in that case. The answer to the posed question is affirmative. In fact, in order to define a new indistinguishability operator for that purpose, we only need to include in the axiomatic in Definition 1 the following axiom:

(ME0) $F(x,y,0) = 0$ for all $x, y \in X$.

Notice that, even in such a case, there exist modular indistinguishability operators that are not fuzzy (pseudo-)metrics. An example of such a kind of operators is given by an easy adaptation of the fuzzy set F_d introduced in Example 3. Indeed, we only need to consider such a fuzzy set defined as in the aforesaid example and, in addition, satisfying $F_d(x,y,0) = 0$ for all $x, y \in X$. Of course, it is easy to check that F_d is a modular indistinguishability operator for the product t-norm $*_P$, which satisfies (**ME0**) but $(F_d, *_P)$ is not a fuzzy (pseudo-)metric.

3. The Duality Relationship

This section is devoted to explore the metric behavior of the new indistinguishability operators. Concretely, we extend, on the one hand, the technique through which a metric can be generated from an indistinguishability operator by means of an additive generator of a t-norm (in Section 3.1) and, on the other hand, the technique that allows for inducing an indistinguishability operator from a metric by means of the pseudo-inverse of the additive generator of a t-norm (in Section 3.2). The same results are also explored when fuzzy (pseudo-)metrics are considered instead of modular indistinguishability operators.

3.1. From Modular Indistinguishability Operators to Metrics

In order to extend Theorem 1 to the modular framework, we need to propose a metric class as a candidate to be induced by a modular indistinguishability operator. We have found that such a candidate is known in the literature as modular metric. Let us recall a few basics about this type of metrics.

According to Chytiakov, V.V. (see [20]), a function $w :]0,\infty[\times X \times X \to [0,\infty]$ is a modular metric on a non-empty set X if, for each $x, y, z \in X$ and each $\lambda, \mu > 0$, the following axioms are fulfilled:

(MM1) $w(\lambda, x, y) = 0$ for all $\lambda > 0$ if and only if $x = y$;
(MM2) $w(\lambda, x, y) = w(\lambda, y, x)$;
(MM3) $w(\lambda + \mu, x, z) \leq w(\lambda, x, y) + w(\mu, y, z)$.

If the axiom (**MM1**) is replaced by the following one,

(MM1′) $w(\lambda, x, x) = 0$ for all $\lambda > 0$,

then w is a called modular pseudo-metric on X.

Of course, the value $w(\lambda, x, y)$ can be understood as a dissimilarity measurement between objects relative to the value λ of a parameter.

Following [20], given $x, y \in X$ and $\lambda > 0$, we will denote from now on the value $w(\lambda, x, y)$ by $w_\lambda(x, y)$.

Notice that, as was pointed out in [20], a (pseudo-)metric is a modular (pseudo-)metric, which is "stationary", i.e., it does not depend on the value t of the parameter. Thus, (pseudo-)metrics on X are modular (pseudo-)metrics $w :]0, \infty[\times X \times X \to [0, \infty]$ such that the assignment $w_{x,y} :]0, \infty[\to [0, \infty]$, given by $w_{x,y}(\lambda) = w_\lambda(x, y)$ is a constant function for each $x, y \in X$.

The following are well-known examples of modular (pseudo-)metrics.

Example 5. *Let d be a (pseudo-)metric on X and let $\varphi :]0, \infty[\to]0, \infty[$ be a non-decreasing function. The functions defined on $]0, \infty[\times X \times X$ as follows:*

(i) $w^1_\lambda(x, y) = \begin{cases} \infty, & \text{if } x \neq y, \\ 0, & \text{if } x = y, \end{cases}$

(ii) $w^2_\lambda(x, y) = \begin{cases} \infty, & \text{if } 0 < \lambda < d(x, y) \text{ and } d(x, y) > 0, \\ 0, & \text{if } \lambda \geq d(x, y) \text{ and } d(x, y) > 0, \\ 0, & \text{if } d(x, y) = 0, \end{cases}$

(iii) $w^3_\lambda(x, y) = \frac{d(x,y)}{\varphi(\lambda)}$,

are modular (pseudo-)metrics on X.

Next, we provide an example of modular metric that will be crucial in Section 3.2.

Proposition 1. *Let d be a metric space on X. Then, the function $w :]0, \infty[\times X \times X \to [0, \infty]$ is a modular metric on X, where*

$$w_\lambda(x, y) = \frac{d^2(x, y)}{\lambda}$$

for each $x, y \in X$ and $\lambda \in]0, \infty[$ (in the last expression, $d^2(x, y)$ denotes $(d(x, y))^2$, as usual).

Proof. It is clear that axioms **(MM1)** and **(MM2)** are satisfied. It remains to show that axiom **(MM3)** holds. Let $x, y, z \in X$ and $\lambda, \mu \in]0, \infty[$. Note that

$$d^2(x, z) \leq \Big(d(x, y) + d(y, z)\Big)^2 = d^2(x, y) + 2d(x, y)d(y, z) + d^2(y, z),$$

since d is a metric and satisfies the triangle inequality.

From the preceding inequality, we deduce the following one:

$$\frac{d^2(x, y)}{\lambda} + \frac{d^2(y, z)}{\mu} - \frac{d^2(x, z)}{\lambda + \mu} = \frac{\mu(\lambda + \mu)d^2(x, y) + \lambda(\lambda + \mu)d^2(y, z) - \lambda\mu d^2(x, z)}{\lambda\mu(\lambda + \mu)} =$$

$$= \frac{\mu\lambda d^2(x, y) + \mu^2 d^2(x, y) + \lambda^2 d^2(y, z) + \lambda\mu d^2(y, z) - \lambda\mu d^2(x, z)}{\lambda\mu(\lambda + \mu)} \geq$$

$$\geq \frac{\mu\lambda d^2(x, y) + \mu^2 d^2(x, y) + \lambda^2 d^2(y, z) + \lambda\mu d^2(y, z) - \lambda\mu(d^2(x, y) + 2d(x, y)d(y, z) + d^2(y, z))}{\lambda\mu(\lambda + \mu)} =$$

$$= \frac{\mu^2 d^2(x, y) + \lambda^2 d^2(y, z) - 2\lambda\mu d(x, y)d(y, z)}{\lambda\mu(\lambda + \mu)} = \frac{(\mu d(x, y) - \lambda d(y, z))^2}{\lambda\mu(\lambda + \mu)} \geq 0.$$

Therefore,

$$w_{\lambda+\mu}(x,z) = \frac{d^2(x,z)}{\lambda+\mu} \leq \frac{d^2(x,y)}{\lambda} + \frac{d^2(y,z)}{\mu} = w_\lambda(x,y) + w_\mu(y,z).$$

Hence, w satisfies **(MM3)**. □

After a brief introduction to modular metric spaces we are able to yield a modular version of Theorem 1.

Theorem 3. *Let X be a non-empty set and let $*$ be a continuous t-norm with additive generator $f_*: [0,1] \to [0,\infty]$. If $\diamond$ is a t-norm, then the following assertions are equivalent:*

(1) $* \leq \diamond$ (i.e., $x * y \leq x \diamond y$ for all $x,y \in [0,1]$).

(2) For any modular indistinguishability operator F on X for $\diamond$, the function $(w^{F,f_*}) :]0,\infty[\times X \times X \to [0,\infty]$ defined by

$$(w^{F,f_*})_\lambda(x,y) = f_*(F(x,y,\lambda)),$$

for each $x,y \in X$ and $\lambda > 0$, is a modular pseudo-metric on X.

(3) For any modular indistinguishability operator F on X for $\diamond$ that separates points, the function $(w^{F,f_*}) :]0,\infty[\times X \times X \to [0,\infty]$ defined by

$$(w^{F,f_*})_\lambda(x,y) = f_*(F(x,y,\lambda)),$$

for each $x,y \in X$ and $\lambda > 0$, is a modular metric on X.

Proof. (1) $\Rightarrow$ (2) Suppose that $* \leq \diamond$ and let F be a modular indistinguishability operator on X for $\diamond$. We will see that (w^{F,f_*}) is a modular pseudo-metric on X.

(MM1') Let $x \in X$. Since $F(x,x,\lambda) = 1$ for each $\lambda > 0$, then $(w^{F,f_*})_\lambda(x,x) = f_*(F(x,x,\lambda)) = f_*(1) = 0$ for each $\lambda > 0$.

(MM2) It is obvious because $F(x,y,\lambda) = F(y,x,\lambda)$ for all $x,y \in X$ and $\lambda > 0$.

(MM3) Let $x,y,z \in X$ and $\lambda,\mu > 0$. We will show that the following inequality

$$(w^{F,f_*})_{\lambda+\mu}(x,z) \leq (w^{F,f_*})_\lambda(x,y) + (w^{F,f_*})_\mu(y,z)$$

holds. First of all, note that F is also a modular indistinguishability operator for $*$ on X due to $\diamond \geq *$. Then, the following inequality is satisfied:

$$F(x,z,\lambda+\mu) \geq F(x,y,\lambda) * F(y,z,\mu) = f_*^{(-1)}\left(f_*(F(x,y,\lambda)) + f_*(F(y,z,\mu))\right).$$

Taking into account that f_* is an additive generator, and thus a decreasing function, we have that

$$f_*(F(x,z,\lambda+\mu)) \leq f_*\left(f_*^{(-1)}\left(f_*(F(x,y,\lambda)) + f_*(F(y,z,\mu))\right)\right).$$

Now, we will distinguish two different cases:

(a) Suppose that $f_*(F(x,y,\lambda)) + f_*(F(y,z,\mu)) \in Ran(f_*)$.
Since f_* is an additive generator of the t-norm $*$, we have that $f_* \circ f_*^{(-1)}|_{Ran(f_*)} = id|_{Ran(f_*)}$. Then,

$$f_*\left(f_*^{(-1)}\left(f_*(F(x,y,\lambda)) + f_*(F(y,z,\mu))\right)\right) = f_*(F(x,y,\lambda)) + f_*(F(y,z,\mu)).$$

It follows that

$$(w^{F,f_*})_{\lambda+\mu}(x,z) = f_*(F(x,z,\lambda+\mu)) \leq f_*(F(x,y,\lambda)) + f_*(F(y,z,\mu)) =$$

$$= (w^{F,f_*})_\lambda(x,y) + (w^{F,f_*})_\mu(y,z).$$

(b) Suppose that $f_*(F(x,y,\lambda)) + f_*(F(y,z,\mu)) \notin Ran(f_*)$. Since f_* is an additive generator of the t-norm $*$, we have that $f_*(a) + f_*(b) \in Ran(f_*) \cup [f_*(0), \infty]$ for each $a, b \in [0,1]$. Then,

$$f_*(F(x,y,\lambda)) + f_*(F(y,z,\mu)) > f_*(0).$$

Thus, we obtain

$$f_*(F(x,z,\lambda+\mu)) \leq f_*(0) < f_*(F(x,y,\lambda)) + f_*(F(y,z,\mu)),$$

where we have that

$$(w^{F,f_*})_{\lambda+\mu}(x,z) \leq (w^{F,f_*})_\lambda(x,y) + (w^{F,f_*})_\mu(y,z),$$

as we claimed.

Therefore, (w^{F,f_*}) is a modular pseudo-metric on X.

(2) $\Rightarrow$ (3) Let F be a modular indistinguishability operator on X for $\diamond$ that separates points. By our assumption, (w^{F,f_*}) is a pseudo-modular metric on X. We will see that (w^{F,f_*}) is a modular metric on X.

Let $x, y \in X$ such that $(w^{F,f_*})_\lambda(x,y) = 0$ for all $\lambda > 0$. By definition, we have that $f_*(F(x,y,\lambda)) = 0$ for all $\lambda > 0$. Then, $F(x,y,\lambda) = 1$ for all $\lambda > 0$, since f_* is an additive generator of $*$. Therefore, $x = y$, since F is a modular indistinguishability operator on X for $\diamond$ that separates points.

(3) $\Rightarrow$ (1) Suppose that, for any modular indistinguishability operator F on X for $\diamond$ that separates points, the function (w^{F,f_*}) is a modular metric on X. We will show that $\diamond \geq *$. To this end, we will prove that $a \diamond b \geq a * b$ provided $a, b \in [0,1[$. Note that the preceding inequality is obvious whenever either $a = 1$ or $b = 1$.

Let $a, b \in [0,1[$. Consider a set constituted by three distinct points $X = \{x, y, z\}$. We define a fuzzy set F on $X \times X \times]0, \infty[$ as follows:

$$F(u,v,t) = F(v,u,t) = \begin{cases} 1, & \text{if } u = v, \\ a \diamond b, & \text{if } u = x \text{ and } v = z, \\ a, & \text{if } u = x \text{ and } v = y, \\ b, & \text{if } u = y \text{ and } v = z, \end{cases}$$

for all $t > 0$.

It is easy to verify, attending to its definition, that F is a modular indistinguishability operator on X for $\diamond$ that separates points. Thus, (w^{F,f_*}) is a modular metric on X. Therefore, given $\lambda > 0$, we have that

$$f_*(a \diamond b) = (w^{F,f_*})_{2\lambda}(x,z) \leq (w^{F,f_*})_\lambda(x,y) + (w^{F,f_*})_\lambda(y,z) = f_*(a) + f_*(b).$$

Notice that, for each $c \in [0,1]$, we have that $(f_*^{(-1)} \circ f_*)(c) = c$, $a * b = f_*^{(-1)}(f_*(a) + f_*(b))$ and that $f_*^{(-1)}$ is decreasing, since f_* is an additive generator of the t-norm $*$. Taking into account the preceding facts and, from the above inequality, we deduce that

$$a \diamond b = f_*^{(-1)}(f_*(a \diamond b)) \geq f_*^{(-1)}(f_*(a) + f_*(b)) = a * b,$$

as we claimed.

This last implication concludes the proof. □

In order to illustrate the technique introduced in the above theorem, we provide two corollaries that establish the particular cases for the Luckasievicz t-norm and the usual product. With this aim, we recall that an additive generator f_{*_L} of $*_L$ and f_{*_P} of $*_P$ is given by

$$\begin{aligned} f_{*_L}(a) &= 1-a, \\ f_{*_P}(a) &= -\log(a), \end{aligned}$$

for each $a \in [0,1]$, respectively. Of course, we have adopted the convention that $\log(0) = -\infty$.

Corollary 1. *Let X be a non-empty set. If $\diamond$ is a t-norm, then the following assertions are equivalent:*

(1) $*_L \leq \diamond$.

(2) *For any modular indistinguishability operator F on X for $\diamond$, the function $(w^{F,f_{*_L}}) :]0,\infty[\times X \times X \to [0,\infty]$ defined by*

$$(w^{F,f_{*_L}})_\lambda(x,y) = 1 - F(x,y,\lambda),$$

for each $x,y \in X$ and $\lambda > 0$, is a modular pseudo-metric on X.

(3) *For any modular indistinguishability operator F on X for $\diamond$ that separates points, the function $(w^{F,f_{*_L}}) :]0,\infty[\times X \times X \to [0,\infty]$ defined by*

$$(w^{F,f_{*_L}})_\lambda(x,y) = 1 - F(x,y,\lambda),$$

for each $x,y \in X$ and $\lambda > 0$, is a modular metric on X.

Corollary 2. *Let X be a non-empty set. If $\diamond$ is a t-norm, then the following assertions are equivalent:*

(1) $*_P \leq \diamond$.

(2) *For any modular indistinguishability operator F on X for $\diamond$, the function $(w^{F,f_{*_P}}) :]0,\infty[\times X \times X \to [0,\infty]$ defined by*

$$(w^{F,f_{*_P}})_\lambda(x,y) = -log(F(x,y,\lambda)),$$

for each $x,y \in X$ and $\lambda > 0$, is a modular pseudo-metric on X.

(3) *For any modular indistinguishability operator F on X for $\diamond$ that separates points, the function $(w^{F,f_{*_P}}) :]0,\infty[\times X \times X \to [0,\infty]$ defined by*

$$(w^{F,f_{*_P}})_\lambda(x,y) = -log(F(x,y,\lambda)),$$

for each $x,y \in X$ and $\lambda > 0$, is a modular metric on X.

Theorem 3 also gives a specific method to generate modular metrics when we focus our attention on fuzzy (pseudo-)metrics instead of modular indistinguishability operators in general.

Corollary 3. *Let X be a non-empty set and let $*$ be a t-norm with additive generator $f_* : [0,1] \to [0,\infty]$. If $\diamond$ is a continuous t-norm, then the following assertions are equivalent:*

(1) $* \leq \diamond$.

(2) *For any fuzzy pseudo-metric $(M,\diamond)$ on X, the function $(w^{M,f_*}) :]0,\infty[\times X \times X \to [0,\infty]$ defined by*

$$(w^{M,f_*})_\lambda(x,y) = f_*(M(x,y,\lambda)),$$

for each $x,y \in X$ and $\lambda > 0$, is a modular pseudo-metric on X.

(3) *For any fuzzy metric* $(M, \diamond)$ *on* X, *the function* $(w^{M,f_*}) :]0,\infty[\times X \times X \to [0,\infty]$ *defined by*

$$(w^{M,f_*})_\lambda(x,y) = f_*(M(x,y,\lambda)),$$

for each $x, y \in X$ *and* $\lambda > 0$, *is a modular metric on* X.

As a consequence of the preceding result, we obtain immediately the following one.

Corollary 4. *Let* X *be a non-empty set and let* $*$ *be a continuous t-norm with additive generator* $f_* : [0,1] \to [0,\infty]$. *Then, the following assertions are equivalent:*

(1) *For any fuzzy pseudo-metric* $(M, *)$ *on* X, *the function* $(w^{M,f_*}) :]0,\infty[\times X \times X \to [0,\infty]$ *defined by*

$$(w^{M,f_*})_\lambda(x,y) = f_*(M(x,y,\lambda)),$$

for each $x, y \in X$ *and* $\lambda > 0$, *is a modular pseudo-metric on* X.

(2) *For any fuzzy metric* $(M, *)$ *on* X, *the function* $(w^{M,f_*}) :]0,\infty[\times X \times X \to [0,\infty]$ *defined by*

$$(w^{M,f_*})_\lambda(x,y) = f_*(M(x,y,\lambda)),$$

for each $x, y \in X$ *and* $\lambda > 0$, *is a modular metric on* X.

It is clear that, when we consider stationary modular indistinguishability operators in the statement of Theorem 3, we obtain as a particular case Theorem 1 and, thus, the classical technique to induce a metric from an indistinguishability operator by means of an additive generator. Clearly, if we replace modular indistinguishability operators by stationary fuzzy metrics, we obtain a more restrictive version of the classical technique, provided by Theorem 3, because it only remains valid for continuous t-norms.

3.2. From Modular (Pseudo-)Metrics to Modular Indistinguishability Operators

As was mentioned above, the main goal of this subsection is to provide a version of Theorem 2 when we consider a modular (pseudo-)metric instead of a (pseudo-)metric. Thus, we give a technique to induce a modular indistinguishability operator from a modular (pseudo-)metric by means of the pseudo-inverse of the additive generator of a t-norm. To this end, let us recall the following representation result, which will be crucial in our subsequent discussion, holds for continuous t-norms:

Theorem 4. *A binary operator* $*$ *in* $[0,1]$ *is a continuous Archimedean t-norm if and only if there exists a continuous additive generator* f_* *such that*

$$x * y = f_*^{(-1)}(f_*(x) + f_*(y)), \tag{1}$$

where the pseudo-inverse $f_*^{(-1)}$ *is given by*

$$f_*^{(-1)}(y) = f^{-1}(\min\{f_*(0), y\}) \tag{2}$$

for all $y \in [0,\infty]$.

In the next result, we introduce the promised technique.

Theorem 5. *Let* $*$ *be a continuous t-norm with additive generator* $f_* : [0,1] \to [0,\infty]$. *If* w *is a modular pseudo-metric on* X, *then the function* $F^{w,f_*} : X \times X \times]0,\infty[\to [0,1]$ *defined, for all* $x, y \in X$ *and* $t > 0$, *by*

$$F^{w,f_*}(x,y,t) = f_*^{(-1)}(w_t(x,y))$$

is a modular indistinguishability operator for $*$. *Moreover, the modular indistinguishability operator* F^{w,f_*} *separates points if and only if* w *is a modular metric on* X.

Proof. Let $*$ be a continuous Archimedean t-norm with additive generator $f_* : [0,1] \to [0,\infty]$ and consider w a modular pseudo-metric on X.

We define the function $F^{w,f_*} : X \times X \times]0,\infty[\to [0,1]$ as follows:

$$F^{w,f_*}(x,y,t) = f_*^{(-1)}(w_t(x,y)),$$

for all $x,y \in X$ and $t > 0$. We will see that F^{w,f_*} is a modular inidistinguishability operator for $*$.

(ME1) Let $x \in X$. Since w is a modular pseudo-metric on X, we have that $w_t(x,x) = 0$ for all $t > 0$. Therefore, $F^{w,f_*}(x,x,t) = f_*^{(-1)}(w_t(x,x)) = f_*^{(-1)}(0) = 1$ for all $t > 0$.

(ME2) Is a consequence of the definition of F^{w,f_*}, since w is a modular pseudo-metric and so it satisfies that $w_t(x,y) = w_t(y,x)$ for each $x,y \in X$ and $t > 0$.

(ME3) Let $x,y,z \in X$ and $t,s > 0$. On the one hand, by expression (2), we deduce that

$$F^{w,f_*}(x,z,t+s) = f_*^{(-1)}(w_{t+s}(x,z)) = f_*^{-1}(\min\{f_*(0), w_{t+s}(x,z)\}).$$

Now, since w is a modular pseudo-metric on X, then

$$w_{t+s}(x,z) \leq w_t(x,y) + w_s(y,z)$$

and, hence,

$$F^{w,f_*}(x,z,t+s) \geq f_*^{-1}(\min\{f_*(0), w_t(x,y) + w_s(y,z)\}).$$

On the other hand, we have that

$$F^{w,f_*}(x,y,t) * F^{w,f_*}(y,z,s) = f_*^{(-1)}\left(f_*\left(F^{w,f_*}(x,y,t)\right) + f_*\left(F^{w,f_*}(y,z,s)\right)\right) =$$

$$= f_*^{-1}\left(\min\left\{f_*(0), f_*\left(F^{w,f_*}(x,y,t)\right) + f_*\left(F^{w,f_*}(y,z,s)\right)\right\}\right).$$

Moreover, by expression (2), we obtain that

$$f_*\left(F^{w,f_*}(x,y,t)\right) = f_*\left(f_*^{(-1)}(w_t(x,y))\right) = \min\{f_*(0), w_t(x,y)\}$$

and

$$f_*\left(F^{w,f_*}(y,z,s)\right) = f_*\left(f_*^{(-1)}(w_s(y,z))\right) = \min\{f_*(0), w_s(y,z)\}.$$

To finish the proof, we will see that

$$\min\{f_*(0), w_t(x,y) + w_s(y,z)\} = \min\{f_*(0), \min\{f_*(0), w_t(x,y)\} + \min\{f_*(0), w_s(y,z)\}\}.$$

To this end, we will distinguish three cases:

Case 1. $f_*(0) \leq w_t(x,y)$ and $f_*(0) \leq w_s(y,z)$. Then, we have that

$$\min\{f_*(0), w_t(x,y) + w_s(y,z)\} = f_*(0)$$

and

$$\min\{f_*(0), \min\{f_*(0), w_t(x,y)\} + \min\{f_*(0), w_s(y,z)\}\} = \min\{f_*(0), f_*(0) + f_*(0)\} = f_*(0).$$

Case 2. $f_*(0) > w_t(x,y)$ and $f_*(0) \le w_s(y,z)$ (the case $f_*(0) \le w_t(x,y)$ and $f_*(0) > w_s(y,z)$ runs following the same arguments). It follows that

$$\min\{f_*(0), w_t(x,y) + w_s(y,z)\} = f_*(0)$$

and

$$\min\{f_*(0), \min\{f_*(0), w_t(x,y)\} + \min\{f_*(0), w_s(y,z)\}\} = \min\{f_*(0), w_t(x,y) + f_*(0)\} = f_*(0).$$

Case 3. $f_*(0) > w_t(x,y)$ and $f_*(0) > w_s(y,z)$. Then, we have that

$$\min\{f_*(0), \min\{f_*(0), w_t(x,y)\} + \min\{f_*(0), w_s(y,z)\}\} = \min\{f_*(0), w_t(x,y) + w_S(y,z)\}.$$

Therefore,

$$F^{w,f_*}(x,z,t+s) \ge f_*^{-1}\left(\min\left\{f_*(0), f_*\left(F^{w,f_*}(x,y,t)\right) + f_*\left(F^{w,f_*}(y,z,s)\right)\right\}\right)$$
$$= F^{w,f_*}(x,y,t) * F^{w,f_*}(y,z,s),$$

whence we deduce that F^{w,f_*} is a modular indistinguishability operator for $*$ on X.

Finally, it is clear that $F^{w,f_*}(x,y,t) = 1$ for all $x, y \in X$ and $t > 0$ if, and only if, $f_*^{(-1)}(w_t(x,y)) = 1$ for all $x, y \in X$ and $t > 0$. Since $f_*^{(-1)}(w_t(x,y)) = 1$ for all $x, y \in X$ and $t > 0$ if, and only if, $w_t(x,y) = 0$ for all $x, y \in X$ and $t > 0$, we immediately obtain that F^{w,f_*} is a modular indistinguishability operator that separates points if, and only if, w is a modular metric on X. □

Next, we specify the method given in Theorem 5 for the t-norms $*_L$ and $*_P$. Note that the pseudo-inverse of the additive generator f_{*_L} and f_{*_P} is given by

$$f_{*_L}^{(-1)}(b) = \begin{cases} 1-b, & \text{if } b \in [0,1[, \\ 0, & \text{if } b \in [1,\infty], \end{cases}$$

and

$$f_{*_P}^{(-1)}(b) = \mathrm{e}^{-b},$$

for each $b \in [0,\infty]$, respectively, where we have adopted the convention that $\mathrm{e}^{-\infty} = 0$.

Corollary 5. *If w is a modular pseudo-metric on X, then the function $F^{w,f_{*_L}} : X \times X \times]0,\infty[\to [0,1]$ defined, for all $x, y \in X$ and $t > 0$, by*

$$F^{w,f_{*_L}}(x,y,t) = \begin{cases} 1 - w_t(x,y), & \text{if } w_t(x,y) \in [0,1[, \\ 0, & \text{if } w_t(x,y) \in [1,\infty], \end{cases}$$

is a modular indistinguishability operator for $*_L$. Moreover, the modular indistinguishability operator $F^{w,f_{*_L}}$ separates points if and only if w is a modular metric on X.

Corollary 6. *If w is a modular pseudo-metric on X, then the function $F^{w,f_{*_P}} : X \times X \times]0,\infty[\to [0,1]$ defined, for all $x, y \in X$ and $t > 0$, by*

$$F^{w,f_{*_P}}(x,y,t) = \mathrm{e}^{-w_t(x,y)},$$

is a modular indistinguishability operator for $*_P$. Moreover, the modular indistinguishability operator $F^{w,f_{*_L}}$ separates points if and only if w is a modular metric on X.

In the light of Theorem 5, it seems natural to ask if the continuity of the t-norm can be eliminated from the assumptions of such a result. The next example gives a negative answer to that question. In particular, it proves that there are fuzzy sets F^{w,f_*}, given by Theorem 5, that are not modular indistinguishability operators when the t-norm $*$ under consideration is not continuous.

Example 6. *Consider the Euclidean metric d_E on $\mathbb{R}$. By Proposition 1, the function w^E is a modular metric on $\mathbb{R}$, where*

$$w^E_\lambda(x,y) = \frac{(d_E(x,y))^2}{\lambda}$$

for all $x,y \in \mathbb{R}$ and $\lambda > 0$. Consider the additive generator f_{*_D} of the non-continuous t-norm $*_D$. Recall that f_{*_D} is given by

$$f_{*_D}(x) = \begin{cases} 0, & \text{if } x = 1, \\ 2-x, & \text{if } x \in [0,1[. \end{cases}$$

An easy computation shows that its pseudo-inverse is given by

$$f^{(-1)}_{*_D}(x) = \begin{cases} 1, & \text{if } x \in [0,1], \\ 2-x, & \text{if } x \in]1,2], \\ 0, & \text{if } x \in]2,\infty[. \end{cases}$$

Next, we show that we can find $x,y,z \in \mathbb{R}$ and $\lambda, \mu \in]0,\infty[$ such that

$$F^{w^E,f_{*_D}}(x,z,\lambda+\mu) < F^{w^E,f_{*_D}}(x,y,\lambda) *_D F^{w^E,f_{*_D}}(y,z,\mu).$$

Let $x=0$, $y=1$ and $z=2$, and consider $\lambda=\mu=1$. Then,

$$w^E_{\lambda+\mu}(x,z,\lambda) = \frac{(d_E(x,z))^2}{\lambda+\mu} = \frac{2^2}{2} = 2,$$

$$w^E_\lambda(x,y) = \frac{(d_E(x,y))^2}{\lambda} = \frac{1^2}{1} = 1,$$

and

$$w^E_\mu(y,z) = \frac{(d_E(y,z))^2}{\mu} = \frac{1^2}{1} = 1.$$

Therefore,

$$0 = f^{(-1)}_{*_D}(2) = F^{w^E,f_{*_D}}(x,z,\lambda+\mu) < F^{w^E,f_{*_D}}(x,y,\lambda) *_D F^{w^E,f_{*_D}}(y,z,\mu) = f^{(-1)}_{*_D}(1) *_D f^{(-1)}_{*_D}(1) = 1.$$

Since the continuity is a necessary hypothesis in the statement of Theorem 5, one could expect that the following result would be true.

"Let $*$ be a continuous Archimedean t-norm with additive generator $f_* : [0,1] \to [0,\infty]$. If w is a modular pseudo-metric on X, then the pair $(M^{w,f_*}, *)$ is a fuzzy (pseudo-)metric, where the fuzzy set $M^{w,f_*} : X \times X \times]0,\infty[$ is given, for all $x,y \in X$ and $t > 0$, by

$$M^{w,f_*}(x,y,t) = f^{(-1)}_*(w_t(x,y)).$$

Moreover, $(M^{w,f_*}, *)$ is a fuzzy metric if and only if w is a modular metric on X."

Nevertheless, the following example proves that such a result does not hold. In fact, the technique provided by Theorem 5 does not give in general a fuzzy (pseudo-)metric.

Example 7. *Let d be a metric on a non-empty set X. Consider the modular metric w^2 on X introduced in Example 5, that is,*

$$w_t^2(x,y) = \begin{cases} \infty, & \text{if} \quad 0 < t < d(x,y) \text{ and } d(x,y) > 0, \\ 0, & \text{if} \quad t \geq d(x,y) \text{ and } d(x,y) > 0, \\ 0, & \text{if} \quad d(x,y) = 0, \end{cases}$$

for all $x, y \in X$ and $t > 0$. Then, it is not hard to check that the pair $(M^{w^2,f_{*P}}, *_P)$ is not a fuzzy (pseudo-)metric, where the fuzzy set $M^{w^2,f_{*P}}$ is given by

$$M^{w^2,f_{*P}}(x,y,t) = f_{*P}^{(-1)}(w_t^2(x,y)) = \begin{cases} 0, & \text{if} \quad 0 < t < d(x,y) \text{ and } d(x,y) > 0, \\ 1, & \text{if} \quad t \geq d(x,y) \text{ and } d(x,y) > 0, \\ 1, & \text{if} \quad d(x,y) = 0, \end{cases}$$

*for all $x, y \in X$ and $t > 0$. Notice that $(M^{w^2,f_{*P}}, *_P)$ fails to fulfill axiom (**KM5**), i.e., the function $M_{x,y}^{w^2,f_{*P}}$: $]0,\infty[\to [0,1]$ is not left-continuous.*

The preceding example suggests the study of those conditions that a modular (pseudo-)metric must satisfy in order to induce a fuzzy (pseudo-) metric by means of the technique exposed in Theorem 5. The following lemma, whose proof was given in [20], will help us to find it.

Lemma 1. *Let w be a modular (pseudo-)metric on X. Then, for each $x, y \in X$, we have that $w_s(x,y) \geq w_t(x,y)$ whenever $s, t \in]0,\infty[$ with $s < t$.*

Taking into account the preceding lemma, the next result provides a condition that is useful for our target.

Proposition 2. *Let w be a modular pseudo-metric on X. The function $\tilde{w} :]0,\infty[\times X \times X \to [0,\infty]$ given, for each $x, y \in X$ and $t > 0$, by*

$$\tilde{w}_\lambda(x,y) = \inf_{0<t<\lambda} w_t(x,y)$$

is a modular pseudo-metric on X such that for each $x, y \in X$ the function $\tilde{w}_{x,y} :]0,\infty[\to]0,\infty[$ is left continuous, where $\tilde{w}_{x,y}(\lambda) = \tilde{w}_\lambda(x,y)$ for each $\lambda \in]0,\infty[$. Furthermore, $\tilde{w}$ is a modular metric on X if and only if w it is so.

Proof. It is obvious that $\tilde{w}$ satisfies axiom (**MM2**). Next, we show that $\tilde{w}$ satisfies axioms (**MM1′**) and (**MM3**).

(MM1′) Fix $x \in X$ and let $\lambda \in]0,\infty[$. Since w is a modular pseudo-metric on X, then $w_t(x,x) = 0$ for each $t > 0$. Therefore,

$$\tilde{w}_\lambda(x,x) = \inf_{0<t<\lambda} w_t(x,x) = 0.$$

(MM3) Let $x, y, z \in X$ and $\lambda, \mu \in]0,\infty[$. Next, we prove that

$$\tilde{w}_{\lambda+\mu}(x,z) \leq \tilde{w}_\lambda(x,y) + \tilde{w}_\mu(y,z).$$

With this aim, note that, given $u, v \in X$ and $\alpha \in]0,\infty[$, we have that, for each $\epsilon \in]0,\infty[$, we can find $t \in]0,\alpha[$ satisfying $w_t(u,v) < \tilde{w}_\alpha(u,v) + \epsilon$.

Fixing an arbitrary $\epsilon \in]0,\infty[$, we can then find $t \in]0,\lambda[$ and $s \in]0,\mu[$ such that $w_t(x,y) < \tilde{w}_\lambda(x,y) + \epsilon/2$ and $w_s(y,z) < \tilde{w}_\mu(y,z) + \epsilon/2$. Therefore,

$$\tilde{w}_{\lambda+\mu}(x,z) \leq w_{t+s}(x,z) \leq w_t(x,y) + w_s(y,z) < \tilde{w}_\lambda(x,y) + \tilde{w}_\mu(y,z) + \epsilon,$$

since w is a pseudo-metric on X. Taking into account that $\epsilon \in]0,\infty[$ is arbitrary, we conclude that

$$\tilde{w}_{\lambda+\mu}(x,z) \leq \tilde{w}_\lambda(x,y) + \tilde{w}_\mu(y,z).$$

Thus, $\tilde{w}$ is a modular pseudo-metric on X.

We will continue showing that, for each $x,y \in X$, the function $\tilde{w}_{x,y} :]0,\infty[\to]0,\infty[$ is left continuous. Fix $x,y \in X$ and consider an arbitrary $\lambda_0 \in]0,\infty[$. Then, given $\epsilon \in]0,\infty[$, we can find $\delta \in]0,\infty[$ such that

$$\tilde{w}_\lambda(x,y) - \tilde{w}_{\lambda_0}(x,y) < \epsilon,$$

for each $\lambda \in]\lambda_0 - \delta, \lambda_0]$ (note that $\tilde{w}_\lambda(x,y) \geq \tilde{w}_{\lambda_0}(x,y)$ for each $\lambda \in]\lambda_0 - \delta, \lambda_0]$ by Lemma 1). Indeed, let $\epsilon \in]0,\infty[$. As before, we can find $t \in]0,\lambda_0[$ such that

$$w_t(x,y) < \tilde{w}_{\lambda_0}(x,y) + \epsilon$$

and, again by Lemma 1, we have that $w_s(x,y) < \tilde{w}_{\lambda_0}(x,y) + \epsilon$ for each $s \in]t,\lambda_0]$. Therefore, taking $\delta = \lambda_0 - t$, we have that

$$\tilde{w}_\lambda(x,y) - \tilde{w}_{\lambda_0}(x,y) \leq w_\lambda(x,y) - \tilde{w}_{\lambda_0}(x,y) < \epsilon,$$

for each $\lambda \in]\lambda_0 - \delta, \lambda_0]$, as we claimed. Thus, $\tilde{w}_{x,y}$ is left-continuous on $]0,\infty[$ since λ_0 is arbitrary.

Finally, it is easy to verify that $\tilde{w}$ is a modular metric on X, if and only if w, it is so. Indeed, $\tilde{w}$ is a modular metric on X if and only if $\tilde{w}_\lambda(x,y) = 0$ for each $\lambda \in]0,\infty[$ implies $x = y$, but $\tilde{w}_\lambda(x,y) = \inf_{0<t<\lambda} w_t(x,y) = 0$ for each $\lambda \in]0,\infty[$ if and only if $w_t(x,y) = 0$ for each $t \in]0,\infty[$, which concludes the proof. □

Observe that, in the preceding result, $\tilde{w}$ coincides with w, whenever $w_{x,y}$ is a left-continuous function, for each $x,y \in X$.

Proposition 2 and Theorem 5 allow us to give the searched method for constructing a fuzzy pseudo-metric from a modular pseudo-metric.

Theorem 6. *Let $*$ be a continuous t-norm with additive generator $f_* : [0,1] \to [0,\infty]$. If w is a modular pseudo-metric on X, then the pair $(M^{w,f_*}, *)$ is a fuzzy pseudo-metric on X, where the fuzzy set $M^{w,f_*} : X \times X \times [0,\infty[$ is defined, for all $x,y \in X$, by*

$$M^{w,f_*}(x,y,t) = f_*^{(-1)}\left(\tilde{w}_t(x,y)\right),$$

where $\tilde{w}_t(x,y) = \inf_{0<\lambda<t} w_\lambda(x,y)$. Moreover, $(M^{w,f_*}, *)$ is a fuzzy metric on X if and only if w is a modular metric on X.

Proof. By Proposition 2, we deduce that $\tilde{w}_{x,y}$ is a modular pseudo-metric on X. Theorem 5 guarantees that M^{w,f_*} is a modular indistinguishability operator for $*$ on X. Moreover, continuity of $f_*^{(-1)}$ and the left-continuity, provided by Proposition 2, of the function $\tilde{w}_{x,y}$ guarantee that axiom **(KM5)** is fulfilled. Thus, the pair $(M^{w,f_*}, *)$ is a fuzzy pseudo-metric on X. Finally, by Proposition 2 and Theorem 5, it is obvious that $(M^{w,f_*}, *)$ is a fuzzy metric on X if and only if w is a modular metric on X. □

4. Discussion

In the literature, there are two tools that allow for measuring the degree of similarity between objects. They are the so-called indistinguishability operators and fuzzy metrics. The former provide the degree up to which two objects are equivalent when there is a limitation on the accuracy of measurement between the objects being compared. The fuzzy metrics provide the degree up to which two objects are equivalent when the measurement is relative to a parameter. Motivated by the fact that none of these types of similarity measurements generalizes the other, we have introduced a new notion of indistinguishability operator that unifies both notions, fuzzy metric and indistinguishability operator under a new one. Moreover, we have explored the metric behavior of this new kind of operator in such a way that the new results extend the classical results to the new framework and, in addition, allow for exploring the aforesaid duality relationship when fuzzy metrics are considered instead of indistinguishability operators. The fact that the new notion of indistinguishability operator does not involve the continuity on the t-norm in their axiomatic presents an advantage with respect to the fuzzy metrics. The assumption of continuity could be limiting the range of applications of fuzzy metrics in those cases where (classical) indistinguishability operators work well. A future work remains open to study which properties of classical indistinguishability operators are also verified in the new framework. In addition, the utility of the new operators in applied problems must be explored.

5. Conclusions

In the present paper we have introduced a new type of indistinguishability operators which generalize the notion of fuzzy metric and the classical notion of indistinguishability operator. Moreover, we have discussed the relationship between the new operators and metrics. Specifically, we have proved the existence of a duality between them and the so-called modular metrics, which provide a dissimilarity measurement between objects relative to a parameter. Furthermore, we have yielded examples that illustrate the new results.

Acknowledgments: Oscar Valero acknowledges the support of the Spanish Ministry of Economy and Competitiveness under Grant TIN2016-81731-REDT (LODISCO II) and Agencia Estatal de Investigación (AEI)/Fondo Europeo de Desarrollo Regional (FEDER), UE funds.

Author Contributions: Juan-José Miñana and Oscar Valero have contributed equally to this work.

Conflicts of Interest: The authors declare no conflict of interest.

References

1. Trillas, E. Assaig sobre les relacions d'indistinguibilitat. In *Proceedings of Primer Congrés Català de Lògica Matemàtica*; Universitat Politècnica de Barcelona: Barcelona, Spain, 1982; pp. 51–54.
2. Recasens, J. Indistinguishability Operators: Modelling Fuzzy Equalities and Fuzzy Equivalence Relations. In *Studies in Fuzziness and Soft Computing*; Kacprzyk, J., Ed.; Springer-Verlag: Berlin/Heidelberg, Germany, 2010; ISBN 978-3-642-16221-3.
3. Klement, E.P.; Mesiar, R.; Pap, E. Triangular norms. In *Trends in Logic—Studia Logica Library*; Wójcicki, R., Mundici, D., Priest, G., Segerberg, K., Urquhart, A., Wansing, H., Malinowski, J., Eds.; Kluwer Academic Publishers: Dordrecht, The Netherlands, 2000; ISBN 0-7923-6416-3.
4. De Baets, B.; Mesiar, R. Metrics and T-equalities. *J. Math. Anal. Appl.* **2002**, *267*, 351–347, doi:10.1006/jmaa.2001.7786.
5. Gottwald, S. On t-norms which are related to distances of fuzzy sets. *BUSEFAL* **1992**, *50*, 25–30.
6. Hohle, U. Fuzzy equalities and indistinguishability. In Proceedings of the First European Congress on Fuzzy and Intelligent Technologies (EUFIT"93), Eurogress Aachen, Germany, 7–10 September 1993; Zimmermann, H.J. Ed.; ELITE: Aachen, Germany; pp. 358–363.
7. Ovchinnikov, S. Representation of transitive fuzzy relations. In *Aspects of Vagueness*; Skala, H., Termini, S., Trillas, E., Eds.; Reidel: Dordrecht, The Netherlands, 1984; pp. 105–118.
8. Valverde, L. On the structure of F-indistinguishability operators. *Fuzzy Sets Syst.* **1988**, *17*, 313–328, doi:10.1016/0165-0114(85)90096-X.
9. Deza, M.M.; Deza, E. *Encyclopedia of Distances*, 1st ed.; Springer: Berlin/Heidelberg, Germany, 2009; ISBN 978-3-642-00234-2.
10. Fuster-Parra, P.; Guerrero, J.; Martín, J.; Valero, O. New results on possibilistic cooperative multi-robot system. *LNCS* **2017**, *10451*, 21–28, doi:10.1007/978-3-319-66805-5_1.
11. Guerrero, J; Miñana, J.J.; Valero, O.; Oliver, G. Indistinguishability operators applied to task allocation problems in multi-agent systems. *Appl. Sci.* **2017**, *7*, 963, doi:10.3390/app7100963.

12. Gerrero, J.; Miñana, J.J.; Valero, O. A comparative analysis of indistinguishability operators applied to swarm multi-robot task allocation problem. *LNCS* **2017**, *10451*,1–9, doi:10.1007/978-3-319-66805-5_3.
13. Gregori, V.; Morillas, S.; Sapena, A. Examples of fuzzy metrics and applications. *Fuzzy Sets Syst.* **2011**, *170*, 95–111.
14. Morillas, S.; Gregori, V.; Peris-Fajarnés, G; Latoree, P. A fast impulsive noise color image filter using fuzzy metrics. *Real-Time Imaging* **2005**, *11*, 417–428.
15. Morillas, S.; Gregori, V.; Peris-Fajarnés, G. New adaptative vector filter using fuzzy metrics. *J. Electron. Imaging* **2007**, *16*, 1–15.
16. Kramosil, I.; Michalek, J. Fuzzy metric and statistical metric spaces. *Kybernetika* **1975**, *11*, 336–344.
17. Grabiec, M. Fixed points in fuzzy metric spaces. *Fuzzy Sets Syst.* **1988**, *27*, 385–389, doi:10.1016/0165-0114(88)90064-4.
18. George, A.; Veeramani, P. On some results in fuzzy metric spaces. *Fuzzy Sets Syst.* **1994**, *64*, 395–399, doi:10.1016/0165-0114(94)90162-7.
19. Gregori, V.; Morillas, S.; Sapena, A. A class of completable fuzzy metric spaces. *Fuzzy Sets Syst.* **2010**, *161*, 2193–2205.
20. Chistyakov, V.V. Modular metric spaces, I: Basic concepts. *Nonlinar Anal-Theor.* **2010**, *72*, 1–14, doi:10.1016/j.na.2009.04.057.

axioms

MDPI

Article

Managing Interacting Criteria: Application to Environmental Evaluation Practices

Teresa González-Arteaga [1], Rocio de Andrés Calle [2,*] and Luis Martínez [3]

[1] BORDA and PRESAD Research Groups and Multidisciplinary Institute of Enterprise (IME), University of Valladolid, E-47011 Valladolid, Spain; teresa.gonzalez.arteaga@uva.es

[2] BORDA Research Unit, PRESAD Research Group and Multidisciplinary Institute of Enterprise (IME), University of Salamanca, E-37007 Salamanca, Spain

[3] Department of Computer Science, University of Jaén, E-23071 Jaén, Spain; martin@ujaen.es

* Correspondence: rocioac@usal.es; Tel.: +34-9232-94500 (ext. 6717)

Received: 5 December 2017; Accepted: 8 January 2018; Published: 16 January 2018

Abstract: The need for organizations to evaluate their environmental practices has been recently increasing. This fact has led to the development of many approaches to appraise such practices. In this paper, a novel decision model to evaluate company's environmental practices is proposed to improve traditional evaluation process in different facets. Firstly, different reviewers' collectives related to the company's activity are taken into account in the process to increase company internal efficiency and external legitimacy. Secondly, following the standard ISO 14031, two general categories of environmental performance indicators, management and operational, are considered. Thirdly, since the assumption of independence among environmental indicators is rarely verified in environmental context, an aggregation operator to bear in mind the relationship among such indicators in the evaluation results is proposed. Finally, this new model integrates quantitative and qualitative information with different scales using a multi-granular linguistic model that allows to adapt diverse evaluation scales according to appraisers' knowledge.

Keywords: environmental performance; criteria relationship; Choquet integral; uncertainty information

1. Introduction

Corporate Social Responsibility has a pivotal role in Business models and although it has been differently defined along years, all definitions have five dimensions on common: The environmental, the social, the economic, the stakeholder and the voluntariness dimension. Focusing on the environmental dimension, in recent years companies and institutions have born environmental pressures from regulators. This fact has forced many corporations to expand their attention to environmental responsible practices and to manage their environmental issues as a strategic competitive issue. Therefore, companies need to implement strategies to reduce their environmental impacts and to contribute to environmental sustainability. As a result, the environmentally conscious practices have emerged in form of systems (environmental management systems) and programs to improve processes and products by means of internal policies, actions and plans like design for environment, recycling, waste management, life cycle analysis, green supply chain management and so on (e.g., see [1,2]). Aside from environmental pressures from regulators, companies implement environmentally conscious practices due to the fact that there is a link between carrying on environmental strategies and improving business performance (see [3–5], among other).

In order to reflect the company's efforts and achievements in relation with its environmental performance, many *environmental performance indicators* (EPI) have been developed. These are used for environmental performance evaluation because they allow communication at different organizational levels and also they provide a basis for making decisions to reach specific goals.

Traditionally, environmental indicators are composite indicators, that is, they are a mathematical combination of a set of indicators or criteria that have no common meaningful unit of measurement. EPIs have taken a great importance and they are being used in different ways like the considerable literature in this area shows (e.g., see [6,7], among other). Moreover, there are many works focused on examining some specific indicators (see [7–15]). From another angle, the existence of intertwined relations among the components and elements of the environmental practices is a relevant aspect of environmental evaluation problem and it has called attention in the technical literature (see [16–19]). Such relationships are not easy to take into account in the evaluation process and they have been partially addressed by hierarchical process (see [1,20]).

From the methodology perspective, the evaluation of companies' environmental performance has been addressed from different approaches (see [12,21–23], among other). This paper follows the viewpoint of the *Multi-Criteria Decision Making Theory* because it facilitates an structured approach to identify solutions to complex evaluation problems (see Figure 1). *Decision Analysis* is a renowned discipline to accomplish this kind of processes since allows to consider different criteria and several experts. In detail, this contribution is influenced by the *Fuzzy Set Theory* because it allows to handle simultaneously the qualitative and quantitative information included in environmental performance evaluation process and it has got good results in similar context (see [24,25]).

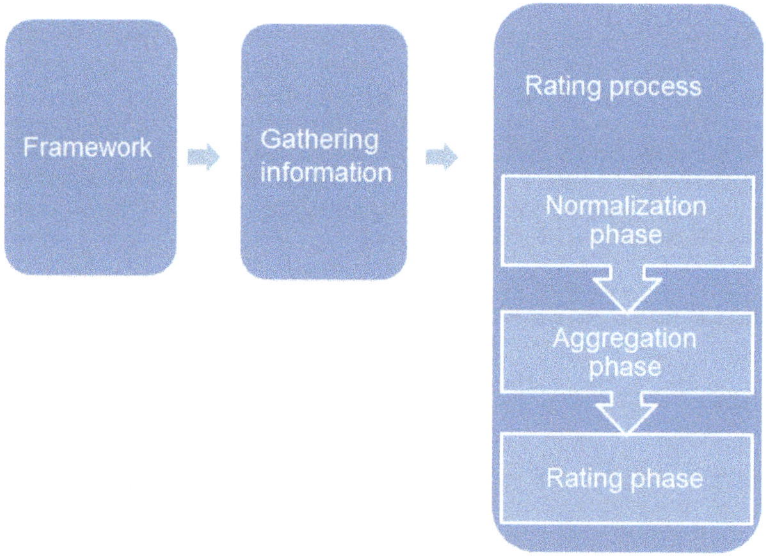

Figure 1. Scheme of the novel evaluation model.

Taking into account the aforementioned wide spectrum of issues, the main aims of this contribution can be summarized as follows. Firstly, proposing a new environmental integral evaluation model for companies, general enough to integrate indicators built over diverse standpoints. Non-focused on selecting a particular environmental indicator set. Secondly, overcoming the limitations associated with some previous models to evaluate environmental practices proposing a model where different sets of reviewers could evaluate company's environmental practices attending to different criteria. Thirdly, presenting a flexible evaluation framework allowing appraisers may express their assessments by means of numerical or linguistic information according to the criteria nature. Finally, for addressing the interdependence among criteria/EPIs in a flexible way, the discrete Choquet integral is adopted in the model-like aggregation operator (see [26]).

The novelty of this model is not only its generality but also its capacity to deal with evaluation frameworks in which different set of reviewers (internal and external, experts and non-experts) can assess criteria of diverse nature conforming a flexible heterogeneous framework where different domains of information (numerical, linguistic) are allowed. Finally another important improvement offered by this novel procedure is the management of interdependence among criteria by using the discrete Choquet integral. Therefore, this proposal for an environmental integral evaluation model is versatile enough to be applied under various settings.

The outline of this paper is as follows. In Section 2, we present a roughly overview of previous works. Section 3 is devoted to detail our proposal of environmental integral evaluation model and Section 4 provides with an illustrative application on a manufacturing company. The paper ends pointing out some concluding remarks.

2. Related Works

Before proposing our new approach to carry out evaluation of company's environmental practices, a short review of the context of environmental evaluation and previous works are included. Initially environmental issues are alluded and then the methodology traditionally used it.

2.1. Evaluating Corporate Environmental Practices

Companies have usually performed different environmental policies from reactive to proactive strategies, some of them react to institutional pressures and stringent environmental regulations (see [9,27]). In many instances, companies respond to influences from large corporate buyers, that demand environmental information to their suppliers. However, a lot of companies voluntarily implement environmental programs, in some cases devoting attention to diverse stakeholder or pursuing environmental reputation (e.g., see [2,28]). For these reasons different authors agree to include stakeholder in the environmental evaluation process (see [1,2,29,30]). Therefore, it is relevant to take into account multiple viewpoints of the concerned parties in the evaluation.

From another point of view, the increasing acceptance of ISO 14001 environmental standards, has caused thousands of facilities worldwide to have adopted Environmental Management System (EMS) that is required to be certified to ISO 14001 (see [31,32]). The environmental standards involve an explicit commitment for improvement of *environmental performance* (EP). Although there is not a compulsory way of building performance indicators, ISO 14001 gives guidance in environmental performance evaluation through environmental performance indicators, defined as the *"specific expression that provides information about an organization's environmental performance"*. In the standard ISO 14031, there are three basic types of EP indicators (EPI): operational (OEPI), management (MEPI) and environmental condition indicators (ECI) (see [33]). The latter intend to provide information about the local, regional, national or global condition of the environment. The MEPIs provide information about management activities in reference to the strategic policy, staffing policy, practices, procedures, decisions and actions at all levels of the organization. The OEPIs provide information about environmental performance of the operations of the organization relative to organization's physical facilities and equipment (material and energy flows).

Although there is not a discussion about this classification, the ECIs are rarely applied by companies while the OEPIs have been reported more widely than the MEPIs, as is shown in different works (see e.g., [16,18,21,34]). It is relevant to point out that those indicators are quite general and they can encompass results of other more specific environmental evaluation tools such as life cycle assessment, environmental risk assessment and so on.

Diverse approaches have been introduced and a great variety of measurement items may be used in the construction of environmental performance indicators (EPIs) (see [16,17,35]). In the literature, a large number of studies have reported a broad range of such indicators (see [7–14], among other).

Another relevant aspect of environmental evaluation context is the existence of intertwined relations among the components and elements of the environmental practices. This is a notable feature and it has deserved certain attention in the literature (see [16–19]). It is not easy to take into account such relationships in the evaluation procedure. One way of addressing partially this issue has been through organizing environmental aspects or factors or components that arise from other subcomponents or elements in a hierarchical way. This common practice has also an inherent interest in the environmental matter (see [1,20]).

2.2. Evaluation Methodologies

Many works have addressed the evaluation of corporate environmental performance with different tools (see e.g., [21,23], among other). This contribution focuses on the use of the *Multi-Criteria Decision Making Theory* to deal with environmental issues. Some approaches are based on traditional methods like Multi-Attribute Analysis technique (MAA) (see e.g., [12]), Analytical Hierarchy Process (AHP) (see e.g., [10,36]), Analytical Network Process (ANP) (see [1,11]) and others specifically developed models (see e.g., [22]). Several references to diverse applications of multi-criteria analysis can be foun to specific issues in corporate environmental evaluation in [15].

Usually it is necessary to handle qualitative information as well as quantitative one in environmental performance evaluation. Given the difficulty of mixing imprecise knowledge in a common way, the *Fuzzy Set Theory* is expanding in this area. Some proposals in this line are the Fuzzy Extended Analytical Hierarchy Process (see e.g., [36]), the Fuzzy TOPSIS (Technique for Order Performance by Similarity to Ideal Solution) (see e.g., [37]), the adaptation of Grey Relational Analysis (see e.g., [38]) and another more general works such as [39].

Some contributions in the environmental evaluation context discuss how to integrate the interdependence relations in the evaluation process (e.g., see [19,20,40,41]).

3. A Multi-Criteria Decision Integral Model for Evaluating Environmental Practices

Taking into account the aforementioned goals of the paper, a new proposal of environmental integral evaluation model is introduced in further detail below and its novelties and improvements.

Generally speaking, the main issue of an evaluation process is to compute a set of overall assessments that summarizes information and provides useful knowledge on evaluated elements. Decision analysis is an excellent field to carried out evaluation processes since includes a wide variety of methods for appraising a set of alternatives considering different criteria and involving several experts. It has got good results as it can be seen in the literature (see [24,25]).

Therefore, in this paper an environmental evaluation process based on a classical decision analysis approach is proposed. It consists of three main stages: establishing a framework suitable for the environmental evaluation, gathering information and rating process. In an heterogeneous information context like environmental evaluation, the rating process begins with a normalization of the information [42]. After that, it is possible to carry out an aggregation in an appropriate way. And, finally, we are in the position to make an assessment of the outcomes. Figure 1 shows a graphical scheme of the proposed model. All this steps are going to be explained in coming subsections.

3.1. Evaluation Framework

The evaluation framework fixes actors and elements that may be considered to evaluate environmental practices of a company with n facility sites, $X = \{x_1, ..., x_n\}$. The type of criteria, appraisers and domains used in our proposal are described below.

3.1.1. Criteria Selection

EPIs depict vast quantity of environmental information, relevant attributes or environmental criteria, in a comprehensive and concise way (see [12,13]).

According to the standard ISO 14031, this contribution considers two general categories of EPIs: *management performance indicators* (MEPI) and *operational performance indicators* (OEPI). Hence, two types of criteria associated to them are distinguished: management performance criteria, $C^M = \{c_1^M, \ldots, c_p^M\}$ and operational performance criteria, $C^O = \{c_1^O, \ldots, c_q^O\}$. Consequently, there are almost $p + q$ criteria.

Both sets of criteria could have a qualitative and quantitative nature. Due to this fact, the set of all criteria C is split into two subsets: one subset of quantitative criteria, C_1, and another one of qualitative, C_2. The criteria structure is then the following one:

- Management performance criteria:

$$C^M = C_1^M + C_2^M, \quad C^M = \{c_k^M \mid c_k^M \in C_1^M \text{ or } c_k^M \in C_2^M\},\ k = 1, \ldots, p.$$

- Operational performance criteria:

$$C^O = C_1^O + C_2^O, \quad C^O = \{c_k^O \mid c_k^O \in C_1^O \text{ or } c_k^O \in C_2^O\},\ k = 1, \ldots, q.$$

Notice $C_1^M \cap C_2^M = \emptyset$ and $C_1^O \cap C_2^O = \emptyset$.

Generally, information provided by reviewers is expressed by means of crisp values (to evaluate quantitative criteria) or by means of words based on *Liker's methodology* (see [43]) (to evaluate qualitative criteria) (see [14,22]). In this novel model the appraisers express their assessments with different linguistic descriptors to asses qualitative criteria. So, our proposal is to manage the qualitative criteria using different scales with a multi-granular linguistic model that allow to adapt the evaluation scales and model the linguistic information by using the fuzzy linguistic approach (see [44]).

3.1.2. Reviewers' Selection

Following the standard ISO 14001, company's environmental practices are assessed from diverse collectives related to the company activity and not only from the top technical environmental managers. This fact increases company internal efficiency and external legitimacy (see [27,30]).

Thus, a method that distinguishes between experts and non-experts reviewers is developed. Moreover, two general set of reviewers, the internal and the external ones are considered due to the importance to include several information sources related to the company environmental practices. These reviewers are classified based on their knowledge and their information about criteria. Therefore, the following collectives of reviewers take part in the evaluation process (see Figure 2 for a symbolic summary):

- A set of *internal reviewers*:
  - A set of company's internal experts: $A^E = \{a_1^E, \ldots, a_m^E\}$.
  - A set of company's internal non-expert (such as managers, staff, employees, etc.): $A^{NE} = \{a_1^{NE}, \ldots, a_r^{NE}\}$.
- A set of *external reviewers*:
  - A set of company's external experts such as auditors: $B^E = \{b_1^E, \ldots, b_s^E\}$.
  - A set of company's external non-experts evaluators, B^{NE}, which is split in two types depending on their relation to the company:
    1. A set of general stakeholders (shareholders, suppliers, government regulators, local communities, intermediate customers, large retailers, final consumers): $B^{NE-G} = \{b_1^{NE-G}, \ldots, b_t^{NE-G}\}$.
    2. A set of social constituents (community groups, trade associations, labor unions, environmental groups): $B^{NE-S} = \{b_1^{NE-S}, \ldots, b_u^{NE-S}\}$.

  Therefore, $B^{NE} = B^{NE-G} \bigcup B^{NE-S}$.

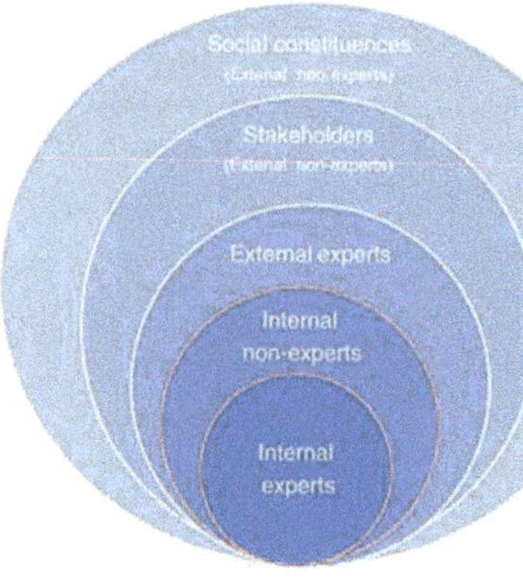

Figure 2. Reviewers of a facility site.

3.2. *Gathering Information*

Once criteria and reviewers have been established, the next step consists of gathering reviewer's assessments about each facility site with regards to each criterion and attending to their knowledge on them. These evaluations are denoted as follows:

- Let $a^E_{i,j,k}$ and $a^{NE}_{i,j,k}$ be the internal reviewers' evaluations, experts and non-experts respectively, on the facility site x_i by the j-th reviewer regarding the criterion c^-_k. Abusing notation, on occasions we refer to criterion k as c^-_k, where superscript denotes M or O, the criterion type.

$$a^E_{i,j,k} \in \begin{cases} \mathbb{R}^+, & \text{if } c^-_k \in C^-_1, \\ S^k_{A^E}, & \text{if } c^-_k \in C^-_2. \end{cases} \qquad a^{NE}_{i,j,k} \in \begin{cases} \mathbb{R}^+, & \text{if } c^-_k \in C^-_1, \\ S^k_{A^{NE}}, & \text{if } c^-_k \in C^-_2. \end{cases}$$

  where $\mathbb{R}^+$ is the set of non-negative real numbers and $S^k_{A^E}$ and $S^k_{A^{NE}}$ are the linguistic term sets used by the internal experts and non-experts reviewers, respectively, to evaluate the criterion k.
- In the same way, let $b^E_{i,j,k}$, $b^{NE-G}_{i,j,k}$, and $b^{NE-S}_{i,j,k}$ be the external reviewers' evaluations, experts and non-experts from stakeholders and social constituencies, respectively, on the facility site x_i by the j-th reviewer with regard to the criterion c^-_k.

$$b^E_{i,j,k} \in \begin{cases} \mathbb{R}^+ \text{ if } c^-_k \in C^-_1, \\ S^k_{B^E} \text{ if } c^-_k \in C^-_2 \end{cases} \qquad b^{NE-G}_{i,j,k} \in \begin{cases} \mathbb{R}^+ \text{ if } c^-_k \in C^-_1, \\ S^k_{B^{NE-G}} \text{ if } c^-_k \in C^-_2 \end{cases} \qquad b^{NE-S}_{i,j,k} \in \begin{cases} \mathbb{R}^+ \text{ if } c^-_k \in C^-_1, \\ S^k_{B^{NE-S}} \text{ if } c^-_k \in C^-_2 \end{cases}$$

  where $\mathbb{R}^+$ is the set of non-negative real numbers and $S^k_{B^E}$, $S^k_{B^{NE-G}}$ and $S^k_{B^{NE-S}}$ are the linguistic term sets used by the external experts and non-experts reviewers, respectively, to evaluate the criterion k.

Note that any appropriate linguistic term set , S^k_-, is characterized by its cardinality $|S^k_-|$. In order to keep the notation as simple as possible we use S^k_-, the subscript denotes the reviewers collective.

3.3. *Rating Process*

To deal with the information gathered, we adapt the proposal presented in [42] to our aim. Each step of this process is described in the following subsections and its scheme is depicted in Figure 1.

3.3.1. Normalization Phase

In order to manage properly the assessments from reviewers, all gathered heterogeneous information should be unified in the normalization phase. Due to the fact that criteria used to evaluate environmental practices could have different domains, it is necessary to achieve the normalization process by means of different ways, depending on the criterion nature.

Firstly, if the compiled information is regarded to a qualitative criterion $c_k^- \in C_2^-$, the registered values are linguistic labels that belong to a linguistic term set. Each collective of reviewers can use different linguistic term sets to evaluate qualitative criteria for each facility site. Therefore, we need to bring all the linguistic labels into a unique expression domain, a common linguistic term set called *Basic Linguistic Term Set* (BLTS). The BLTS is selected with the aim of keeping as much knowledge as possible. It is denoted by $\overline{S} = \{\overline{s}_0, \ldots, \overline{s}_g\}$ and its granularity $|\overline{S}|$ is g (see [42]). This procedure conducts the heterogeneous information into fuzzy sets and later into linguistic 2-tuples in the BLTS (see [44,45] for more details). The definition of the 2-tuples is recalled below.

Definition 1. *([44]) Let $S = \{s_0, \ldots, s_g\}$ be a set of linguistic terms. The 2-tuple set associated with S is determined for $\langle S \rangle = S \times [-0.5, 0.5)$. The function $\Delta_S : [0, g] \longrightarrow \langle S \rangle$ is given by,*

$$\Delta_S(\beta) = (s_i, \alpha), \ \ with \ \begin{cases} i = round\,(\beta), \\ \alpha = \beta - i, \end{cases}$$

where round assigns to β the integer number $i \in \{0, 1, \ldots, g\}$ closest to β.

Secondly, if gathered information is on a quantitative criterion $c_k^- \in C_1^-$, we have to distinguish between benefit or cost criterion to accomplish accurately the normalization process.

Let $y_{i,j,k}$ be the assessment to the facility site x_i of the reviewer j, on the criterion k, $c_k^- \in C_1^-$. This assessment is normalized in $[0, 1]$:

$$\widetilde{y}_{i,j,k} = \begin{cases} \frac{y_{i,j,k}}{y_{k\,max}}, & \text{if } c_k^- \in C_1^- \text{ is a benefit criterion,} \\ \left(1 - \frac{y_{i,j,k}}{y_{k\,max}}\right), & \text{if } c_k^- \in C_1^- \text{ is a cost criterion,} \end{cases}$$

where $y_{k\,max}$ is the maximum assessment expressed for all reviewers over all facility sites attending to the k-th criterion, that is, $\widetilde{y}_{i,j,k}$.

Then, $\widetilde{y}_{i,j,k}$ is conducted it into the BLTS by means of linguistic 2-tuples following [42,46].

Once the information is normalized and unified, we proceed to the next step of the rating process, the aggregation.

3.3.2. Aggregation Phase

In this phase, the individual assessments are aggregated by means of aggregation operators according to the step of the aggregation phase and considering their properties. We focus especially on proposing the OWA and discrete Choquet integral as aggregation operators. The former does not distinguish the origin of the values (they are *anonymous*). This allows to manage properly information from different reviewers. The latter operator is useful to manage interdependence among criteria which are commonly found in environmental evaluation.

In the environmental evaluation literature the existence of relations among environmental indicators is well-known but it is rarely explicitly defined. Obviously, not to take into account these relations it could have consequences over outcomes and it could disturb company's environmental final decisions. With the aim of overcoming this drawback, our proposal manages such a relationship by means of discrete Choquet integral (see [26,47,48]). This aggregation operator allows us to take into account interactions among criteria or EPIs and, also, the global importance of each criteria or EPI.

Then, each stage of the aggregation process is developed. The Figure 3 displays an outline of this phase.

1. *Computing EPIs for each reviewers' collective and each criterion.*

   Since we assume that the reviewers give their evaluations individually, we propose to use the 2-tuple OWA operator from [46]. We reproduce its definition below.

   Definition 2. *Let* $((l_1, \alpha_1), \ldots, (l_m, \alpha_m)) \in \langle \overline{S} \rangle^m$ *be a vector of linguistic 2-tuples and* $W = (w_1, \ldots, w_m) \in [0,1]^m$ *be a weighting vector such that* $\sum_{i=1}^{m} w_i = 1$. *The 2-tuple OWA operator associated with* w *is the function* $G^{\mathbf{w}} : \langle \overline{S} \rangle^m \longrightarrow \langle \overline{S} \rangle$ *defined by*

$$G^{\mathbf{w}}((l_1, \alpha_1), \ldots, (l_m, \alpha_m)) = \Delta_{\overline{S}} \left(\sum_{i=1}^{m} \mathrm{w}_i \, \beta_i^* \right),$$

   where β_i^* *is the i-th largest element of* $\left\{ \Delta_{\overline{S}}^{-1}(l_1, \alpha_1), \ldots, \Delta_{\overline{S}}^{-1}(l_m, \alpha_m) \right\}$.

   In order to apply this operator, the weighting vector can be computed using the well-known non-decreasing quantifiers proposed by Yager (see [49]). It is important to note that each concrete aggregation procedure with OWA operators can use a different quantifier, in other words, a different weighting vector. This adds flexibility to the model.

   The reviewers' assessments are aggregated for each criterion and each collective (see Figure 3) by means of a 2-tuple OWA operator, G_{-}^{-}. Then, for each collective and for every criterion c_k^{-}, the process is conducted as follows:

   - For *internal reviewers* (experts and non-experts, respectively):

$$\begin{aligned} I_k^{A^E}(x_j) &= G_k^{W_{A^E}}(\widetilde{a}_{1,j,k}^{E}, \ldots, \widetilde{a}_{m,j,k}^{E}), \\ I_k^{A^{NE}}(x_j) &= G_k^{W_{A^{NE}}}(\widetilde{a}_{1,j,k}^{NE}, \ldots, \widetilde{a}_{r,j,k}^{NE}). \end{aligned}$$

   - For *external reviewers* (experts and non-experts, respectively):

$$\begin{aligned} I_k^{B^E}(x_j) &= G_k^{W_{B^E}}(\widetilde{b}_{1,j,k}^{E}, \ldots, \widetilde{b}_{s,j,k}^{E}), \\ I_k^{B^{NE}}(x_j) &= G_k^{W_{B^{NE}}}\left(I_k^{B^{NE-G}}(x_j), I_k^{B^{NE-S}}(x_j)\right), \end{aligned}$$

     where $I_k^{B^{NE-G}}(x_j)$ is the environmental performance indicator for stakeholder reviewers:

$$I_k^{B^{NE-G}}(x_j) = G_k^{W_{B^{NE-G}}}(\widetilde{b}_{1,j,k}^{NE-G}, \ldots, \widetilde{b}_{t,j,k}^{NE-G}),$$

     and $I_k^{B^{NE-S}}(x_j)$ is the environmental performance indicator for social constituents reviewers:

$$I_k^{B^{NE-S}}(x_j) = G_k^{W_{B^{NE-S}}}(\widetilde{b}_{1,j,k}^{NE-S}, \ldots, \widetilde{b}_{u,j,k}^{NE-S}).$$

2. *Computing EPIs for experts/non-experts reviewers and each criterion.*

   As in the preceding step, the OWA operator is used. The previous environmental performance indicators for the x_j facility site: $I_k^{A^E}(x_j)$, $I_k^{A^{NE}}(x_j)$, $I_k^{B^E}(x_j)$ and $I_k^{B^{NE}}(x_j)$ are aggregated for each criterion taking into account if the reviewers are experts or not (see Figure 3). The previous indicators belonging to the experts reviewers are then aggregated by means of an OWA operator for each criterion c_k^-.

$$I_k^E(x_j) = G_k^{W_E}\left(I_k^{A^E}(x_j), I_k^{B^E}(x_j)\right).$$

Analogously to the experts reviewers, an environmental performance indicator is computed for each criterion c_k^- by aggregating the opinions of all non-experts reviewers.

$$I_k^{NE}(x_j) = G_k^{W_{NE}}\left(I_k^{A^{NE}}(x_j), I_k^{B^{NE}}(x_j)\right).$$

3. *Computing global EPIs.*

The proposed environmental integral evaluation model puts forward the computation of three global environmental performance indicators: an overall that includes all the issues, a management one relative to management issues and an operational one for the operational issues considered. In this way, the major recommendations issued by the ISO 14001 is followed and the model takes the most advantage of the gathered information.

From the previous step, there are two values for every single criterion, one from experts reviewers and another from non-experts reviewers for each facility site.

Now to aggregate the values corresponding to different criteria, we propose to use the discrete Choquet integral as aggregation operator. It allows for consideration of the interrelations among criteria through the choice of an specific fuzzy measure.

(a) *An overall global EPI.*

In order to cover better the possible interdependences among criteria when we compute an overall global EPI, we do not distinguish between management and operational criteria because we could have interdependences among some management criteria with some operational criteria. Therefore, we do not limit to consider the interdependence among management criteria on the one hand, or only among operational criteria on the another hand.

We propose to determine a fuzzy measure, μ, over the set of all criteria,

$$C = C^M \bigcup C^O = \{c_1^M, \ldots, c_p^M, c_1^O, \ldots, c_q^O\}.$$

And then, this fuzzy measure is used to calculate the associated Choquet integrals. Theoretical aspects of Choquet Integral are included in Appendix A. We must do twice, one for the values coming from the experts reviewers and another one for the values coming from the non-experts reviewers. Afterwards, a convex linear combination of both is computed which is the overall global EPI. The parameter $\beta \in (0,1)$ is chosen arbitrary taking into account the interest of the company. It stands for the relative importance assigned for the expert reviewers assessments versus the non-expert reviewers assessments.

$$I^E(x_j) = {}_{\mu}IC\left(I_{1M}^E(x_j), \ldots, I_{pM}^E(x_j), I_{1O}^E(x_j), \ldots, I_{qO}^E(x_j)\right).$$

$$I^{NE}(x_j) = {}_{\mu}IC\left(I_{1M}^{NE}(x_j), \ldots, I_{pM}^{NE}(x_j), I_{1O}^{NE}(x_j), \ldots, I_{qO}^{NE}(x_j)\right).$$

$$I(x_j) = \beta I^E(x_j) + (1-\beta) I^{NE}(x_j)).$$

(b) *Management and operational global EPIS.*

At this point we are interested in calculating two more global indicators, a management global EPI and an operational global EPI, with the aim of conforming to the standard ISO 14001. The criteria are divided according to their type, management and operational. Each type of criteria is aggregated separately by means of two discrete Choquet integrals because it can have interrelations (see Figure 3).

In this case, we take the fuzzy measure previously built on the set of all criteria, C, and we derive from it two new fuzzy measures, one on the set of management criteria, C^M, and another on the set of operational criteria, C^O, called μ^M and μ^O, respectively. So, they are coherent to the previous one, μ. These are defined as follows

$$\mu^M : \mathcal{P}(C^M) \longrightarrow [0,1] \qquad \mu^M(T) = \frac{\mu(T \cap C^M)}{\mu(C^M)} \quad \text{where} \quad T \subseteq C^M$$

$$\mu^O : \mathcal{P}(C^O) \longrightarrow [0,1] \qquad \mu^O(S) = \frac{\mu(S \cap C^O)}{\mu(C^O)} \quad \text{where} \quad S \subseteq C^O$$

Now, from these fuzzy measures we proceed to calculate the mentioned indicators using a discrete Choquet integral for each one.

- *Management global environmental performance indicator (MEPI).* In order to calculate this indicator we aggregate the experts and non-experts indicators for management criteria, $c^M = \{c_1^M, \ldots, c_p^M\}$ by means of a Choquet integral based on the fuzzy measure μ^M previously defined. We then compute a convex linear combination of both of them. $\gamma \in (0,1)$ is selected depending on the interest of the company (see Figure 3).

$$I_M^E(x_j) = {}_{\mu^M}IC^M\left(I_{1M}^E(x_j), \ldots, I_{pM}^E(x_j)\right).$$

$$I_M^{NE}(x_j) = {}_{\mu^M}IC^M\left(I_{1M}^{NE}(x_j), \ldots, I_{pM}^{NE}(x_j)\right).$$

$$I^M(x_j) = \gamma I_M^E(x_j) + (1-\gamma) I_M^{NE}(x_j)).$$

- *Operational global environmental performance indicator (OEPI).* Analogously to management global EPI, an operational global environmental performance indicator is computed for operational criteria, $c^O = \{c_1^O, \ldots, c_q^O\}$ adopting the same strategy to aggregate the experts and no-experts indicators for such criteria. We use the Choquet integral based on the fuzzy measure μ^O defined above and a convex linear combination of both with a constant $\delta \in (0,1)$ chosen according of the interest of the company (see Figure 3).

$$I_O^E(x_j) = {}_{\mu^O}IC^O\left(I_{1O}^E(x_j), \ldots, I_{qO}^E(x_j)\right).$$

$$I_O^{NE}(x_j) = {}_{\mu^O}IC^O\left(I_{1O}^{NE}(x_j), \ldots, I_{qO}^{NE}(x_j)\right).$$

$$I^O(x_j) = \delta I_O^E(x_j) + (1-\delta) I_O^{NE}(x_j)).$$

All these indicators obtained in each step of the aggregation process, $I_k^{A^E}(x_j)$, $I_k^{A^{NE}}(x_j)$, $I_k^{B^E}(x_j)$, $I_k^{B^{NE-G}}(x_j)$, $I_k^{B^{NE-S}}(x_j)$, $I_k^E(x_j)$, $I_k^{NE}(x_j)$, $I^M(x_j)$, $I^O(x_j)$ and $I(x_j)$ are used for evaluating company environmental practices, not only the overall global EPI for each facility site.

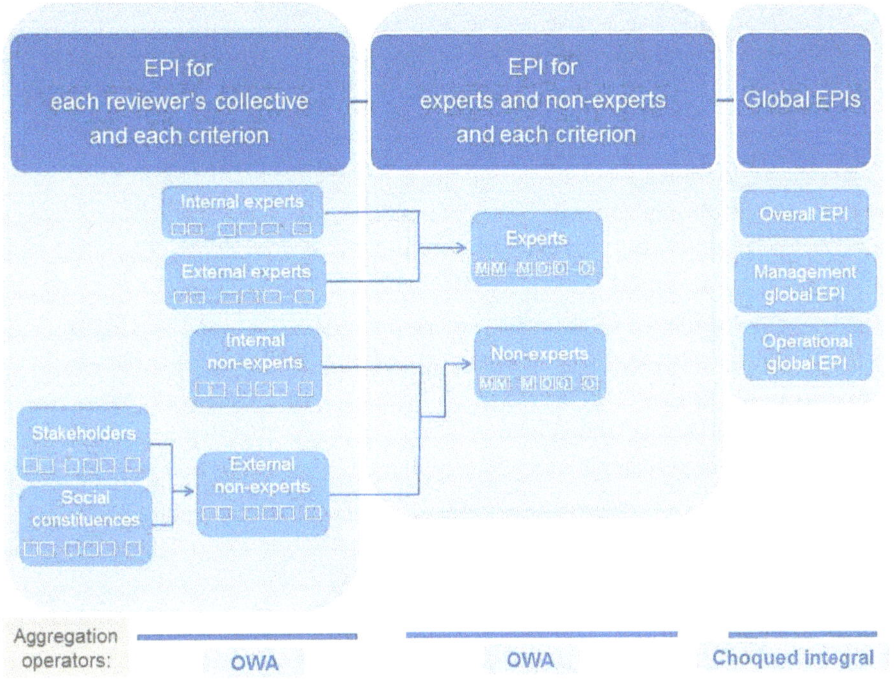

Figure 3. Multi-step aggregation scheme.

3.3.3. Rating Phase

In the rating phase, the management team shall classify and order facility sites according to the environmental performance indicators obtained in the previous phase. Since those values are linguistic 2-tuples over the BLTS, the sorting and ranking of facility sites are carried out according to the ordinary lexicographic order presented in [46].

One of the most important points of an environmental practices evaluation process is providing feedback to facility sites. With this environmental integral evaluation model, several indicators are obtained for each facility site. Among them are the MEPI and OEPI recommended in the standard ISO14001. In this way, manager team know comprehensive results in the process evaluation and they can use them to formulate environmental policy and programs in order to improve company's environmental practices.

4. An Illustrative Application

A manufacturing company which is carrying out an evaluation on their environmental practices is considered. This company involves in the evaluation process assessments from internal and external reviewers.

To begin with, the evaluation framework is presented. Without loss of generality a company with two facility sites to be evaluated: $X = \{x_1, x_2\}$ is considered according to the criteria described below.

- c_1^M: Extension in pollution control initiatives.
- c_2^M: Green purchasing. Assessing how far they incorporate environmental considerations in the purchasing process.
- c_3^M: Proportion of research and development funds applied to projects with environmental significance.

- c_1^O: The amount of CO_2 emissions measured in Kg of CO_2/m^2.
- c_2^O: The scope of use of renewable power sources.
- c_3^O: The electric power consumption in KWh/m^2.

The criteria can be sorted out according to different issues. We can classify them by the criteria nature and based on the type of information gathered (see Section 3.1.1). In this case, these classifications of the cited criteria are shown in Table 1.

Table 1. Criteria along with their classifications considered in the example.

	Criteria	Classification		
c_1^M	Pollution control initiatives	M	Qualitative	Granularity 7
c_2^M	Green purchasing	M	Qualitative	Granularity 7
c_3^M	Funds in research projects with environmental significance	M	Quantitative	Benefit
c_1^O	CO_2 emissions	O	Quantitative	Cost
c_2^O	The scope of renewable power source	O	Qualitative	Granularity 5
c_3^O	The electric power consumption	O	Quantitative	Cost

Abbreviations: M: Management, O: Operational.

These criteria are assessed by diverse reviewers' collectives:

- Internal reviewers. This reviewers' collective is made up of:
  - A set of three company's internal experts: $A^E = \{a_1^E, a_2^E, a_3^E\}$.
  - A set of two company's internal non-experts: $A^{NE} = \{a_1^{NE}, a_2^{NE}\}$.
- External reviewers. This collective consists of:
  - A set of two company's external experts such as auditors: $B^E = \{b_1^E, b_2^E\}$.
  - A set of five company's external non-experts evaluators which is made up of two different reviewers' collectives:
    1. A set of three general stakeholders: $B^{NE-G} = \{b_1^{NE-G}, b_2^{NE-G}, b_3^{NE-G}\}$.
    2. A set of two social constituents: $B^{NE-S} = \{b_1^{NE-S}, b_2^{NE-S}\}$.

It is worth noticing that each group of reviewers express their appraisals on the facility sites, but not necessarily attending to the same criteria. For example, the non-experts do not assess the criterion c_3^M relative to funds in research projects because they do not have enough knowledge.

In relation with the expression domains used specifically in the assessments, there are three quantitative and three qualitative criteria. The term set referred to criteria c_1^M, c_2^M ($S^1_{A^E}$, $S^1_{A^{NE}}$, $S^1_{B^E}$, $S^1_{B^{NE-G}}$, $S^1_{B^{NE-S}}$) have seven linguistic terms whose semantics $\{$*Very poor (VP), Medium poor (MP), Fair(F), Medium good (MG), Very good (VG)*$\}$ is shown in Figure 4 (left). For the criterion c_2^O the term sets ($S^2_{A^E}$, $S^2_{A^{NE}}$, $S^2_{B^E}$, $S^2_{B^{NE-G}}$, $S^2_{B^{NE-S}}$) have five linguistic terms $\{$*Very poor (VP), Medium poor(MP), Fair(F), Medium good (MG), Very good (VG)*$\}$ whose semantics is included in Figure 4 (right).

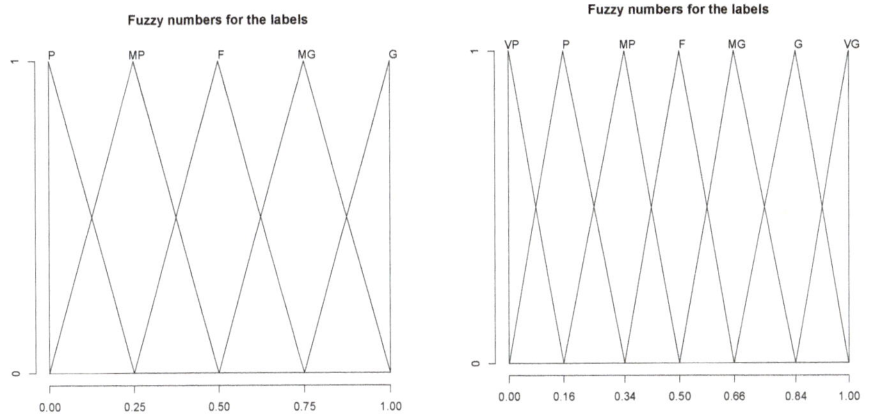

Figure 4. Left: label set for criteria c_2^O and its associated fuzzy numbers; **Right**: label set for criteria c_1^M and c_2^M and its associated fuzzy numbers.

When the evaluation framework has been fixed, the reviewers express their opinions about each facility site in relation to each criterion. The values are shown in Tables 2–7.

Table 2. Appraisals from the reviewers for criterion c_1^M.

	a_1^E	a_2^E	a_3^E	a_1^{NE}	a_2^{NE}	b_1^E	b_2^E	b_1^{NE-G}	b_2^{NE-G}	b_3^{NE-G}	b_1^{NE-S}	b_2^{NE-S}
Site x_1	P	VP	P	VP	P	MP	VP	VP	P	P	P	P
Site x_2	P	P	VP	VP	P	P	VP	P	P	VP	MP	P

Table 3. Appraisals from the reviewers for criterion c_2^M.

	a_1^E	a_2^E	a_3^E	a_1^{NE}	a_2^{NE}	b_1^E	b_2^E	b_1^{NE-G}	b_2^{NE-G}	b_3^{NE-G}	b_1^{NE-S}	b_2^{NE-S}
Site x_1	MG	G	MG	MG	MG	F	MG	G	F	MG	G	G
Site x_2	MG	G	G	G	MG	MG	F	MG	MG	F	MG	G

Table 4. Appraisals from the reviewers for criterion c_3^M.

	a_1^E	a_2^E	a_3^E	b_1^E	b_2^E
Site x_1	0.150	0.135	0.156	0.105	0.087
Site x_2	0.291	0.300	0.300	0.243	0.219

Table 5. Appraisals from the reviewers for criterion c_1^O.

	a_1^E	a_2^E	a_3^E	b_1^E	b_2^E
Site x_1	134	120	134	112	88
Site x_2	88	112	156	146	200

Table 6. Appraisals from the reviewers for criterion c_2^O.

	a_1^E	a_2^E	a_3^E	a_1^{NE}	a_2^{NE}	b_1^E	b_2^E	b_1^{NE-G}	b_2^{NE-G}	b_3^{NE-G}	b_1^{NE-S}	b_2^{NE-S}
Site x_1	MP	MP	F	MP	F	MG	F	MP	F	F	F	F
Site x_2	VP	MP	VP	VP	MP	MP	VP	MP	MP	VP	MP	MP

Table 7. Appraisals from the reviewers for criterion c_3^O.

	a_1^E	a_2^E	a_3^E	b_1^E	b_2^E
Site x_1	182.4	163.4	186.2	167.2	133.0
Site x_2	357.2	372.4	380.0	326.8	330.6

After gathering the appraisals, the rating process begins. According to the proposed environmental evaluation process in Section 3, the three phases of the rating process are carried out.

The first phase of the rating process is the normalization phase, whose target is unified all the information gathered from reviewers. We begin by transforming the numerical values into the interval $[0,1]$ as it was stated in Section 3.3.1. Let us recall that c_1^O and c_3^O are cost criteria and c_3^M is a benefit criterion. After that, both numerical and linguistic values will be conducted into a unique linguistic term set called BLTS by means of fuzzy sets. In this case the BLTS is $\overline{S} = \{\overline{VP}, \ldots, \overline{VG}\}$. Following the procedure described previously, all the values are conducted into linguistic 2-tuples in $\langle\overline{S}\rangle$. The results are not shown here because of the length (see complementary material).

Now, a global assessment for each facility site in a multi-step aggregation process is computed like it was proposed in Section 3.3.2 (see Figure 3).

1. *Computing environmental performance indicators for each reviewers' collective and each criterion.*

   In the first step of this process, we apply the 2-tuple OWA operator, which requires a weighting vector. This vector can be chosen in different ways. Particularly, in this example, we use the weighting vector determined by a fuzzy linguistic quantifier (see [50]), the quantifier "most" whose parameters are $(0.3, 0.8)$.

   The aggregation value with these OWA operators is computed for each criterion for the collectives of internal expert and internal non-experts and for each site ($I_k^{A^E}(x_j)$ and $I_k^{A^{NE}}(x_j)$), see Table 8.

Table 8. Environmental indicators for internal reviewers for each criterion.

	c_1^M	c_2^M	c_3^M	c_1^O	c_2^O	c_3^O
$I^{A^E}(x_1)$	(P, 0)	(MG, 0.27)	(F, 0.03)	(MP, 0.06)	(MP, −0.09)	(F, 0.21)
$I^{A^{NE}}(x_1)$	(P, −0.4)	(MG, 0)			(MP, 0.4)	
$I^{A^E}(x_2)$	(P, 0)	(G, 0)	(VG, 0)	(F, −0.17)	(P, −0.18)	(VP, 0.19)
$I^{A^{NE}}(x_2)$	(P, −0.4)	(G, −0.4)			(P, 0.13)	

The environmental performance indicators for general stakeholder reviewers and for social constituents reviewers are obtained by aggregating the values associated to them for each criterion and for each site $I_k^{B^{NE-S}}(x_j)$ and $I_k^{B^{NE-G}}(x_j)$). They are used to compute the indicators for each criterion for external non-experts and for each site ($I_k^{B^{NE}}(x_j)$. In addition, the aggregation value is obtained for each criterion for the external experts and for each site ($I_k^{B^E}(x_j)$). As noted above, all these calculations have been made using OWA operators with the appropriated weights. The results are showed in Table 9.

Table 9. Environmental indicators for external reviewers for each criterion.

	c_1^M	c_2^M	c_3^M	c_1^O	c_2^O	c_3^O
$I^{B^E}(x_1)$	(P, 0.2)	(MG, −0.4)	(MP, −0.07)	(F, 0.08)	(MG, −0.1)	(MG, −0.29)
$I^{B^{NE}}(x_1)$	(P, 0)	(G, −0.29)			(F, 0)	
$I^{B^E}(x_2)$	(P, −0.4)	(MG, −0.4)	(G, −0.34)	(P, −0.03)	(P, 0.13)	(P, −0.15)
$I^{B^{NE}}(x_2)$	(P, 0.36)	(MG, 0.36)			(MP, −0.5)	

2. *Computing environmental performance indicators for experts/non-experts reviewers and each criterion.*

In the second step of the process, the 2-tuple OWA operator is also applied using the weighting vector calculated before. It is worth pointing out that there is the possibility to use another one attending to the specific characteristics of a particular case study. After these calculations, an environmental performance indicator is computed for experts and non-experts, for each criterion about each facility site. These results are shown in Table 10. The Figure 5a, displays those values for each criteria.

Table 10. EPIs for experts and non experts for each criterion.

	c_1^M	c_2^M	c_3^M	c_1^O	c_2^O	c_3^O
$I^E(x_1)$	(P, 0.12)	(MG, 0)	(F, −0.41)	(F, −0.33)	(F, 0.1)	(MG, −0.49)
$I^{NE}(x_1)$	(P, −0.16)	(MG, 0.42)			(F, −0.24)	
$I^E(x_2)$	(P, −0.16)	(MG, 0.44)	(G, 0.46)	(MP, 0.08)	(P, 0)	(P, −0.41)
$I^{NE}(x_2)$	(P, 0.06)	(G, −0.5)			(P, 0.35)	

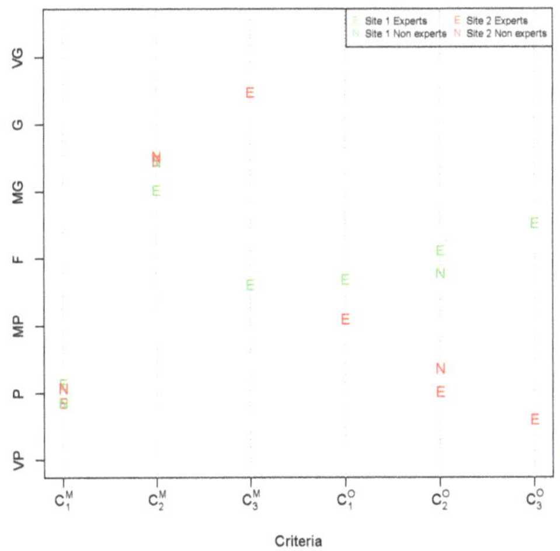

(a)

Figure 5. *Cont.*

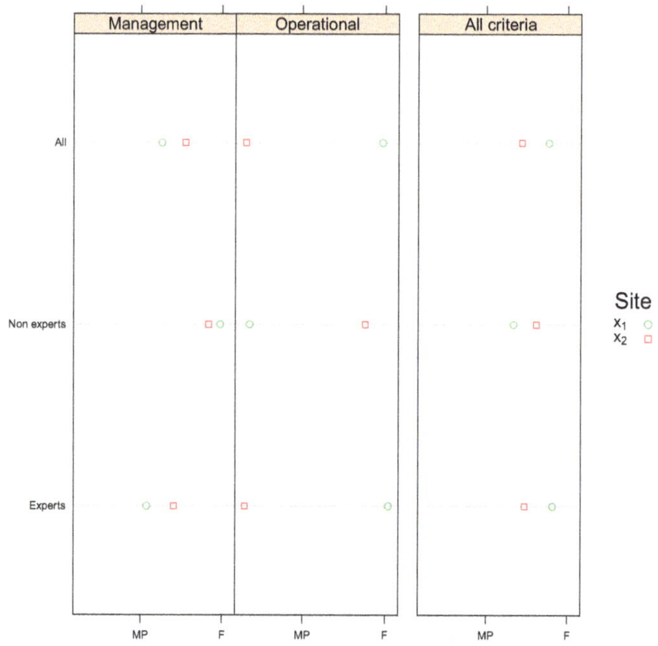

(b)

Figure 5. EPIs for site x_1 and x_2. (**a**) EPIs for each criteria for experts and non-experts; (**b**) Partial and global EPIs.

3. *Computing global environmental performance indicators.*

   In this step, the values in Table 10 are aggregated for each facility site. At this point, it is needed to aggregate the indicators from different criteria. Therefore, according to the proposed environmental integral evaluation model, the 2-tuple Choquet integral operator is used, which is able to cope with interaction among criteria. Consequently, specific fuzzy measures suited for this example are required and included in Appendix B.

   (a) *An overall global environmental performance indicator.*

   Now, the fuzzy measures from Table A2 and from Table A4 are employed to calculate the associated 2-tuple Choquet integral for experts and for non experts, respectively, in each site (I^E and I^{NE}). These computed values are shown in Table 11.

Table 11. Experts and non-experts EPIs for sites x_1 and x_2.

	Experts	Non Experts
Site x_1	(F, −0.18)	(F, −0.38)
Site x_2	(MP, 0.47)	(MP, 0.34)

Finally, the overall global environmental performance indicator for each facility site is computed from previous indicators as it was established in the Section 3.3.2, using for example $\beta = 0.75$. They are shown in Table 12.

Table 12. Overall global EPI for sites x_1 and x_2.

	Overall EPI
Site x_1	(F, −0.23)
Site x_2	(MP, 0.44)

(b) *Management and operational global environmental performance indicators.*

In the last step of the process, a management global indicator and an operational global indicator for each facility site are derived. In a similar way to the previous step, the values for several criteria are aggregated using a 2-tuple Choquet integral separately for the management criteria and for operational criteria. In the latter calculation it is used both the fuzzy measure for management μ^M and the fuzzy measure for operational criteria μ^O, respectively. The results are in Table 13. In order to compute the final values by the proposed procedure, both types of criteria are aggregated for experts and non experts using $\gamma = \delta = 0.75$. The resultant management and operational global performance indicators are shown in Table 14.

Table 13. Partial EPIs for sites x_1 and x_2.

	Site x_1		Site x_2	
	Experts	Non Experts	Experts	Non Experts
Manag.	(MP, 0.08)	(F, −0.16)	(MP, 0.42)	(F, −0.02)
Opera.	(F, 0.04)	(F, −0.24)	(P, 0.29)	(P, 0.35)

Table 14. Management and operational global EPIs for sites x_1 and x_2.

	MEPI	OEPI
Site x_1	(MP, 0.27)	(F, −0.03)
Site x_2	(F, −0.44)	(P, 0.31)

In order to ease the evaluation of the outcomes, the partial and global results obtained in the aggregation process are displayed in Figure 5b.

Finally, a classification between the facility sites is established with the purpose of identifying the best one, both overall as global management and operational issues. Using the ordinary lexicographic order on $\langle \overline{S} \rangle$, the facility site x_1, with a value $(Fair, -0.4) \in \langle \overline{S} \rangle$, is the best globally carrying out environmental practices with this evaluation. As well, the results show the facility site x_2 is slightly better than facility site x_1 in management issues, but not in operational ones. These indicators make visible different facets of the environmental strategies, as suggested by the standard ISO 14001 guidelines. Also note that both experts and non-experts consider better the site x_1. The company now knows the partial and global results which indicate the weak points of the facility sites and it allows to synchronize environmental facility site goals with company's environmental goals.

5. Conclusions

The evaluation of environmental practices allows companies and organizations to determine their efficiency and effectiveness of their efforts concerning environmental aspects. In this paper, an environmental integral evaluation model has been presented facilitating the comprehensive analysis. This new approach ensures the most relevant issues are regarded in the evaluation process. So, it can deal with different reviewers' collectives related to the company's activity. These reviewers express their objective and subjective perceptions presenting different degrees of knowledge about evaluated facility sites. Thus in our proposal, appraisers could express their assessments in different scales according to their knowledge and criteria nature. In addition, this model handles the interdependence among environmental criteria by means of the use of the discrete Choquet integral. It is worth emphasizing that the proposed model is quite flexible to be adapted easily to different settings. A computational experiment for making sensitivity analysis on fuzzy measures could be considered as future research.

Finally, the proposed model not only produces an overall global environmental performance indicator for each facility site, but also a management global and an operational global environmental performance indicators. In addition, it generates intermediate environmental performance indicators according to the opinions of each set of reviewers and each criterion. These are useful for the organization in its environmental improvement effort.

In the following appendixes some definitions and explanations are included to better understand our approach and our illustrative application.

Acknowledgments: The authors thank the anonymous reviewers and Academic Editor for their valuable comments and recommendations. The authors acknowledge financial support by the Spanish Ministerio de Economía y Competitividad under Projects ECO2016-77900-P (R. de Andrés Calle and T. González-Arteaga) and TIN2015-66524-P (L. Martínez).

Author Contributions: Teresa González-Arteaga, Rocio de Andrés Calle and Luis Martínez have contributed equally to this work.

Conflicts of Interest: The authors declare no conflict of interest.

Appendix A. A Short Survey on Discrete Choquet Integral for Dealing with Dependencies of EPIs

The application of the Choquet integral in multiple criteria decision making problems has been expanding and the main advances can be seen in [26,47]. When criteria have inter-dependent characteristics, it is not suitable to aggregate them by traditional aggregation operators based on additive measures because it would cause some bias effects in results. Then, we can use alternative operators that has been developed for this distinctive feature. These operators are based on fuzzy measures (or capacities), which only fulfil a monotonicity instead of additivity property. In this appendix, some definitions and important needed results are included for this application.

Definition A1. *Let $C = \{c_1, \ldots, c_n\}$ be a finite universe. A fuzzy measure or capacity is a set function $\mu : \mathcal{P}(C) \longrightarrow [0,1]$ which satisfies:*

1. $\mu(\emptyset) = 0$ *and* $\mu(C) = 1$,
2. $A \subseteq B \Rightarrow \mu(A) \leq \mu(B)$,

where $\mathcal{P}(C)$ is the set of all subsets of C.

Definition A2. *Let μ a fuzzy measure on C, the Möbius representation of μ is a set function $m_\mu : \mathcal{P}(C) \longrightarrow \Re$ given by*

$$m_\mu(S) = \sum_{T \subseteq S} (-1)^{|S|-|T|} \mu(T), \qquad \forall S \subseteq C$$

We note that if it is known the *Möbius* representation, it is easy to recover the fuzzy measure from which was derived $\mu(T) = \sum_{S \subseteq T} m_{\mu(S)}, \forall T \subseteq C$.

Definition A3. *Let $f : C \longrightarrow \mathbb{R}^+$ be a function with $x_k = f(c_k)$ for $k \in \{1, \ldots, n\}$. The discrete Choquet integral of f with respect to a fuzzy measure μ is given by*

$${}_\mu IC\,(x_1, \ldots, x_n) = \sum_{T \subseteq C} \left(m_\mu(T) \bigwedge_{i \in T} x_i \right)$$

where $\bigwedge$ denote the minimum.

Remark A1. *From a mathematical point of view, the discrete Choquet integral is a linear expression up to a reordering of the elements. In addition, as it has been proved in several references such as [26,51,52], among others, the Choquet integral fulfills very significant mathematical properties like idempotence, continuity, non-decreasing monotonicity, stability under the same positive linear transformation, decomposability and compensativeness.*

In this paper, we underscore the importance of the latter because it is essential for our application. Compensativeness indicates that:

$$\min(x_1,\ldots,x_n) \leq \ {}_{\mu}IC(x_1,\ldots,x_n) \leq \max(x_1,\ldots,x_n).$$

This allows us to transform the results of the aggregation by means of the Choquet integral into 2-tuples in the linguistic term set using the results in [46], faciliting understanding. It proves suitable the following definition.

Definition A4. *[53] Let $(l_1,\alpha_1),\ldots,(l_m,\alpha_n)$ be a vector of linguistic 2-tuple, $C = \{1,2,\ldots,n\}$ a finite universe and μ a fuzzy measure on C, then the 2-tuples Choquet integral operator is defined as:*

$${}_{\mu}IC((l_1,\alpha_1),\ldots,(l_n,\alpha_n)) = \Delta\left(\sum_{T\subseteq C}\left(m_{\mu}(T)\Delta^{-1}(\bigwedge_{i\in T}(l_i,\alpha_i)\right)\right)$$

here $\bigwedge$ denote the minimum.

It is relevant to point out that the fuzzy measure plays a key role in the Choquet integral comparable with the choice of weights in the weighted mean. Moreover, the weighted mean and the OWA aggregation operator are special cases of the Choquet integral for specific fuzzy measures which are connected with the weighting vector (see [52]).

The value, which the fuzzy measure assigns to a set A of criteria, $\mu(A)$ defines the weight of one combination of criteria, but it does not give the global importance of the criteria not either the degree of interaction among them. The overall importance of one or more criteria is given through all the measures $\mu(A)$ where A contains such criteria. On the basis of this idea, it has been proposed different indexes being the most common the *Shapley index*, which expresses the relative importance of a single criterion or a number of criteria into the decision problem. Thus, those indexes help to analyse the behavioral properties of the aggregation operator (see [51]).

In order to use the Choquet integral in our approach, we need to have at our disposal a fuzzy measure. Since it is convenient to choose one adjusted to our decision problem, we put forward identifying an appropriate fuzzy measure from its concrete characteristics. The main possibilities to do that are enclosed below. Once a fuzzy measure is identified the associated Choquet integral can be used as an aggregation tool.

Identification of the Fuzzy Measure

The identification of a suitable fuzzy measure to a particular Multi-Criteria Decision problem is the great importance since all the knowledge concerning the criteria is embedded into the fuzzy measure. As we manage a problem with n criteria, $2^n - 2$ real numbers in $[0,1]$ is needed to be completely determined. It seems obvious that these coefficients are hard to be provided by the decision makers. In order to address this issue a number of strategies have been considered in the literature. The fuzzy measure identification methods have been studied by many authors (see [26,47,54,55]). The key point is to use optimization methods to build the fuzzy measure belonging to a particular family with the aim of reducing the initial requirements. The most relevant family is the *k-additive fuzzy measures*. This allows us to model interaction among at most k criteria.

A fuzzy measure μ is *k-additive* if its *Möbius* representation verifies for each $A \subseteq C$ with $|A| > k$ $m_{\mu}(A) = 0$ and there exist at least one subset A of cardinality k such that $m_{\mu}(A) \neq 0$.

An optimization problem in order to obtain an specific fuzzy measure is posed for our problem. The alternative optimization methods differ among them with respect to the objective function and with respect to the input information used to establish constrains. The latter usually come from the monotonicity of fuzzy measures, the particular characteristics of the family of fuzzy measures used. It is also usual to use learning data that may include semantical considerations. An overview is provided in [47].

From a practical point of view these methods can be applied by means of the *Kappalab R package* (see [56]). This package contains high-level routines for R, which is free software environment release under GNU (GNU is a recursive acronym for "GNU is Not Unix") for statistical computing and graphics [57]. As a consequence, this software smooths the way for the applications of the Choquet integral, in particular, in the environmental context.

Finally, it is important to emphasize the advantage of allowing decision makers the flexibility to incorporate major interdependencies among criteria through the formulation of the fuzzy measure identification problem.

Appendix B. Computing Some Fuzzy Measures for the Illustrative Example

This appendix is devoted to computing the needed fuzzy measures used in the illustrative application of Section 4.

As we stated in Appendix A, there are several methods to achieve a suited fuzzy measure which collects the distinctive features of the evaluation context. In order to build a particular fuzzy measure for the application, we look for a 2-additive fuzzy measures using the minimum variance method (see [47]). Therefore, some initial requirements are considered. The choice made is just to show some potentials that may be interesting, but there are plenty of possibilities.

- Firstly, we have a clear orientation to criteria relative to practices in outputs of the company (c_1^M, c_1^O, c_3^O). We translate this through some constrains in the optimization problem. Let us consider Table A1 that contains a learning set of possible normalized criteria values. We demand that the aggregated values, with Choquet integral for its rows, satisfy this order

$$_{\mu}IC(\text{Row }1) \geq {}_{\mu}IC(\text{Row }2) \geq {}_{\mu}IC(\text{Row }3) \geq {}_{\mu}IC(\text{Row }4) \geq {}_{\mu}IC(\text{Row }5)$$

Here the aggregation values for rows 1 and 2 are greater than for rows 3, 4, and 5 because the outputs criteria have bigger values whatever are the rest of criteria. Besides, when the output criteria have equal values we prefer to have a greater value in criteria c_2^M, c_2^O, relative to practices in inputs of the company than in c_3^M. This implies that Row 1 has an aggregation value greater than Row 2 and the same happens with Rows 4 and 5.

Table A1. A learning set used for determining the fuzzy measure (criteria reordered).

	Output			Input		
	c_1^M	c_1^O	c_3^O	c_2^M	c_2^O	c_3^M
Row 1	G	MP	MP	G	MP	MP
Row 2	G	MP	MP	MP	MP	G
Row 3	MP	MP	MP	G	MP	G
Row 4	MP	MP	MP	G	MP	MP
Row 5	MP	MP	MP	MP	MP	G

This means the following constraints:

$$_{\mu}IC(Row\,1) - {}_{\mu}IC(Row\,2) \geq 0, \;\; {}_{\mu}IC(Row\,2) - {}_{\mu}IC(Row\,3) \geq 0,$$
$$_{\mu}IC(Row\,3) - {}_{\mu}IC(Row\,4) \geq 0, \;\; {}_{\mu}IC(Row\,4) - {}_{\mu}IC(Row\,5) \geq 0.$$

- Secondly, the decision maker brings some ideas about the interactions of some criteria. The pairs (c_1^M, c_1^O) and (c_1^M, c_3^O) have to interact in a redundancy way. That is, the contribution of the pair should be inferior to the sum of the contribution of the criterion into it. However, the pair (c_1^M, c_2^O) has to interact in a complementary way, i.e., it has a contribution superior to the sum of the contributions of the criteria in it. This means the following constraints on the Shapley indexes:

  $I_\mu(c_1^M, c_1^O) < 0$, $I_\mu(c_1^M, c_3^O) < 0$ and $I_\mu(c_1^M, c_2^O) > 0$
- Thirdly, due to the fact that the CO_2 emissions are the great importance for the company, it is imposed that the overall importance of this criterion should be at least 0.15 over 1. This adds a new constraint through the Shapley value for this criterion.

  This implies the constraint on the Shapley value $I_\mu(c_1^O) \geq 0.15$.

In order to get the fuzzy measure we are looking for we incorporate all the previous requirements in the routine for minimum variance method implemented in the kappalab R package. As the resulting fuzzy measure μ has $2^6 = 64$ values we show its Mobius transformation, which is shorter, in Table A2. Analogously, Table A3 contains the Mobius transformation of the associated fuzzy measures μ^M and μ^O.

Table A2. Möbius transformation for the fuzzy measure μ for expert reviewers.

Subset	$\{c_1^M\}$	$\{c_2^M\}$	$\{c_3^M\}$	$\{c_1^O\}$	$\{c_2^O\}$	$\{c_3^O\}$
$m_\mu(subset)$	0.503	0.203	0.036	0.266	0.109	0.222
Subset	$\{c_1^M, c_2^M\}$	$\{c_1^M, c_3^M\}$	$\{c_1^M, c_1^O\}$	$\{c_1^M, c_2^O\}$	$\{c_1^M, c_3^O\}$	$\{c_2^M, c_3^M\}$
$m_\mu(subset)$	−0.003	−0.003	−0.219	0.100	−0.197	0.131
Subset	$\{c_2^M, c_1^O\}$	$\{c_2^M, c_2^O\}$	$\{c_2^M, c_3^O\}$	$\{c_3^M, c_1^O\}$	$\{c_3^M, c_2^O\}$	$\{c_3^M, c_3^O\}$
$m_\mu(subset)$	−0.047	−0.099	−0.025	0.000	−0.010	0.000
Subset	$\{c_1^O, c_2^O\}$	$\{c_1^O, c_3^O\}$	$\{c_2^O, c_3^O\}$			
$m_\mu(subset)$	0.006	0.028	0.000			

Table A3. Mobius for μ^M and μ^O for management and for operational criteria, respectively (for experts).

Subset	$\{c_1^M\}$	$\{c_2^M\}$	$\{c_3^M\}$	$\{c_1^M, c_2^M\}$	$\{c_1^M, c_3^M\}$	$\{c_2^M, c_3^M\}$
$m_{\mu^M}(subset)$	0.58	0.23	0.04	−0.00	−0.00	0.15
Subset	$\{c_1^O\}$	$\{c_2^O\}$	$\{c_3^O\}$	$\{c_1^O, c_2^O\}$	$\{c_1^O, c_3^O\}$	$\{c_2^O, c_3^O\}$
$m_{\mu^O}(subset)$	0.42	0.17	0.35	0.01	0.04	0.00

Besides, since some of the criteria cannot be assessed by non-experts, we need to compute another fuzzy measure suited for the aggregation step that implies those criteria appraised by non-experts (c_1^M, c_2^M, c_2^O). Hence, we adapt the previous requirements in appropriate way. The results are shown in Table A4

Table A4. Mobius transformation for the fuzzy measure μ for non-expert reviewers.

Subset	$\{c_1^M\}$	$\{c_2^M\}$	$\{c_2^O\}$	$\{c_1^M, c_2^M\}$	$\{c_1^M, c_2^O\}$	$\{c_2^M, c_2^O\}$
$m_\mu(subset)$	0.29	0.35	0.29	−0.01	0.10	−0.01
$m_{\mu^M}(subset)$	0.47	0.56	-	−0.02	-	-
$m_{\mu^O}(subset)$	-	-	1	-	-	-

In order to understand better the role of each criteria in the aggregation operator we obtain the Shapley values associated to those previous fuzzy measures (see Table A5). These values shows the global involvement of each criteria in the results obtained with the Choquet integral. So, the criterion c_1^M (pollution control initiatives) is the more relevant for the expert reviewers but not for the non experts.

Table A5. Shapley values for fuzzy measure μ, $\mu^M$and μ^M for experts (right) and non-experts (left).

	Non-Experts							Experts		
	c_1^M	c_2^M	c_3^M	c_1^O	c_2^O	c_3^O		c_1^M	c_2^M	c_2^O
μ	0.34	0.18	0.10	0.15	0.11	0.12	μ	0.33	0.33	0.33
μ^M	0.58	0.31	0.12				μ^M	0.45	0.55	
μ^O				0.45	0.18	0.37	μ^O			1

References

1. Sarkis, J. Evaluating environmentally conscious business practices. *Eur. J. Oper. Res.* **1998**, *107*, 159–174.
2. Darnall, N.; Henriques, I.; Sadorsky, P. Do environmental management systems improve business performance in an international setting? *J. Int. Manag.* **2008**, *14*, 364–376.
3. Stanwick, P.; Stanwick, S. CEO compensation: Does it pay to be green? *Bus. Strategy Environ.* **2001**, *10*, 176–182.
4. Rivera, J. Assessing a voluntary environmental initiative in the developing world: The Costa Rican Certification for Sustainable Tourism. *Policy Sci.* **2002**, *35*, 333–360.
5. Iraldo, F.; Testa, F.; Frey, M. Is an environmental management system able to influence environmental and competitive performance? The case of the eco-management and audit scheme (EMAS) in the European union. *J. Clean. Prod.* **2009**, *17*, 1444–1452.
6. Jørgensen, S.; Burkhard, B.; Müller, F. Twenty volumes of ecological indicators: An accounting short review. *Ecol. Indic.* **2013**, *28*, 4–9.
7. Rahdari, A.; Rostamy, A. Designing a general set of sustainability indicators at the corporate level. *J. Clean. Prod.* **2015**, *108*, 757–771.
8. Jasch, C. Environmental performance evaluation and indicators. *J. Clean. Prod.* **2000**, *8*, 79–88.
9. Lin, B.; Jones, C.; Hsieh, C. Environmental practices and assessment: A process perspective. *Ind. Manag. Data Syst.* **2001**, *101*, 71–79.
10. Handfield, R.; Walton, S.; Sroufe, R.; Melnyk, S. Applying environmental criteria to supplier assessment: A study in the application of the Analytical Hierarchy Process. *Eur. J. Oper. Res.* **2002**, *141*, 70–87.
11. Sarkis, J. A strategic decision framework for green supply chain management. *J. Clean. Prod.* **2003**, *11*, 397–409.
12. Humphreys, P.; Wong, Y.; Chan, F. Integrating environmental criteria into the supplier selection process. *J. Mater. Process. Technol.* **2003**, *138*, 349–356.
13. Zhu, Q.; Sarkis, J.; Lai, K.H. Confirmation of a measurement model for green supply chain management practices implementation. *Int. J. Prod. Econ.* **2008**, *111*, 261–273.
14. Jabbour, A.B.L.S.; Jabbour, C.J.C. Are supplier selection criteria going green? Case studies of companies in Brazil. *Ind. Manag. Data Syst.* **2009**, *109*, 477–495.
15. Herva, M.; Roca, E. Review of combined approaches and multi-criteria analysis for corporate environmental evaluation. *J. Clean. Prod.* **2013**, *39*, 355–371.
16. Olsthoorn, X.; Tyteca, D.; Wehrmeyer, W.; Wagner, M. Environmental indicators for business: A review of the literature and standardisation methods. *J. Clean. Prod.* **2001**, *9*, 453–463.
17. Niemeijer, D.; de Groot, R.S. A conceptual framework for selecting environmental indicator sets. *Ecol. Indic.* **2008**, *8*, 14–25.
18. Brunklaus, B.; Malmqvist, T.; Baumann, H. Managing Stakeholders or the Environment? The Challenge of Relating Indicators in Practice. *Corp. Soc. Responsib. Environ. Manag.* **2009**, *16*, 27–37.
19. Gandhi, S.; Mangla, S.; Kumar, P.; Kumar, D. Evaluating factors in implementation of successful green supply chain management using DEMATEL: A case study. *Int. Strateg. Manag. Rev.* **2015**, *3*, 96–109.
20. Tseng, M.L. Using a hybrid MCDM model to evaluate firm environmental knowledge management in uncertainty. *Appl. Soft Comput.* **2011**, *11*, 1340–1352.
21. Xie, S.; Hayase, K. Corporate Environmental Performance Evaluation: A Measurement Model and a New Concept. *Bus. Strategy Environ.* **2007**, *16*, 148–168.
22. Tsoulfas, G.T.; Pappis, C.P. A model for supply chains environmental performance analysis and decision making. *J. Clean. Prod.* **2008**, *16*, 1647–1657.

23. Ho, W.; Xu, X.; Dey, P.K. Multi-criteria decision making approaches for supplier evaluation and selection: A literature review. *Eur. J. Oper. Res.* **2010**, *202*, 16–24.
24. Arfi, B. Fuzzy decision making in politics: A linguistic fuzzy-set approach. *Political Anal.* **2005**, *13*, 23–56.
25. Martínez, L. Sensory evaluation based on linguistic decision analysis. *Int. J. Approx. Reason.* **2007**, *44*, 148–164.
26. Grabisch, M. The application of fuzzy integrals in multicriteria decision making. *Eur. J. Oper. Res.* **1996**, *89*, 445–456.
27. Darnall, N.; Seol, I.; Sarkis, J. Perceived stakeholder influences and organizations' use of environmental audits. *Acc. Organ. Soc.* **2009**, *34*, 170–187.
28. King, A.; Lenox, M.; Terlaak, A. The strategic use of decentralized institutions: Exploring certification with the ISO 14001 management standard. *Acad. Manag. J.* **2005**, *48*, 1091–1106.
29. Kain, J.H.; Saderberg, H. Management of complex knowledge in planning for sustainable development: The use of multi-criteria decision aids. *Environ. Impact Assess. Rev.* **2008**, *28*, 7–21.
30. Nawrocka, D.; Parker, T. Finding the connection: Environmental management systems and environmental performance. *J. Clean. Prod.* **2009**, *17*, 601–607.
31. International Organization for Standardization (ISO). *Environmental Management Systems–Requirements with Guidance for Use (ISO 14001:2004)*; ISO: Geneva, Switzerland, 2004.
32. To, W.; Lee, P. Diffusion of ISO 14001 environmental management system: Global, regional and country-level analyses. *J. Clean. Prod.* **2014**, *66*, 489–498.
33. International Organization for Standardization (ISO). *Standards 14031. Environmental Management Environmental Performance Evaluation: Guidelines*; ISO: Geneva, Switzerland, 1999.
34. Henri, J.F.; Journeault, M. Environmental performance indicators: An empirical study of Canadian manufacturing firms. *J. Environ. Manag.* **2008**, *87*, 165–176.
35. Thoresen, J. Environmental performance evaluation—A tool for industrial improvement. *J. Clean. Prod.* **1999**, *7*, 365–370.
36. Lee, A.H.; Kang, H.Y.; Hsu, C.F.; Hung, H.C. A green supplier selection model for high-tech industry. *Expert Syst. Appl.* **2009**, *36*, 7917–7927.
37. Awasthi, A.; Chauhan, S.S.; Goyal, S. A fuzzy multicriteria approach for evaluating environmental performance of suppliers. *Int. J. Prod. Econ.* **2010**, *126*, 370–378.
38. Tseng, M.L.; Chiu, A.S. Evaluating firm's green supply chain management in linguistic preferences. *J. Clean. Prod.* **2013**, *40*, 22–31.
39. Espinilla, M.; de Andres, R.; Martinez, F.J.; Martinez, L. *A Heterogeneous Evaluation Model for the Assessment of Sustainable Energy Policies*; MeloPinto, P., Couto, P., Serodio, C., Fodor, J., DeBaets, B., Eds.; Advances in Intelligent and Soft Computing; Springer: Berlin/Heidelberg, Germany, 2011; Volume 107, pp. 209–220.
40. Tseng, M.L.; Chiang, J.H.; Lan, L.W. Selection of optimal supplier in supply chain management strategy with analytic network process and choquet integral. *Comput. Ind. Eng.* **2009**, *57*, 330–340.
41. Buyukozkan, G.; Cifci, G. A novel fuzzy multi-criteria decision framework for sustainable supplier selection with incomplete information. *Comput. Ind.* **2011**, *62*, 164–174.
42. Herrera, F.; Martínez, L.; Sánchez, P. Managing non-homogeneous information in group decision making. *Eur. J. Oper. Res.* **2005**, *166*, 115–132.
43. Likert, R. A Technique for the measumerement of attitudes. *Arch. Psychol.* **1932**, *140*, 1–55.
44. Herrera, F.; Herrera-Viedma, E.; Martínez, L. A fusion approach for managing multi-granularity linguistic term sets in decision making. *Fuzzy Sets Syst.* **2000**, *114*, 43–58.
45. Martínez, L.; Herrera, F. An overview on the 2-tuple linguistic model for computing with words in decision making: Extensions, applications and challenges. *Inf. Sci.* **2012**, *207*, 1–18.
46. Herrera, F.; Martínez, L. A 2-tuple fuzzy linguistic representation model for computing with words. *IEEE Trans. Fuzzy Syst.* **2000**, *8*, 746–752.
47. Grabisch, M.; Kojadinovic, I.; Meyer, P. A review of methods for capacity identification in Choquet integral based multi-attribute utility theory: Applications of the Kappalab R package. *Eur. J. Oper. Res.* **2008**, *186*, 766–785.
48. Klement, E.; Mesiar, R. Discrete Integrals and Axiomatically Defined Functionals. *Axioms* **2012**, *1*, 9–20.
49. Yager, R. On Ordered Weighted Averaging Operators in Multicriteria Decision Making. *IEEE Trans. Syst. Man Cybern.* **1988**, *18*, 183–190.

50. Zadeh, L. A Computational Approach to Fuzzy Quantifiers in Natural Languages. *Comput. Math. Appl.* **1983**, *9*, 149–184.
51. Marichal, J.L. Behavioral analysis of aggregation in multicriteria decision aid. In *Preferences and Decisions under Incomplete Knowledge*; Fodor, J., De Baets, B., Perny, P., Eds.; Studies in Fuzziness and Soft Computing; Physica Verlag: Heidelberg, Germany, 2000; Volume 51, pp. 153–178.
52. Beliakov, G.; Pradera, A.; Calvo, T. *Aggregation Functions: A Guide for Practitioners*; Springer: Berlin, Germany, 2007.
53. Yang, W.; Chen, Z. New aggregation operators based on the Choquet integral and 2-tuple linguistic information. *Expert Syst. Appl.* **2012**, *39*, 2662–2668.
54. Meyer, P.; Roubens, M. Choice, Ranking and Sorting in Fuzzy Multiple Criteria Decision Aid. In *Multiple Criteria Decision Analysis: State of the Art Surveys*; Figueira, J., Greco, S., Ehrgott, M., Eds.; International Series in Operations Research and Management Science; Springer: New York, NY, USA, 2005; Volume 78, pp. 471–503.
55. Wu, J.Z.; Zhang, Q. 2-order additive fuzzy measure identification method based on diamond pairwise comparison and maximum entropy principle. *Fuzzy Optim. Decis. Mak.* **2010**, *9*, 435–453.
56. Grabisch, M.; Kojadinovic, I.; Meyer, P. *Kappalab: Non-Additive Measure and Integral Manipulation Functions*, R package version 0.4-7; Repository CRAN, 2015. Available online: https://cran.r-project.org/web/packages/kappalab/index.html (accessed on 20 September 2017).
57. R Development Core Team. *R: A Language and Environment for Statistical Computing;* R Foundation for Statistical Computing: Vienna, Austria, 2017; ISBN 3-900051-07-0.

axioms

MDPI

Article

Cubic Interval-Valued Intuitionistic Fuzzy Sets and Their Application in *BCK/BCI*-Algebras

Young Bae Jun [1], Seok-Zun Song [2,*] and Seon Jeong Kim [3]

1 Department of Mathematics Education, Gyeongsang National University, Jinju 52828, Korea; skywine@gmail.com
2 Department of Mathematics, Jeju National University, Jeju 63243, Korea
3 Department of Mathematics, Natural Science of College, Gyeongsang National University, Jinju 52828, Korea; skim@gnu.ac.kr
* Correspondence: szsong@jejunu.ac.kr; Tel.: +82-64-754-3562

Received: 28 December 2017; Accepted: 15 January 2018; Published: 23 January 2018

Abstract: As a new extension of a cubic set, the notion of a cubic interval-valued intuitionistic fuzzy set is introduced, and its application in BCK/BCI-algebra is considered. The notions of α-internal, β-internal, α-external and β-external cubic IVIF set are introduced, and the P-union, P-intersection, R-union and R-intersection of α-internal and α-external cubic IVIF sets are discussed. The concepts of cubic IVIF subalgebra and ideal in BCK/BCI-algebra are introduced, and related properties are investigated. Relations between cubic IVIF subalgebra and cubic IVIF ideal are considered, and characterizations of cubic IVIF subalgebra and cubic IVIF ideal are discussed.

Keywords: α-internal; β-internal; α-external; β-external cubic interval-valued intuitionistic fuzzy set; cubic interval-valued intuitionistic fuzzy subalgebra; cubic interval-valued intuitionistic fuzzy ideal

JEL Classification: MSC; 06F35; 03B60; 03B52

---

1. Introduction

In 2012, Jun et al. [1] introduced cubic sets, and then this notion is applied to several algebraic structures (see [2–11]).

The aim of this paper is to introduce the notion of a cubic IVIF set which is an extended concept of a cubic set. We introduce cubic interval-valued intuitionistic fuzzy set which is an extension of cubic set, and apply it to BCK/BCI-algebra. We investigate P-union, P-intersection, R-union and R-intersection of α-internal and α-external cubic IVIF sets.

We define cubic IVIF subalgebra and ideal in BCK/BCI-algebra, and investigate related properties. We consider relations between cubic IVIF subalgebra and cubic IVIF ideal. We discuss characterizations of cubic IVIF subalgebra and cubic IVIF ideal.

2. Preliminaries

A *fuzzy set* in a set X is defined to be a function $\mu : X \to [0,1]$. For $I = [0,1]$, denote by I^X the collection of all fuzzy sets in a set X. Define a relation $\leq$ on I^X as follows:

$$(\forall \mu, \lambda \in I^X)\ (\mu \leq \lambda \iff (\forall x \in X)(\mu(x) \leq \lambda(x))). \tag{1}$$

The complement of $\mu \in I^X$, denoted by μ^c, is defined by

$$(\forall x \in X)\ (\mu^c(x) = 1 - \mu(x)). \tag{2}$$

For a family $\{\mu_i \mid i \in \Lambda\}$ of fuzzy sets in X, we define the join ($\vee$) and meet ($\wedge$) operations as follows:

$$\bigvee_{i\in\Lambda} \mu_i : X \to [0,1],\ x \mapsto \sup\{\mu_i(x) \mid i \in \Lambda\},$$
$$\bigwedge_{i\in\Lambda} \mu_i : X \to [0,1]\ x \mapsto \inf\{\mu_i(x) \mid i \in \Lambda\}.$$

An *interval-valued intuitionistic fuzzy set* (briefly, IVIF set) A over a set X (see [12]) is an object having the form

$$\tilde{A} = \{\langle x, \alpha[\tilde{A}](x), \beta[\tilde{A}](x)\rangle \mid x \in X\}$$

where $\alpha[\tilde{A}](x) \subseteq [0,1]$ and $\beta[\tilde{A}](x) \subseteq [0,1]$ are intervals and for every $x \in X$,

$$\sup \alpha[\tilde{A}](x) + \sup \beta[\tilde{A}](x) \leq 1$$

Especially, if

$$\sup \alpha[\tilde{A}](x) = \inf \alpha[\tilde{A}](x) \text{ and } \sup \beta[\tilde{A}](x) = \inf \beta[\tilde{A}](x), \tag{3}$$

then the IVIF set $\tilde{A}$ is reduced to an intuitionistic fuzzy set (see [13]).

Given two closed subintervals $D_1 = [D_1^-, D_1^+]$ and $D_2 = [D_2^-, D_2^+]$ of $[0,1]$, we define the order "$\ll$" as follows:

$$D_1 \ll D_2 \Leftrightarrow D_1^- \leq D_2^- \text{ and } D_1^+ \leq D_2^+.$$

We also define the refined minimum (briefly, rmin) and refined maximum (briefly, rmax) as follows:

$$\mathrm{rmin}\{D_1, D_2\} = \left[\min\{D_1^-, D_2^-\}, \min\{D_1^+, D_2^+\}\right],$$
$$\mathrm{rmax}\{D_1, D_2\} = \left[\max\{D_1^-, D_2^-\}, \max\{D_1^+, D_2^+\}\right].$$

For a family $\{D_i = [D_i^-, D_i^+] \mid i \in \Lambda\}$ of closed subintervals of $[0,1]$, we define rinf (refined infimum) and rsup (refined supermum) as follows:

$$\underset{i\in\Lambda}{\mathrm{rinf}} D_i = \left[\inf_{i\in\Lambda} D_i^-, \inf_{i\in\Lambda} D_i^+\right] \text{ and } \underset{i\in\Lambda}{\mathrm{rsup}} D_i = \left[\sup_{i\in\Lambda} D_i^-, \sup_{i\in\Lambda} D_i^+\right].$$

In this paper we use the interval-valued intuitionistic fuzzy set

$$\tilde{A} = \{\langle x, \alpha[\tilde{A}](x), \beta[\tilde{A}](x)\rangle \mid x \in X\}$$

over X in which $\alpha[\tilde{A}](x)$ and $\beta[\tilde{A}](x)$ are closed subintervals of $[0,1]$ for all $x \in X$, that is, $\alpha[\tilde{A}] : X \to D[0,1]$ and $\beta[\tilde{A}] : X \to D[0,1]$ are interval-valued fuzzy (briefly, IVF) sets in X where $D[0,1]$ is the set of all closed subintervals of $[0,1]$. Also, we use the notations $\alpha[\tilde{A}]^-(x)$ and $\alpha[\tilde{A}]^+(x)$ to mean the left end point and the right end point of the interval $\alpha[\tilde{A}](x)$, respectively, and so we have $\alpha[\tilde{A}](x) = [\alpha[\tilde{A}]^-(x), \alpha[\tilde{A}]^+(x)]$. The interval-valued intuitionistic fuzzy set

$$\tilde{A} = \{\langle x, \alpha[\tilde{A}](x), \beta[\tilde{A}](x)\rangle \mid x \in X\}$$

over X is simply denoted by $\tilde{A}(x) = (\alpha[\tilde{A}](x), \beta[\tilde{A}](x))$ for $x \in X$ or $\tilde{A} = (\alpha[\tilde{A}], \beta[\tilde{A}])$.

An algebra $(X; *, 0)$ of type $(2,0)$ is called a *BCI-algebra* if it satisfies the following axioms:

(I) $(\forall x, y, z \in X)\ (((x * y) * (x * z)) * (z * y) = 0)$,
(II) $(\forall x, y \in X)\ ((x * (x * y)) * y = 0)$,
(III) $(\forall x \in X)\ (x * x = 0)$,
(IV) $(\forall x, y \in X)\ (x * y = 0, y * x = 0 \Rightarrow x = y)$.

If a BCI-algebra X satisfies the following identity:

(V) $(\forall x \in X)\ (0 * x = 0)$,

then X is called a *BCK-algebra*. We can define a partial ordering $\leq$ on X by $x \leq y$ if and only if $x * y = 0$. Any BCK/BCI-algebra X satisfies the following conditions:

$$(\forall x \in X)\ (x * 0 = x), \tag{4}$$
$$(\forall x, y, z \in X)\ (x \leq \ \Rightarrow x * z \leq y * z,\ z * y \leq z * x), \tag{5}$$
$$(\forall x, y, z \in X)\ ((x * y) * z = (x * z) * y), \tag{6}$$
$$(\forall x, y, z \in X)\ ((x * z) * (y * z) \leq x * y). \tag{7}$$

A nonempty subset S of a BCK/BCI-algebra X is called a *subalgebra* of X if $x * y \in S$ for all $x, y \in S$. A subset I of a BCK/BCI-algebra X is called an *ideal* of X if $0 \in I$ and the following condition is valid.

$$(\forall x, y \in X)\ (x * y \in I,\ y \in I \Rightarrow x \in I). \tag{8}$$

We refer the reader to the books [14,15] and the paper [16] for further information regarding BCK/BCI-algebras.

3. Cubic Interval-Valued Intuitionistic Fuzzy Sets

Definition 1. *Let X be a nonempty set. By a cubic interval-valued intuitionistic fuzzy set (briefly, cubic IVIF set) in X we mean a structure*

$$\mathcal{A} = \{\langle x, \mu(x), \tilde{A}(x)\rangle \mid x \in X\}$$

in which μ is a fuzzy set in X and $\tilde{A}$ is an interval-valued intuitionistic fuzzy set in X.

A cubic IVIF set $\mathcal{A} = \{\langle x, \mu(x), \tilde{A}(x)\rangle \mid x \in X\}$ is simply denoted by $\mathcal{A} = \langle \mu, \tilde{A}\rangle$.

Let $\mathcal{A} = \langle \mu, \tilde{A}\rangle$ be a cubic IVIF set in a nonempty set X. Given $\varepsilon \in [0,1]$ and $([s_\alpha, t_\alpha], [s_\beta, t_\beta]) \in D[0,1] \times D[0,1]$, we consider the sets

$$\mu[\varepsilon] := \{x \in X \mid \mu(x) \geq \varepsilon\},$$
$$\alpha[\tilde{A}][s_\alpha, t_\alpha] := \{x \in X \mid \alpha[\tilde{A}](x) \ll [s_\alpha, t_\alpha]\},$$
$$\beta[\tilde{A}][s_\beta, t_\beta] := \{x \in X \mid \beta[\tilde{A}](x) \gg [s_\beta, t_\beta]\}.$$

Definition 2. *Let X be a nonempty set. A cubic IVIF set $\mathcal{A} = \langle \mu, \tilde{A}\rangle$ is said to be*

- *α-internal if $\mu(x) \in \alpha[\tilde{A}](x)$ for all $x \in X$,*
- *β-internal if $\mu(x) \in \beta[\tilde{A}](x)$ for all $x \in X$,*
- *internal if it is both α-internal and β-internal.*
- *α-external if $\mu(x) \notin (\alpha[\tilde{A}]^-(x), \alpha[\tilde{A}]^+(x))$ for all $x \in X$.*
- *β-external if $\mu(x) \notin (\beta[\tilde{A}]^-(x), \beta[\tilde{A}]^+(x))$ for all $x \in X$.*
- *external if it is both α-external and β-external.*

It is clear that if $\mathcal{A} = \langle \mu, \tilde{A}\rangle$ is a cubic IVIF set in X with $\alpha[\tilde{A}](x) \cap \beta[\tilde{A}](x) = \emptyset$ for some $x \in X$, then $\mathcal{A} = \langle \mu, \tilde{A}\rangle$ cannot be internal.

Example 1. *Let $\mathcal{A} = \langle \mu, \tilde{A} \rangle$ be a cubic IVIF set in I in which $\tilde{A}(x) = ([0.3, 0.5], [0.1, 0.4])$ and $\mu(x) = k$ for all $x \in X$.*

(1) *If $k \in (0.4, 0.5]$, then $\mathcal{A} = \langle \mu, \tilde{A} \rangle$ is α-internal which is not β-internal.*
(2) *If $k \in [0.1, 0.3)$, then $\mathcal{A} = \langle \mu, \tilde{A} \rangle$ is β-internal which is not α-internal.*
(3) *If $k \in [0.3, 0.4]$, then $\mathcal{A} = \langle \mu, \tilde{A} \rangle$ is internal.*
(4) *If either $k \geq 0.5$ or $k \leq 0.1$, then $\mathcal{A} = \langle \mu, \tilde{A} \rangle$ is external.*
(5) *If either $k \geq 0.5$ or $k \leq 0.3$, then $\mathcal{A} = \langle \mu, \tilde{A} \rangle$ is α-external but may not be β-external.*
(6) *If either $k \geq 0.4$ or $k \leq 0.1$, then $\mathcal{A} = \langle \mu, \tilde{A} \rangle$ is β-external but may not be α-external.*

Proposition 1. *Let $\mathcal{A} = \langle \mu, \tilde{A} \rangle$ be a cubic IVIF set in a nonempty set X in which $\beta[\tilde{A}]^-(x) \leq \alpha[\tilde{A}]^-(x) < \alpha[\tilde{A}]^+(x) \leq \beta[\tilde{A}]^+(x)$ for all $x \in X$. If $\mathcal{A} = \langle \mu, \tilde{A} \rangle$ is α-internal, then it is β-internal and so internal. Also, if $\mathcal{A} = \langle \mu, \tilde{A} \rangle$ is β-external, then it is α-external and so external.*

Proof. Straightforward. □

Proposition 2. *If $\mathcal{A} = \langle \mu, \tilde{A} \rangle$ is a cubic IVIF set in X which is not external, then there exist $x, y \in X$ such that $\mu(x) \in \alpha[\tilde{A}](x)$ or $\mu(y) \in \beta[\tilde{A}](y)$.*

Proof. Straightforward. □

Theorem 1. *Let $\mathcal{A} = \langle \mu, \tilde{A} \rangle$ be a cubic IVIF set in X in which the right end point of $\alpha[\tilde{A}](x)$ (or $\beta[\tilde{A}](x)$) is equal to the left end point of $\beta[\tilde{A}](x)$ (or $\alpha[\tilde{A}](x)$) for all $x \in X$. If we define μ by*

$$(\forall x \in X) \left(\mu(x) = \alpha[\tilde{A}]^+(x) = \beta[\tilde{A}]^-(x) \text{ or } \mu(x) = \alpha[\tilde{A}]^-(x) = \beta[\tilde{A}]^+(x) \right),$$

then $\mathcal{A} = \langle \mu, \tilde{A} \rangle$ is external.

Proof. Straightforward. □

For any cubic IVIF set $\mathcal{A} = \langle \mu, \tilde{A} \rangle$ in X, let

$$U_\alpha(\tilde{A}) := \{\alpha[\tilde{A}]^+(x) \mid x \in X\}, \; L_\alpha(\tilde{A}) := \{\alpha[\tilde{A}]^-(x) \mid x \in X\},$$
$$U_\beta(\tilde{A}) := \{\beta[\tilde{A}]^+(x) \mid x \in X\}, \; L_\beta(\tilde{A}) := \{\beta[\tilde{A}]^-(x) \mid x \in X\}.$$

Theorem 2. *Let $\mathcal{A} = \langle \mu, \tilde{A} \rangle$ be a cubic IVIF set in X. If $\mathcal{A} = \langle \mu, \tilde{A} \rangle$ is both α-internal and α-external (resp., both β-internal and β-external), then $\mu(x) \in U_\alpha(\tilde{A}) \cup L_\alpha(\tilde{A})$ (resp., $\mu(x) \in U_\beta(\tilde{A}) \cup L_\beta(\tilde{A})$) for all $x \in X$.*

Proof. Assume that $\mathcal{A} = \langle \mu, \tilde{A} \rangle$ is both α-internal and α-external. Then $\mu(x) \in \alpha[\tilde{A}](x)$ and $\mu(x) \notin (\alpha[\tilde{A}]^-(x), \alpha[\tilde{A}]^+(x))$ for all $x \in X$. It follows that $\mu(x) = \alpha[\tilde{A}]^-(x)$ or $\mu(x) = \alpha[\tilde{A}]^+(x)$, that is, $\mu(x) \in L_\alpha(\tilde{A})$ or $\mu(x) \in U_\alpha(\tilde{A})$. Hence $\mu(x) \in U_\alpha(\tilde{A}) \cup L_\alpha(\tilde{A})$ for all $x \in X$. Similarly, if $\mathcal{A} = \langle \mu, \tilde{A} \rangle$ is both β-internal and β-external, then $\mu(x) \in U_\alpha(\tilde{A}) \cup L_\alpha(\tilde{A})$ for all $x \in X$. □

Given an IVIF set $\tilde{A} = (\alpha[\tilde{A}], \beta[\tilde{A}])$ in X, the *complement* of $\tilde{A}$ is denoted by $\tilde{A}^c$ and is defined as follows:

$$\tilde{A}^c : X \to D[0,1] \times D[0,1], \; x \mapsto \langle \alpha[\tilde{A}^c], \beta[\tilde{A}^c] \rangle \tag{9}$$

where

$$\alpha[\tilde{A}^c] : X \to D[0,1], \; x \mapsto [\alpha[\tilde{A}^c]^-(x), \alpha[\tilde{A}^c]^+(x)] \tag{10}$$

and

$$\beta[\tilde{A}^c] : X \to D[0,1],\ x \mapsto [\beta[\tilde{A}^c]^-(x), \beta[\tilde{A}^c]^+(x)] \tag{11}$$

with $\alpha[\tilde{A}^c]^-(x) = 1 - \alpha[\tilde{A}]^+(x)$, $\alpha[\tilde{A}^c]^+(x) = 1 - \alpha[\tilde{A}]^-(x)$, $\beta[\tilde{A}^c]^-(x) = 1 - \beta[\tilde{A}]^+(x)$, and $\beta[\tilde{A}^c]^+(x) = 1 - \beta[\tilde{A}]^-(x)$.

Definition 3. *For any cubic IVIF sets $\mathcal{A} = \langle \mu, \tilde{A} \rangle$ and $\mathcal{B} = \langle \lambda, \tilde{B} \rangle$ in X, we define*

$$\text{(Equality) } \mathcal{A} = \mathcal{B} \Leftrightarrow \tilde{A} = \tilde{B} \text{ and } \mu = \lambda, \tag{12}$$

$$\text{(P-order) } \mathcal{A} \Subset_P \mathcal{B} \Leftrightarrow \tilde{A} \subseteq \tilde{B} \text{ and } \mu \leq \lambda, \tag{13}$$

$$\text{(R-order) } \mathcal{A} \Subset_R \mathcal{B} \Leftrightarrow \tilde{A} \subseteq \tilde{B} \text{ and } \mu \geq \lambda, \tag{14}$$

where $\tilde{A} \subseteq \tilde{B}$ means that $\alpha[\tilde{A}] \ll \alpha[\tilde{B}]$ and $\beta[\tilde{A}] \gg \beta[\tilde{B}]$, that is,

$$(\forall x \in X) \begin{pmatrix} \alpha[\tilde{A}]^-(x) \leq \alpha[\tilde{B}]^-(x),\ \alpha[\tilde{A}]^+(x) \leq \alpha[\tilde{B}]^+(x) \\ \beta[\tilde{A}]^-(x) \geq \beta[\tilde{B}]^-(x),\ \beta[\tilde{A}]^+(x) \geq \beta[\tilde{B}]^+(x) \end{pmatrix}.$$

It is clear that the set of all cubic IVIF sets in X forms a poset under the P-order $\Subset_P$ and the R-order $\Subset_R$.

For any $\tilde{A}_i = \{\langle x, \alpha[\tilde{A}_i](x), \beta[\tilde{A}_i](x) \rangle\}$ where $i \in \Lambda$, we define

$$\bigcup_{i\in\Lambda} \tilde{A}_i = \left\{ \langle x, \alpha[\bigcup_{i\in\Lambda} \tilde{A}_i](x), \beta[\bigcup_{i\in\Lambda} \tilde{A}_i](x) \rangle \mid x \in X \right\} \tag{15}$$

and

$$\bigcap_{i\in\Lambda} \tilde{A}_i = \left\{ \langle x, \alpha[\bigcap_{i\in\Lambda} \tilde{A}_i](x), \beta[\bigcap_{i\in\Lambda} \tilde{A}_i](x) \rangle \mid x \in X \right\} \tag{16}$$

where $\alpha[\bigcup_{i\in\Lambda} \tilde{A}_i](x) = \underset{i\in\Lambda}{\mathrm{rsup}}\alpha[\tilde{A}_i](x)$, $\beta[\bigcup_{i\in\Lambda} \tilde{A}_i](x) = \underset{i\in\Lambda}{\mathrm{rinf}}\beta[\tilde{A}_i](x)$, $\alpha[\bigcap_{i\in\Lambda} \tilde{A}_i](x) = \underset{i\in\Lambda}{\mathrm{rinf}}\alpha[\tilde{A}_i](x)$, and $\beta[\bigcap_{i\in\Lambda} \tilde{A}_i](x) = \underset{i\in\Lambda}{\mathrm{rsup}}\beta[\tilde{A}_i](x)$.

Definition 4. *Given a family $\{\mathcal{A}_i = \langle \mu_i, \tilde{A}_i \rangle \mid i \in \Lambda\}$ of cubic IVIF sets in X, we define*

(1) $\underset{i\in\Lambda}{\Cup_P} \mathcal{A}_i := \left\langle \bigvee_{i\in\Lambda} \mu_i, \bigcup_{i\in\Lambda} \tilde{A}_i \right\rangle$ *(P-union)*

(2) $\underset{i\in\Lambda}{\Cup_R} \mathcal{A}_i := \left\langle \bigwedge_{i\in\Lambda} \mu_i, \bigcup_{i\in\Lambda} \tilde{A}_i \right\rangle$ *(R-union)*

(3) $\underset{i\in\Lambda}{\Cap_P} \mathcal{A}_i := \left\langle \bigwedge_{i\in\Lambda} \mu_i, \bigcap_{i\in\Lambda} \tilde{A}_i \right\rangle$ *(P-intersection)*

(4) $\underset{i\in\Lambda}{\Cap_R} \mathcal{A}_i := \left\langle \bigvee_{i\in\Lambda} \mu_i, \bigcap_{i\in\Lambda} \tilde{A}_i \right\rangle$ *(R-intersection)*

Proposition 3. *For any cubic IVIF sets $\mathcal{A} = \langle \mu, \tilde{A} \rangle$, $\mathcal{B} = \langle \lambda, \tilde{B} \rangle$, $\mathcal{C} = \langle \nu, \tilde{C} \rangle$ and $\mathcal{D} = \langle \rho, \tilde{D} \rangle$ in X, we have*

(1) $\mathcal{A} \Subset_P \mathcal{B},\ \mathcal{A} \Subset_P \mathcal{C} \Rightarrow \mathcal{A} \Subset_P \mathcal{B} \Cap_P \mathcal{C}$.

(2) $\mathcal{A} \Subset_R \mathcal{B},\ \mathcal{A} \Subset_R \mathcal{C} \Rightarrow \mathcal{A} \Subset_R \mathcal{B} \Cap_R \mathcal{C}$.

(3) $\mathcal{A} \Subset_P \mathcal{C},\ \mathcal{B} \Subset_P \mathcal{C} \Rightarrow \mathcal{A} \Cup_P \mathcal{B} \Subset_P \mathcal{C}$.

(4) $\mathcal{A} \Subset_R \mathcal{C},\ \mathcal{B} \Subset_R \mathcal{C} \Rightarrow \mathcal{A} \Cup_R \mathcal{B} \Subset_R \mathcal{C}$.

(5) $\mathcal{A} \Subset_P \mathcal{B},\ \mathcal{C} \Subset_P \mathcal{D} \Rightarrow \mathcal{A} \Cup_P \mathcal{C} \Subset_P \mathcal{B} \Cup_P \mathcal{D},\ \mathcal{A} \Cap_P \mathcal{C} \Subset_P \mathcal{B} \Cap_P \mathcal{D}$.

(6) $\mathcal{A} \Subset_R \mathcal{B},\ \mathcal{C} \Subset_R \mathcal{D} \Rightarrow \mathcal{A} \Cup_R \mathcal{C} \Subset_R \mathcal{B} \Cup_R \mathcal{D},\ \mathcal{A} \Cap_R \mathcal{C} \Subset_R \mathcal{B} \Cap_R \mathcal{D}$.

Proof. Straightforward. □

Theorem 3. *Let $\mathcal{A} = \langle \mu, \tilde{A} \rangle$ be a cubic IVIF set in X. If $\mathcal{A} = \langle \mu, \tilde{A} \rangle$ is α-internal (resp., α-external), then so is the complement of $\mathcal{A} = \langle \mu, \tilde{A} \rangle$.*

Proof. Assume that $\mathcal{A} = \langle \mu, \tilde{A} \rangle$ is α-internal. Then $\mu(x) \in \alpha[\tilde{A}](x)$, that is, $\alpha[\tilde{A}]^-(x) \leq \mu(x) \leq \alpha[\tilde{A}]^+(x)$ for all $x \in X$. Thus $1 - \alpha[\tilde{A}]^+(x) \leq 1 - \mu(x) \leq 1 - \alpha[\tilde{A}]^-(x)$, that is, $\mu^c(x) \in \alpha[\tilde{A}^c](x)$ for all $x \in X$. Hence $\mathcal{A}^c = \langle \mu^c, \tilde{A}^c \rangle$ is α-internal. If $\mathcal{A} = \langle \mu, \tilde{A} \rangle$ is α-external, then $\mu(x) \notin (\alpha[\tilde{A}]^-(x), \alpha[\tilde{A}]^+(x))$ for all $x \in X$, and so $\mu(x) \in [0, \alpha[\tilde{A}]^-(x)]$ or $\mu(x) \in [\alpha[\tilde{A}]^+(x), 1]$. It follows that $1 - \alpha[\tilde{A}]^-(x) \leq 1 - \mu(x) \leq 1$ or $0 \leq 1 - \mu(x) \leq 1 - \alpha[\tilde{A}]^+(x)$, and so that

$$\begin{aligned} \mu^c(x) &= 1 - \mu(x) \notin (1 - \alpha[\tilde{A}]^+(x), 1 - \alpha[\tilde{A}]^-(x)) \\ &= (\alpha[\tilde{A}^c]^-(x), \alpha[\tilde{A}^c]^+(x)) \end{aligned}$$

Therefore $\mathcal{A}^c = \langle \mu^c, \tilde{A}^c \rangle$ is α-external. □

Theorem 4. *Let $\mathcal{A} = \langle \mu, \tilde{A} \rangle$ be a cubic IVIF set in X. If $\mathcal{A} = \langle \mu, \tilde{A} \rangle$ is β-internal (resp., β-external), then so is the complement of $\mathcal{A} = \langle \mu, \tilde{A} \rangle$.*

Proof. It is similar to the proof of Theorem 3. □

Corollary 1. *If a cubic IVIF set $\mathcal{A} = \langle \mu, \tilde{A} \rangle$ in X is internal, then so is $\mathcal{A}^c = \langle \mu^c, \tilde{A}^c \rangle$.*

Theorem 5. *If $\{\mathcal{A}_i = \langle \mu_i, \tilde{A}_i \rangle \mid i \in \Lambda\}$ is a family of α-internal cubic IVIF sets in X, then the P-union and the P-intersection of $\{\mathcal{A}_i = \langle \mu_i, \tilde{A}_i \rangle \mid i \in \Lambda\}$ are also α-internal cubic IVIF sets in X.*

Proof. Since $\mathcal{A}_i$ is α-internal, we have $\mu_i(x) \in \alpha[\tilde{A}_i](x)$, that is,

$$\alpha[\tilde{A}_i]^-(x) \leq \mu_i(x) \leq \alpha[\tilde{A}_i]^+(x)$$

for all $x \in X$ and $i \in \Lambda$. It follows that

$$\alpha[\bigcup_{i\in\Lambda} \tilde{A}_i]^-(x) \leq \left(\bigvee_{i\in\Lambda} \mu_i \right)(x) \leq \alpha[\bigcup_{i\in\Lambda} \tilde{A}_i]^+(x)$$

and

$$\alpha[\bigcap_{i\in\Lambda} \tilde{A}_i]^-(x) \leq \left(\bigwedge_{i\in\Lambda} \mu_i \right)(x) \leq \alpha[\bigcap_{i\in\Lambda} \tilde{A}_i]^+(x).$$

Therefore $\Cup_{P\, i\in\Lambda} \mathcal{A}_i$ and $\Cap_{P\, i\in\Lambda} \mathcal{A}_i$ are α-internal cubic IVIF sets in X. □

The following example shows that the R-union and R-intersection of α-internal cubic IVIF sets need not be α-internal.

Example 2. *Let $\mathcal{A} = \langle \mu, \tilde{A} \rangle$ and $\mathcal{B} = \langle \lambda, \tilde{B} \rangle$ be cubic IVIF sets in $I = [0, 1]$ with $\tilde{A}(x) = ([0.2, 0.45], [0.3, 0.5])$ and $\tilde{B}(x) = ([0.6, 0.8], [0.1, 0.2])$ for all $x \in I$. Then $\tilde{A} \cup \tilde{B} = (\alpha[\tilde{A} \cup \tilde{B}], \beta[\tilde{A} \cup \tilde{B}])$ and $\tilde{A} \cap \tilde{B} = (\alpha[\tilde{A} \cap \tilde{B}], \beta[\tilde{A} \cap \tilde{B}])$, where*

$$\begin{aligned} \alpha[\tilde{A} \cup \tilde{B}](x) &= \text{rmax}\{\alpha[\tilde{A}](x), \beta[\tilde{B}](x)\} \\ &= \text{rmax}\{[0.2, 0.45], [0.6, 0.8]\} = [0.6, 0.8] \end{aligned}$$

$$\beta[\tilde{A} \cup \tilde{B}](x) = \text{rmin}\{\beta[\tilde{A}](x), \beta[\tilde{B}](x)\}$$
$$= \text{rmin}\{[0.3, 0.5], [0.1, 0.2]\} = [0.1, 0.2]$$

$$\alpha[\tilde{A} \cap \tilde{B}](x) = \text{rmin}\{\alpha[\tilde{A}](x), \alpha[\tilde{B}](x)\}$$
$$= \text{rmin}\{[0.2, 0.45], [0.6, 0.8]\} = [0.2, 0.45]$$

$$\beta[\tilde{A} \cap \tilde{B}](x) = \text{rmax}\{\beta[\tilde{A}](x), \beta[\tilde{B}](x)\}$$
$$= \text{rmin}\{[0.3, 0.5], [0.1, 0.2]\} = [0.3, 0.5].$$

If $\mu(x) = 0.4$ and $\lambda(x) = 0.7$ for all $x \in I$, then $\mathcal{A}$ and $\mathcal{B}$ are α-internal. The R-union of $\mathcal{A}$ and $\mathcal{B}$ is

$$\mathcal{A} \Cup_R \mathcal{B} = \{\langle x, \mu(x), \tilde{B}(x)\rangle \mid x \in I\}$$

and it is not α-internal. The R-intersection of $\mathcal{A}$ and $\mathcal{B}$ is

$$\mathcal{A} \Cap_R \mathcal{B} = \{\langle x, \lambda(x), \tilde{A}(x)\rangle \mid x \in I\}$$

which is not α-internal.

We provide a condition for the R-union of two α-internal cubic IVIF sets to be α-internal.

Theorem 6. *Let $\mathcal{A} = \langle \mu, \tilde{A}\rangle$ and $\mathcal{B} = \langle \lambda, \tilde{B}\rangle$ be cubic IVIF sets in X such that*

$$(\forall x \in X)\left(\max\{\alpha[\tilde{A}]^-(x), \alpha[\tilde{B}]^-(x)\} \leq (\mu \wedge \lambda)(x)\right). \quad (17)$$

If $\mathcal{A} = \langle \mu, \tilde{A}\rangle$ and $\mathcal{B} = \langle \lambda, \tilde{B}\rangle$ are α-internal, then so is the R-union of $\mathcal{A} = \langle \mu, \tilde{A}\rangle$ and $\mathcal{B} = \langle \lambda, \tilde{B}\rangle$.

Proof. Assume that $\mathcal{A} = \langle \mu, \tilde{A}\rangle$ and $\mathcal{B} = \langle \lambda, \tilde{B}\rangle$ are α-internal cubic IVIF sets in X that satisfies the condition (17). Then

$$\alpha[\tilde{A}]^-(x) \leq \mu(x) \leq \alpha[\tilde{A}]^+(x)$$

and

$$\alpha[\tilde{B}]^-(x) \leq \lambda(x) \leq \alpha[\tilde{B}]^+(x)$$

for all $x \in X$, which imply that $(\mu \wedge \lambda)(x) \leq \alpha[\tilde{A} \cup \tilde{B}]^+(x)$. It follows from (17) that

$$\alpha[\tilde{A} \cup \tilde{B}]^-(x) = \max\{\alpha[\tilde{A}]^-(x), \alpha[\tilde{B}]^-(x)\} \leq (\mu \wedge \lambda)(x) \leq \alpha[\tilde{A} \cup \tilde{B}]^+(x)$$

for all $x \in X$. Therefore

$$\mathcal{A} \Cup_R \mathcal{B} = \{\langle x, (\mu \wedge \lambda)(x), (\tilde{A} \cup \tilde{B})(x)\rangle \mid x \in X\}$$

is an α-internal cubic IVIF set in X. □

We provide a condition for the R-intersection of two α-internal cubic IVIF sets to be α-internal.

Theorem 7. *Let $\mathcal{A} = \langle \mu, \tilde{A}\rangle$ and $\mathcal{B} = \langle \lambda, \tilde{B}\rangle$ be cubic IVIF sets in X such that*

$$(\forall x \in X)\left(\min\{\alpha[\tilde{A}]^+(x), \alpha[\tilde{B}]^+(x)\} \geq (\mu \vee \lambda)(x)\right). \quad (18)$$

If $\mathcal{A} = \langle \mu, \tilde{A}\rangle$ and $\mathcal{B} = \langle \lambda, \tilde{B}\rangle$ are α-internal, then so is the R-intersection of $\mathcal{A} = \langle \mu, \tilde{A}\rangle$ and $\mathcal{B} = \langle \lambda, \tilde{B}\rangle$.

Proof. Let $\mathcal{A} = \langle \mu, \tilde{A} \rangle$ and $\mathcal{B} = \langle \lambda, \tilde{B} \rangle$ be α-internal cubic IVIF sets in X which satisfy the condition (18). Then $\alpha[\tilde{A}]^-(x) \leq \mu(x) \leq \alpha[\tilde{A}]^+(x)$ and $\alpha[\tilde{B}]^-(x) \leq \lambda(x) \leq \alpha[\tilde{B}]^+(x)$ for all $x \in X$. It follows from (18) that

$$\alpha[\tilde{A} \cap \tilde{B}]^-(x) \leq (\mu \vee \lambda)(x) \leq \min\{\alpha[\tilde{A}]^+(x), \alpha[\tilde{B}]^+(x)\} = \alpha[\tilde{A} \cap \tilde{B}]^+(x)$$

for all $x \in X$. Therefore

$$\mathcal{A} \Cap_R \mathcal{B} = \{\langle x, (\mu \vee \lambda)(x), (\tilde{A} \cap \tilde{B})(x) \rangle \mid x \in X\}$$

is an α-internal cubic IVIF set in X. □

The following example shows that the P-union and the P-intersection of α-external cubic IVIF sets need not be α-external.

Example 3. *Let $\mathcal{A} = \langle \mu, \tilde{A} \rangle$ and $\mathcal{B} = \langle \lambda, \tilde{B} \rangle$ be cubic IVIF sets in X which are given in Example 2. If $\mu(x) = 0.7$ and $\lambda(x) = 0.4$ for all $x \in I$, then $\mathcal{A}$ and $\mathcal{B}$ are α-external. The P-union of $\mathcal{A}$ and $\mathcal{B}$ is*

$$\mathcal{A} \Cup_P \mathcal{B} = \{\langle x, \mu(x), \tilde{B}(x) \rangle \mid x \in I\}$$

and it is not α-external. The P-intersection of $\mathcal{A}$ and $\mathcal{B}$ is

$$\mathcal{A} \Cap_P \mathcal{B} = \{\langle x, \lambda(x), \tilde{A}(x) \rangle \mid x \in I\}$$

which is not α-external.

We consider conditions for the P-union and P-intersection of two α-external cubic IVIF sets to be α-external.

Theorem 8. *Let $\mathcal{A} = \langle \mu, \tilde{A} \rangle$ and $\mathcal{B} = \langle \lambda, \tilde{B} \rangle$ be cubic IVIF sets in X such that*

$$\{x \in X \mid \mu(x) \leq \alpha[\tilde{A}]^-(x),\ \lambda(x) \geq \alpha[\tilde{B}]^+(x)\} = \emptyset \tag{19}$$

and

$$\{x \in X \mid \mu(x) \geq \alpha[\tilde{A}]^+(x),\ \lambda(x) \leq \alpha[\tilde{B}]^-(x)\} = \emptyset. \tag{20}$$

If $\mathcal{A} = \langle \mu, \tilde{A} \rangle$ and $\mathcal{B} = \langle \lambda, \tilde{B} \rangle$ are α-external, then so are the P-union and P-intersection of $\mathcal{A} = \langle \mu, \tilde{A} \rangle$ and $\mathcal{B} = \langle \lambda, \tilde{B} \rangle$.

Proof. If $\mathcal{A} = \langle \mu, \tilde{A} \rangle$ and $\mathcal{B} = \langle \lambda, \tilde{B} \rangle$ are α-external, then $\mu(x) \notin (\alpha[\tilde{A}]^-(x), \alpha[\tilde{A}]^+(x))$ and $\lambda(x) \notin (\alpha[\tilde{B}]^-(x), \alpha[\tilde{B}]^+(x))$, that is,

$$\mu(x) \leq \alpha[\tilde{A}]^-(x) \text{ or } \mu(x) \geq \alpha[\tilde{A}]^+(x)$$

and

$$\lambda(x) \leq \alpha[\tilde{B}]^-(x) \text{ or } \lambda(x) \geq \alpha[\tilde{B}]^+(x)$$

for all $x \in X$. It follows from conditions (19) and (20) that

$$(\forall x \in X)\ (\mu(x) \leq \alpha[\tilde{A}]^-(x),\ \lambda(x) \leq \alpha[\tilde{B}]^-(x)) \tag{21}$$

or

$$(\forall x \in X)\ \left(\mu(x) \geq \alpha[\tilde{A}]^+(x),\ \lambda(x) \geq \alpha[\tilde{B}]^+(x)\right). \tag{22}$$

The conditions (21) and (22) induce

$$(\mu \vee \lambda)(x) \leq \max\{\alpha[\tilde{A}]^-(x), \alpha[\tilde{B}]^-(x)\}$$

and

$$(\mu \vee \lambda)(x) \geq \max\{\alpha[\tilde{A}]^+(x), \alpha[\tilde{B}]^+(x)\}$$

respectively, for all $x \in X$. Hence

$$(\mu \vee \lambda)(x) \notin \left(\alpha[\tilde{A} \cup \tilde{B}]^-(x), \alpha[\tilde{A} \cup \tilde{B}]^+(x)\right)$$

for all $x \in X$, and therefore

$$\mathcal{A} \Cup_P \mathcal{B} = \{\langle x, (\mu \vee \lambda)(x), (\tilde{A} \cup \tilde{B})(x)\rangle \mid x \in I\}$$

is an α-external cubic IVIF set in X. Also, (21) and (22) imply that

$$(\mu \wedge \lambda)(x) \leq \min\{\alpha[\tilde{A}]^-(x), \alpha[\tilde{B}]^-(x)\}$$

and

$$(\mu \wedge \lambda)(x) \geq \min\{\alpha[\tilde{A}]^+(x), \alpha[\tilde{B}]^+(x)\}$$

respectively, for all $x \in X$. Therefore

$$\mathcal{A} \Cap_P \mathcal{B} = \{\langle x, (\mu \wedge \lambda)(x), (\tilde{A} \cap \tilde{B})(x)\rangle \mid x \in I\}$$

is an α-external cubic IVIF set in X. □

The following example shows that the R-union and R-intersection of α-external cubic IVIF sets in X may not be α-external.

Example 4. *Let $\mathcal{A} = \langle \mu, \tilde{A}\rangle$ and $\mathcal{B} = \langle \lambda, \tilde{B}\rangle$ be cubic IVIF sets in $I = [0,1]$ with $\tilde{A}(x) = ([0.2, 0.4], [0.3, 0.5])$ and $\tilde{B}(x) = ([0.3, 0.6], [0.2, 0.4])$ for all $x \in I$. Then $\tilde{A} \cup \tilde{B} = (\alpha[\tilde{A} \cup \tilde{B}], \beta[\tilde{A} \cup \tilde{B}])$ and $\tilde{A} \cap \tilde{B} = (\alpha[\tilde{A} \cap \tilde{B}], \beta[\tilde{A} \cap \tilde{B}])$, where*

$$\begin{aligned}\alpha[\tilde{A} \cup \tilde{B}](x) &= \text{rmax}\{\alpha[\tilde{A}](x), \alpha[\tilde{B}](x)\}\\ &= \text{rmax}\{[0.2, 0.4], [0.3, 0.6]\} = [0.3, 0.6]\end{aligned}$$

$$\begin{aligned}\beta[\tilde{A} \cup \tilde{B}](x) &= \text{rmin}\{\beta[\tilde{A}](x), \beta[\tilde{B}](x)\}\\ &= \text{rmin}\{[0.3, 0.5], [0.2, 0.4]\} = [0.2, 0.4]\end{aligned}$$

$$\begin{aligned}\alpha[\tilde{A} \cap \tilde{B}](x) &= \text{rmin}\{\alpha[\tilde{A}](x), \alpha[\tilde{B}](x)\}\\ &= \text{rmin}\{[0.2, 0.4], [0.3, 0.6]\} = [0.2, 0.4]\end{aligned}$$

$$\begin{aligned}\beta[\tilde{A} \cap \tilde{B}](x) &= \text{rmax}\{\beta[\tilde{A}](x), \beta[\tilde{B}](x)\}\\ &= \text{rmin}\{[0.3, 0.5], [0.2, 0.4]\} = [0.3, 0.5].\end{aligned}$$

If $\mu(x) = 0.5$ *and* $\lambda(x) = 0.7$ *for all* $x \in I$*, then* $\mathcal{A}$ *and* $\mathcal{B}$ *are both* α*-external. The R-union of* $\mathcal{A}$ *and* $\mathcal{B}$ *is*

$$\mathcal{A} \Cup_R \mathcal{B} = \{\langle x, \mu(x), \tilde{B}(x)\rangle \mid x \in I\}$$

and it is not α*-external. If* $\mu(x) = 0.1$ *and* $\lambda(x) = 0.3$ *for all* $x \in I$*, then* $\mathcal{A}$ *and* $\mathcal{B}$ *are* α*-external. The R-intersection of* $\mathcal{A}$ *and* $\mathcal{B}$ *is*

$$\mathcal{A} \Cap_R \mathcal{B} = \{\langle x, \lambda(x), \tilde{A}(x)\rangle \mid x \in I\}$$

which is not α*-external.*

We discuss conditions for the R-union and R-intersection of two cubic IVIF sets to be α-external. Let $\mathcal{A} = \langle \mu, \tilde{A}\rangle$ and $\mathcal{B} = \langle \lambda, \tilde{B}\rangle$ be cubic IVIF sets in X such that

$$(\forall x \in X)\ (\alpha[\tilde{A}]^-(x) \leq \alpha[\tilde{B}]^-(x) \leq \alpha[\tilde{A}]^+(x) \leq \alpha[\tilde{B}]^+(x)). \tag{23}$$

If $\mu(x) \leq \alpha[\tilde{A}]^-(x)$ for all $x \in X$, then

$$(\mu \wedge \lambda)(x) \leq \alpha[\tilde{A}]^-(x) \leq \max\{\alpha[\tilde{A}]^-(x), \alpha[\tilde{B}]^-(x)\}$$

and so $(\mu \wedge \lambda)(x) \notin (\alpha[\tilde{A} \cup \tilde{B}]^-(x), \alpha[\tilde{A} \cup \tilde{B}]^+(x))$.

If $(\mu \wedge \lambda)(x) \geq \alpha[\tilde{B}]^+(x)$ for all $x \in X$, then

$$(\mu \wedge \lambda)(x) \geq \alpha[\tilde{B}]^+(x) = \max\{\alpha[\tilde{A}]^+(x), \alpha[\tilde{B}]^+(x)\}$$

and thus $(\mu \wedge \lambda)(x) \notin (\alpha[\tilde{A} \cup \tilde{B}]^-(x), \alpha[\tilde{A} \cup \tilde{B}]^+(x))$.

If $(\mu \vee \lambda)(x) \leq \alpha[\tilde{A}]^-(x)$ for all $x \in X$, then

$$(\mu \vee \lambda)(x) \leq \alpha[\tilde{A}]^-(x) = \min\{\alpha[\tilde{A}]^-(x), \alpha[\tilde{B}]^-(x)\}$$

which implies that $(\mu \vee \lambda)(x) \notin (\alpha[\tilde{A} \cap \tilde{B}]^-(x), \alpha[\tilde{A} \cap \tilde{B}]^+(x))$.

If $\mu(x) \geq \alpha[\tilde{B}]^+(x)$ for all $x \in X$, then

$$(\mu \vee \lambda)(x) \geq \mu(x) \geq \alpha[\tilde{B}]^+(x) \geq \min\{\alpha[\tilde{A}]^+(x), \alpha[\tilde{B}]^+(x)\}$$

which induces $(\mu \vee \lambda)(x) \notin (\alpha[\tilde{A} \cap \tilde{B}]^-(x), \alpha[\tilde{A} \cap \tilde{B}]^+(x))$.

Therefore we have the following theorem.

Theorem 9. *Let* $\mathcal{A} = \langle \mu, \tilde{A}\rangle$ *and* $\mathcal{B} = \langle \lambda, \tilde{B}\rangle$ *be cubic IVIF sets in X satisfying the condition* (23)*. If* μ *and* λ *satisfies any one of the following conditions.*

$$(\forall x \in X)\ (\mu(x) \leq \alpha[\tilde{A}]^-(x)), \tag{24}$$

$$(\forall x \in X)\ ((\mu \wedge \lambda)(x) \geq \alpha[\tilde{B}]^+(x)), \tag{25}$$

then the R-union of $\mathcal{A} = \langle \mu, \tilde{A}\rangle$ *and* $\mathcal{B} = \langle \lambda, \tilde{B}\rangle$ *is* α*-external. If* μ *and* λ *satisfies any one of the following conditions.*

$$(\forall x \in X)\ ((\mu \vee \lambda)(x) \leq \alpha[\tilde{A}]^-(x)), \tag{26}$$

$$(\forall x \in X)\ (\mu(x) \geq \alpha[\tilde{B}]^+(x)), \tag{27}$$

then the R-intersection of $\mathcal{A} = \langle \mu, \tilde{A}\rangle$ *and* $\mathcal{B} = \langle \lambda, \tilde{B}\rangle$ *is* α*-external.*

Let $\mathcal{A} = \langle \mu, \tilde{A} \rangle$ and $\mathcal{B} = \langle \lambda, \tilde{B} \rangle$ be cubic IVIF sets in X such that

$$(\forall x \in X)\ (\alpha[\tilde{B}]^-(x) \leq \alpha[\tilde{A}]^-(x) \leq \alpha[\tilde{A}]^+(x) \leq \alpha[\tilde{B}]^+(x)). \tag{28}$$

If $\mu(x) \leq \alpha[\tilde{B}]^-(x)$ for all $x \in X$, then

$$(\mu \wedge \lambda)(x) \leq \alpha[\tilde{B}]^-(x) \leq \max\{\alpha[\tilde{A}]^-(x), \alpha[\tilde{B}]^-(x)\}$$

and so $(\mu \wedge \lambda)(x) \notin (\alpha[\tilde{A} \cup \tilde{B}]^-(x), \alpha[\tilde{A} \cup \tilde{B}]^+(x))$.

If $(\mu \wedge \lambda)(x) \geq \alpha[\tilde{B}]^+(x)$ for all $x \in X$, then

$$(\mu \wedge \lambda)(x) \geq \alpha[\tilde{B}]^+(x) = \max\{\alpha[\tilde{A}]^+(x), \alpha[\tilde{B}]^+(x)\}$$

and thus $(\mu \wedge \lambda)(x) \notin (\alpha[\tilde{A} \cup \tilde{B}]^-(x), \alpha[\tilde{A} \cup \tilde{B}]^+(x))$.

If $\{x \in X \mid \mu(x) \geq \alpha[\tilde{B}]^+(x)\} = X$, then

$$(\mu \vee \lambda)(x) \geq \mu(x) \geq \alpha[\tilde{B}]^+(x) \geq \min\{\alpha[\tilde{A}]^+(x), \alpha[\tilde{B}]^+(x)\}$$

which induces $(\mu \vee \lambda)(x) \notin (\alpha[\tilde{A} \cap \tilde{B}]^-(x), \alpha[\tilde{A} \cap \tilde{B}]^+(x))$ for all $x \in X$.

If $\{x \in X \mid (\mu \vee \lambda)(x) \leq \alpha[\tilde{B}]^-(x)\} = X$, then

$$(\mu \vee \lambda)(x) \leq \alpha[\tilde{B}]^-(x) = \min\{\alpha[\tilde{A}]^-(x), \alpha[\tilde{B}]^-(x)\}$$

and so $(\mu \vee \lambda)(x) \notin (\alpha[\tilde{A} \cap \tilde{B}]^-(x), \alpha[\tilde{A} \cap \tilde{B}]^+(x))$ for all $x \in X$.

Therefore we have the following theorem.

Theorem 10. *Let $\mathcal{A} = \langle \mu, \tilde{A} \rangle$ and $\mathcal{B} = \langle \lambda, \tilde{B} \rangle$ be cubic IVIF sets in X satisfying the condition (28). If μ and λ satisfies*

$$\{x \in X \mid \mu(x) \leq \alpha[\tilde{B}]^-(x)\} = X$$

or

$$\{x \in X \mid (\mu \wedge \lambda)(x) \geq \alpha[\tilde{B}]^+(x)\} = X,$$

then the R-union of $\mathcal{A} = \langle \mu, \tilde{A} \rangle$ and $\mathcal{B} = \langle \lambda, \tilde{B} \rangle$ is α-external. Also, if

$$\{x \in X \mid \mu(x) \geq \alpha[\tilde{B}]^+(x)\} = X$$

or

$$\{x \in X \mid (\mu \vee \lambda)(x) \leq \alpha[\tilde{B}]^-(x)\} = X,$$

then the R-intersection of $\mathcal{A} = \langle \mu, \tilde{A} \rangle$ and $\mathcal{B} = \langle \lambda, \tilde{B} \rangle$ is α-external.

4. Cubic IVIF Subalgebras and Ideals

In what follows, let X be a BCK/BCI-algebra unless otherwise specified.

Definition 5. *A cubic IVIF set $\mathcal{A} = \langle \mu, \tilde{A} \rangle$ in X is called a cubic IVIF subalgebra of X if the following conditions are valid.*

$$(\forall x, y \in X)\ (\mu(x * y) \geq \min\{\mu(x), \mu(y)\}), \tag{29}$$

$$(\forall x, y \in X) \begin{pmatrix} \alpha[\tilde{A}](x * y) \ll \mathrm{rmax}\{\alpha[\tilde{A}](x), \alpha[\tilde{A}](y)\} \\ \beta[\tilde{A}](x * y) \gg \mathrm{rmin}\{\beta[\tilde{A}](x), \beta[\tilde{A}](y)\} \end{pmatrix}. \tag{30}$$

Example 5. *Let $X = \{0, a, b, c\}$ be a BCK-algebra with the Cayley table (Table 1).*

Table 1. Tabular representation of the binary operation $*$.

$*$	0	a	b	c
0	0	0	0	0
a	a	0	0	a
b	b	a	0	b
c	c	c	c	0

Define a cubic IVIF set $\mathcal{A} = \langle \mu, \tilde{A} \rangle$ in X by the tabular representation in Table 2.

Table 2. Tabular representation of $\mathcal{A} = \langle \mu, \tilde{A} \rangle$.

X	μ	$\tilde{A} = (\alpha[\tilde{A}], \beta[\tilde{A}])$
0	0.8	$([0.1, 0.4], [0.3, 0.5])$
a	0.3	$([0.2, 0.4], [0.2, 0.4])$
b	0.6	$([0.2, 0.5], [0.1, 0.3])$
c	0.5	$([0.3, 0.5], [0.2, 0.5])$

It is routine to verify that $\mathcal{A} = \langle \mu, \tilde{A} \rangle$ is a cubic IVIF subalgebra of X.

Proposition 4. *If $\mathcal{A} = \langle \mu, \tilde{A} \rangle$ is a cubic IVIF subalgebra of X, then $\mu(0) \geq \mu(x)$, $\alpha[\tilde{A}](0) \ll \alpha[\tilde{A}](x)$ and $\beta[\tilde{A}](0) \gg \beta[\tilde{A}](x)$ for all $x \in X$.*

Proof. Since $x * x = 0$ for all $x \in X$, we have

$$\mu(0) = \mu(x * x) \geq \min\{\mu(x), \mu(x)\} = \mu(x),$$
$$\alpha[\tilde{A}](0) = \alpha[\tilde{A}](x * x) \ll \text{rmax}\{\alpha[\tilde{A}](x), \alpha[\tilde{A}](x)\} = \alpha[\tilde{A}](x),$$
$$\beta[\tilde{A}](0) = \beta[\tilde{A}](x * x) \gg \text{rmin}\{\beta[\tilde{A}](x), \beta[\tilde{A}](x)\} = \beta[\tilde{A}](x).$$

This completes the proof. □

Proposition 5. *If $\mathcal{A} = \langle \mu, \tilde{A} \rangle$ is a cubic IVIF subalgebra of a BCI-algebra X, then $\mu(0 * x) \geq \mu(x)$, $\alpha[\tilde{A}](0 * x) \ll \alpha[\tilde{A}](x)$ and $\beta[\tilde{A}](0 * x) \gg \beta[\tilde{A}](x)$ for all $x \in X$.*

Proof. It follows from Definition 5 and Proposition 4. □

Proposition 6. *For a cubic IVIF subalgebra $\mathcal{A} = \langle \mu, \tilde{A} \rangle$ of X, if there exists a sequence $\{x_n\}$ in X such that $\lim_{n\to\infty} \mu(x_n) = 1$, $\lim_{n\to\infty} \alpha[\tilde{A}](x_n) = [0, 0]$ and $\lim_{n\to\infty} \beta[\tilde{A}](x_n) = [1, 1]$, then $\mu(0) = 1$, $\alpha[\tilde{A}](0) = [0, 0]$ and $\beta[\tilde{A}](0) = [1, 1]$.*

Proof. Using Proposition 4, we have $\mu(0) \geq \mu(x_n)$, $\alpha[\tilde{A}](0) \ll \alpha[\tilde{A}](x_n)$ and $\beta[\tilde{A}](0) \gg \beta[\tilde{A}](x_n)$. It follows from hypothesis that

$$1 \geq \mu(0) \geq \lim_{n\to\infty} \mu(x_n) = 1,$$
$$[0, 0] \ll \alpha[\tilde{A}](0) \ll \lim_{n\to\infty} \alpha[\tilde{A}](x_n) = [0, 0],$$
$$[1, 1] \gg \beta[\tilde{A}](0) \gg \lim_{n\to\infty} \beta[\tilde{A}](x_n) = [1, 1],$$

Hence $\mu(0) = 1$, $\alpha[\tilde{A}](0) = [0, 0]$ and $\beta[\tilde{A}](0) = [1, 1]$. □

Theorem 11. *If $\mathcal{A} = \langle \mu, \tilde{A} \rangle$ is a cubic IVIF subalgebra of X, then the sets $\mu[\varepsilon]$, $\alpha[\tilde{A}][s,t]$ and $\beta[\tilde{A}][s,t]$, are subalgebras of X for all $\varepsilon \in [0,1]$ and $[s,t] \in D[0,1]$.*

Proof. For any $\varepsilon \in [0,1]$ and $[s,t] \in D[0,1]$, let $x, y \in X$ be such that

$$x, y \in \mu[\varepsilon] \cap \alpha[\tilde{A}][s,t] \cap \beta[\tilde{A}][s,t].$$

Then $\mu(x) \geq \varepsilon$, $\mu(y) \geq \varepsilon$, $\alpha[\tilde{A}](x) \ll [s,t]$, $\alpha[\tilde{A}](y) \ll [s,t]$, $\beta[\tilde{A}](x) \gg [s,t]$ and $\beta[\tilde{A}](y) \gg [s,t]$. It follows that

$$\begin{aligned} &\mu(x*y) \geq \min\{\mu(x), \mu(y)\} \geq \min\{\varepsilon, \varepsilon\} = \varepsilon, \\ &\alpha[\tilde{A}](x*y) \ll \text{rmax}\{\alpha[\tilde{A}](x), \alpha[\tilde{A}](y)\} \ll \text{rmax}\{[\text{s,t}], [\text{s,t}]\} = [\text{s,t}], \\ &\beta[\tilde{A}](x*y) \gg \text{rmin}\{\beta[\tilde{A}](x), \beta[\tilde{A}](y)\} \gg \text{rmin}\{[\text{s,t}], [\text{s,t}]\} = [\text{s,t}], \end{aligned}$$

that is, $x*y \in \mu[\varepsilon]$, $x*y \in \alpha[\tilde{A}][s,t]$ and $x*y \in \beta[\tilde{A}][s,t]$. Therefore $\mu[\varepsilon]$, $\alpha[\tilde{A}][s,t]$ and $\beta[\tilde{A}][s,t]$ are subalgebras of X for all $\varepsilon \in [0,1]$ and $[s,t] \in D[0,1]$. □

Corollary 2. *If $\mathcal{A} = \langle \mu, \tilde{A} \rangle$ is a cubic IVIF subalgebra of X, then $\mu[\varepsilon] \cap \alpha[\tilde{A}][s,t] \cap \beta[\tilde{A}][s,t]$ is a subalgebra of X for all $\varepsilon \in [0,1]$ and $[s,t] \in D[0,1]$.*

The following example shows that the converse of Corollary 2 is not true in general.

Example 6. *Let $X = \{0,1,2,3,4\}$ be a set with the Cayley table (Table 3).*

Table 3. Tabular representation of the binary operation $*$.

$*$	0	1	2	3	4
0	0	0	0	0	0
1	1	0	0	0	0
2	2	1	0	1	0
3	3	3	3	0	0
4	4	4	4	4	0

Then X is a BCK-algebra (see [15]). Define a cubic IVIF set $\mathcal{A} = \langle \mu, \tilde{A} \rangle$ in X by the tabular representation in Table 4.

Table 4. Tabular representation of $\mathcal{A} = \langle \mu, \tilde{A} \rangle$.

X	μ	$\tilde{A} = (\alpha[\tilde{A}], \beta[\tilde{A}])$
0	0.8	$([0.1, 0.3], [0.3, 0.6])$
1	0.3	$([0.2, 0.4], [0.2, 0.4])$
2	0.8	$([0.2, 0.5], [0.1, 0.3])$
3	0.5	$([0.3, 0.5], [0.2, 0.5])$
4	0.3	$([0.3, 0.5], [0.2, 0.5])$

It is routine to verify that $\mu[\varepsilon] \cap \alpha[\tilde{A}][s,t] \cap \beta[\tilde{A}][s,t]$ is a subalgebra of X for all $\varepsilon \in [0,1]$ and $[s,t] \in D[0,1]$. But, $\mathcal{A} = \langle \mu, \tilde{A} \rangle$ is not a cubic IVIF subalgebra of X since $\mu(2*3) = \mu(1) = 0.3 < 0.5 = \min\{\mu(2), \mu(3)\}$.

We provide conditions for a cubic IVIF set $\mathcal{A} = \langle \mu, \tilde{A} \rangle$ in X to be a cubic IVIF subalgebra of X.

Theorem 12. *Let $\mathcal{A} = \langle \mu, \tilde{A} \rangle$ be a cubic IVIF set in X such that $\mu[\varepsilon]$, $\alpha[\tilde{A}][s_\alpha, t_\alpha]$ and $\beta[\tilde{A}][s_\beta, t_\beta]$ are subalgebras of X for all $\varepsilon \in [0,1]$, and $([s_\alpha, t_\alpha], [s_\beta, t_\beta]) \in D[0,1] \times D[0,1]$. Then $\mathcal{A} = \langle \mu, \tilde{A} \rangle$ is a cubic IVIF subalgebra of X.*

Proof. For any $x, y \in X$, putting

$$\begin{aligned}\min\{\mu(x),\mu(y)\} &:= \varepsilon\\ \text{rmax}\{\alpha[\tilde{A}](x),\alpha[\tilde{A}](y)\} &:= [s_\alpha,t_\alpha],\\ \text{rmin}\{\beta[\tilde{A}](x),\beta[\tilde{A}](y)\} &:= [s_\beta,t_\beta],\end{aligned}$$

imply that $\mu(x) \geq \varepsilon$, $\mu(y) \geq \varepsilon$, $\alpha[\tilde{A}](x) \ll [s_\alpha, t_\alpha]$, $\alpha[\tilde{A}](y) \ll [s_\alpha, t_\alpha]$, and $\beta[\tilde{A}](x) \gg [s_\beta, t_\beta]$, $\beta[\tilde{A}](y) \gg [s_\beta, t_\beta]$. Hence $x \in \mu[\varepsilon]$, $y \in \mu[\varepsilon]$, $x \in \alpha[\tilde{A}][s_\alpha, t_\alpha]$, $y \in \alpha[\tilde{A}][s_\alpha, t_\alpha]$, $x \in \beta[\tilde{A}][s_\beta, t_\beta]$, and $y \in \beta[\tilde{A}][s_\beta, t_\beta]$. It follows from hypothesis that $x * y \in \mu[\varepsilon]$, $x * y \in \alpha[\tilde{A}][s_\alpha, t_\alpha]$, $x * y \in \beta[\tilde{A}][s_\beta, t_\beta]$, and so that

$$\begin{aligned}&\mu(x*y) \geq \varepsilon = \min\{\mu(x),\mu(y)\}\\ &\alpha[\tilde{A}](x*y) \ll [s_\alpha,t_\alpha] = \text{rmax}\{\alpha[\tilde{A}](x),\alpha[\tilde{A}](y)\},\\ &\beta[\tilde{A}](x*y) \gg [s_\beta,t_\beta] = \text{rmin}\{\beta[\tilde{A}](x),\beta[\tilde{A}](y)\}.\end{aligned}$$

Therefore $\mathcal{A} = \langle \mu, \tilde{A} \rangle$ is a cubic IVIF subalgebra of X. □

The following theorem gives us a way to establish a new cubic IVIF subalgebra from old one in BCI-algebras.

Theorem 13. *Given a cubic IVIF subalgebra $\mathcal{A} = \langle \mu, \tilde{A} \rangle$ of a BCI-algebra X, let $\mathcal{A}^* = \langle \mu^*, A^* \rangle$ be a cubic IVIF set in X defined by $\mu^*(x) = \mu(0 * x)$, $\alpha[\tilde{A}]^*(x) = \alpha[\tilde{A}](0 * x)$ and $\beta[\tilde{A}]^*(x) = \beta[\tilde{A}](0 * x)$ for all $x \in X$. Then $\mathcal{A}^* = \langle \mu^*, A^* \rangle$ is a cubic IVIF subalgebra of X.*

Proof. Since $0 * (x * y) = (0 * x) * (0 * y)$ for all $x, y \in X$, it follows that

$$\begin{aligned}\mu^*(x*y) &= \mu(0*(x*y)) = \mu((0*x)*(0*y))\\ &\geq \min\{\mu(0*x),\mu(0*y)\}\\ &= \min\{\mu^*(x),\mu^*(y)\},\end{aligned}$$

$$\begin{aligned}\alpha[\tilde{A}]^*(x*y) &= \alpha[\tilde{A}](0*(x*y)) = \alpha[\tilde{A}]((0*x)*(0*y))\\ &\ll \text{rmax}\{\alpha[\tilde{A}](0*x),\alpha[\tilde{A}](0*y)\}\\ &= \text{rmax}\{\alpha[\tilde{A}]^*(x),\alpha[\tilde{A}]^*(y)\},\end{aligned}$$

and

$$\begin{aligned}\beta[\tilde{A}]^*(x*y) &= \beta[\tilde{A}](0*(x*y)) = \beta[\tilde{A}]((0*x)*(0*y))\\ &\gg \text{rmin}\{\beta[\tilde{A}](0*x),\beta[\tilde{A}](0*y)\}\\ &= \text{rmin}\{\beta[\tilde{A}]^*(x),\beta[\tilde{A}]^*(y)\}\end{aligned}$$

for all $x, y \in X$. Therefore $\mathcal{A}^* = \langle \mu^*, A^* \rangle$ is a cubic IVIF subalgebra of X. □

Definition 6. *A cubic IVIF set $\mathcal{A} = \langle \mu, \tilde{A} \rangle$ is called a cubic IVIF ideal of X if the following conditions are valid.*

$$(\forall x \in X)\ (\mu(0) \geq \mu(x)), \tag{31}$$

$$(\forall x \in X)\ (\alpha[\tilde{A}](0) \ll \alpha[\tilde{A}](x),\ \beta[\tilde{A}](0) \gg \beta[\tilde{A}](x)) \tag{32}$$

$$(\forall x, y \in X)\ (\mu(x) \geq \min\{\mu(x * y), \mu(y)\}) \tag{33}$$

$$(\forall x, y \in X) \left(\begin{array}{c} \alpha[\tilde{A}](x) \ll \mathrm{rmax}\{\alpha[\tilde{A}](x * y), \alpha[\tilde{A}](y)\} \\ \beta[\tilde{A}](x) \gg \mathrm{rmin}\{\beta[\tilde{A}](x * y), \beta[\tilde{A}](y)\} \end{array} \right) \tag{34}$$

Example 7. *Let $X = \{0, a, 1, 2, 3\}$ be a BCI-algebra with the Cayley table in Table 5.*

Table 5. Tabular representation of the binary operation $*$.

$*$	0	a	1	2	3
0	0	0	3	2	1
a	a	0	3	2	1
1	1	1	0	3	2
2	2	2	1	0	3
3	3	3	2	1	0

Define a cubic IVIF set $\mathcal{A} = \langle \mu, \tilde{A} \rangle$ in X by Table 6.

Table 6. Tabular representation of $\mathcal{A} = \langle \mu, \tilde{A} \rangle$.

X	μ	$\tilde{A} = (\alpha[\tilde{A}], \beta[\tilde{A}])$
0	0.7	([0.1, 0.3], [0.5, 0.6])
a	0.7	([0.2, 0.4], [0.4, 0.5])
1	0.2	([0.3, 0.6], [0.2, 0.3])
2	0.5	([0.3, 0.6], [0.3, 0.4])
3	0.2	([0.3, 0.6], [0.2, 0.3])

Then $\mathcal{A} = \langle \mu, \tilde{A} \rangle$ is a cubic IVIF ideal of X.

Proposition 7. *Every cubic IVIF ideal $\mathcal{A} = \langle \mu, \tilde{A} \rangle$ of X satisfies:*

$$\begin{aligned} &\mu(x) \geq \min\{\mu(y), \mu(z)\}. \\ &\alpha[\tilde{A}](x) \ll \mathrm{rmax}\{\alpha[\tilde{A}](y), \alpha[\tilde{A}](z)\}, \\ &\beta[\tilde{A}](x) \gg \mathrm{rmin}\{\beta[\tilde{A}](y), \beta[\tilde{A}](z)\}, \end{aligned} \tag{35}$$

for all $x, y, z \in X$ with $x * y \leq z$.

Proof. Let $x, y, z \in X$ be such that $x * y \leq z$. Then $(x * y) * z = 0$, and so

$$\begin{aligned} &\mu(x * y) \geq \min\{\mu((x * y) * z), \mu(z)\} = \min\{\mu(0), \mu(z)\} = \mu(z), \\ &\alpha[\tilde{A}](x * y) \ll \mathrm{rmax}\{\alpha[\tilde{A}]((x * y) * z), \alpha[\tilde{A}](z)\} = \mathrm{rmax}\{\alpha[\tilde{A}](0), \alpha[\tilde{A}](z)\} = \alpha[\tilde{A}](z), \\ &\beta[\tilde{A}](x * y) \gg \mathrm{rmin}\{\beta[\tilde{A}]((x * y) * z), \beta[\tilde{A}](z)\} = \mathrm{rmin}\{\beta[\tilde{A}](0), \beta[\tilde{A}](z)\} = \beta[\tilde{A}](z). \end{aligned}$$

It follows that

$$\begin{aligned} &\mu(x) \geq \min\{\mu(x * y), \mu(y)\} \geq \min\{\mu(y), \mu(z)\}, \\ &\alpha[\tilde{A}](x) \ll \mathrm{rmax}\{\alpha[\tilde{A}](x * y), \alpha[\tilde{A}](y)\} \ll \mathrm{rmax}\{\alpha[\tilde{A}](y), \alpha[\tilde{A}](z)\}, \\ &\beta[\tilde{A}](x) \gg \mathrm{rmin}\{\beta[\tilde{A}](x * y), \beta[\tilde{A}](y)\} \gg \mathrm{rmin}\{\beta[\tilde{A}](y), \beta[\tilde{A}](z)\}. \end{aligned}$$

This completes the proof. □

We provide conditions for a cubic IVIF set to be a cubic IVIF ideal.

Theorem 14. *Let $\mathcal{A} = \langle \mu, \tilde{A} \rangle$ be a cubic IVIF set in X in which conditions (31) and (32) are valid. If $\mathcal{A} = \langle \mu, \tilde{A} \rangle$ satisfies the condition (35) for all $x, y, z \in X$ with $x * y \leq z$, then $\mathcal{A} = \langle \mu, \tilde{A} \rangle$ is a cubic IVIF ideal of X.*

Proof. Since $x * (x * y) \leq y$ for all $x, y \in X$, it follows from (35) that

$$\begin{aligned}&\mu(x) \geq \min\{\mu(x * y), \mu(y)\},\\ &\alpha[\tilde{A}](x) \ll \text{rmax}\{\alpha[\tilde{A}](x * y), \alpha[\tilde{A}](y)\},\\ &\beta[\tilde{A}](x) \gg \text{rmin}\{\beta[\tilde{A}](x * y), \beta[\tilde{A}](y)\},\end{aligned}$$

Therefore $\mathcal{A} = \langle \mu, \tilde{A} \rangle$ is a cubic IVIF ideal of X. □

Lemma 1. *Every cubic IVIF ideal $\mathcal{A} = \langle \mu, \tilde{A} \rangle$ in X satisfies the following condition.*

$$(\forall x, y \in X)\ (x \leq y \Rightarrow \mu(x) \geq \mu(y),\ \alpha[\tilde{A}](x) \ll \alpha[\tilde{A}](y),\ \beta[\tilde{A}](x) \gg \beta[\tilde{A}](y)). \tag{36}$$

Proof. Assume that $x \leq y$ for all $x, y \in X$. Then $x * y = 0$, and so

$$\begin{aligned}&\mu(x) \geq \min\{\mu(x * y), \mu(y)\} = \min\{\mu(0), \mu(y)\} = \mu(y),\\ &\alpha[\tilde{A}](x) \ll \text{rmax}\{\alpha[\tilde{A}](x * y), \alpha[\tilde{A}](y)\} = \text{rmax}\{\alpha[\tilde{A}](0), \alpha[\tilde{A}](y)\} = \alpha[\tilde{A}](y),\\ &\beta[\tilde{A}](x) \gg \text{rmin}\{\beta[\tilde{A}](x * y), \beta[\tilde{A}](y)\} = \text{rmin}\{\beta[\tilde{A}](0), \beta[\tilde{A}](y)\} = \beta[\tilde{A}](y)\end{aligned}$$

by (31)–(34). □

Theorem 15. *In a BCK-algebra X, every cubic IVIF ideal is a cubic IVIF subalgebra.*

Proof. Let $\mathcal{A} = \langle \mu, \tilde{A} \rangle$ be a cubic IVIF ideal of a BCK-algebra X. Since $x * y \leq x$ for all $x, y \in X$, we have $\mu(x * y) \geq \mu(x)$, $\alpha[\tilde{A}](x * y) \ll \alpha[\tilde{A}](x)$, $\beta[\tilde{A}](x * y) \gg \beta[\tilde{A}](x)$ by Lemma 1. It follows from (33) and (34) that

$$\begin{aligned}&\mu(x * y) \geq \mu(x) \geq \min\{\mu(x * y), \mu(y)\} \geq \min\{\mu(x), \mu(y)\},\\ &\alpha[\tilde{A}](x * y) \ll \alpha[\tilde{A}](x) \ll \text{rmax}\{\alpha[\tilde{A}](x * y), \alpha[\tilde{A}](y)\} \ll \text{rmax}\{\alpha[\tilde{A}](x), \alpha[\tilde{A}](y)\},\\ &\beta[\tilde{A}](x * y) \gg \beta[\tilde{A}](x) \gg \text{rmin}\{\beta[\tilde{A}](x * y), \beta[\tilde{A}](y)\} \gg \text{rmin}\{\beta[\tilde{A}](x), \beta[\tilde{A}](y)\}.\end{aligned}$$

Therefore $\mathcal{A} = \langle \mu, \tilde{A} \rangle$ is a cubic IVIF subalgebra of X. □

The converse of Theorem 15 is not true in general as seen in the following example.

Example 8. *The cubic IVIF subalgebra $\mathcal{A} = \langle \mu, \tilde{A} \rangle$ in Example 5 is not a cubic IVIF ideal of X since $\mu(a) = 0.3 < 0.6 = \min\{\mu(a * b), \mu(b)\}$.*

We establish a characterization of a cubic IVIF ideal in a BCK-algebra.

Theorem 16. *For a cubic IVIF set $\mathcal{A} = \langle \mu, \tilde{A} \rangle$ in a BCK-algebra X, the following assertions are equivalent.*

(1) *$\mathcal{A} = \langle \mu, \tilde{A} \rangle$ is a cubic IVIF ideal of X.*
(2) *$\mathcal{A} = \langle \mu, \tilde{A} \rangle$ is a cubic IVIF subalgebra of X satisfying (35) for all $x, y, z \in X$ with $x * y \leq z$.*

Proof. (1) ⇒ (2). It follows from Proposition 7 and Theorem 15.
(2) ⇒ (1). It is by Proposition 4 and Theorem 14. □

5. Conclusions

We have introduced a cubic interval-valued intuitionistic fuzzy set which is an extension of a cubic set, and have applied it to BCK/BCI-algebra. We have investigated P-union, P-intersection, R-union and R-intersection of α-internal and α-external cubic IVIF sets. We have defined cubic IVIF subalgebra and ideal in BCK/BCI-algebra, and have investigated related properties. We have considered relations between cubic IVIF subalgebra and cubic IVIF ideal. We have discussed characterizations of cubic IVIF subalgebra and cubic IVIF ideal. In the consecutive research, we will discuss P-union, P-intersection, R-union and R-intersection of β-internal and β-external cubic IVIF sets. We will apply this notion to several kind of ideals in BCK/BCI-algebras, for example, (positive) implicative ideal, commutative ideal, associative ideal, q-ideal, etc. We will also apply this notion to several algebraic substructures in various algebraic structures, for example, MV-algebra, BL-algebra, R_0-algebra, residuated lattice, MTL-algebra, semigroup, semiring, near-ring, etc. We will try to study this notion in the neutrosophic environment.

Acknowledgments: The authors thank the anonymous reviewers for their valuable comments and suggestions.

Author Contributions: Young Bae Jun initiated the main idea of the work and wrote the paper. Seok-Zun Song and Young Bae Jun conceived and designed the new definitions and results. Seok-Zun Song and Seon Jeong Kim performed finding examples and checking contents. All authors have read and approved the final manuscript for submission.

Conflicts of Interest: The authors declare no conflict of interest.

References

1. Jun, Y.B.; Kim, C.S.; Yang, K.O. Cubic sets. *Ann. Fuzzy Math. Inform.* **2012**, *4*, 83–98.
2. Ahn, S.S.; Kim, Y.H.; Ko, J.M. Cubic subalgebras and filters of CI-algebras. *Honam Math. J.* **2014**, *36*, 43–54.
3. Akram, M.; Yaqoob, N.; Gulistan, M. Cubic KU-subalgebras. *Int. J. Pure Appl. Math.* **2013**, *89*, 659–665.
4. Jun, Y.B.; Khan, A. Cubic ideals in semigroups. *Honam Math. J.* **2013**, *35*, 607–623.
5. Jun, Y.B.; Kim, C.S.; Kang, J.G. Cubic q-ideals of BCI-algebras. *Ann. Fuzzy Math. Inform.* **2011**, *1*, 25–34.
6. Jun, Y.B.; Lee, K.J. Closed cubic ideals and cubic ∘-subalgebras in BCK/BCI-algebras. *Appl. Math. Sci.* **2010**, *4*, 3395–3402.
7. Jun, Y.B.; Lee, K.J.; Kang, M.S. Cubic structures applied to ideals of BCI-algebras. *Comput. Math. Appl.* **2011**, *62*, 3334–3342.
8. Khan, M.; Jun, Y.B.; Gulistan, M.; Yaqoob, N. The generalized version of Jun's cubic sets in semigroups. *J. Intell. Fuzzy Syst.* **2015**, *28*, 947–960.
9. Muhiuddin, G.; Al-roqi, A.M. Cubic soft sets with applications in BCK/BCI-algebras. *Ann. Fuzzy Math. Inform.* **2014**, *8*, 291–304.
10. Senapati, T.; Kim, C.S.; Bhowmik, M.; Pal, M. Cubic subalgebras and cubic closed ideals of B-algebras. *Fuzzy Inf. Eng.* **2015**, *7*, 129–149.
11. Yaqoob, N.; Mostafa, S.M.; Ansari, M.A. On cubic KU-ideals of KU-algebras. *ISRN Algebra* **2013**, doi:10.1155/2013/935905.
12. Atanassov, K.; Gargov, G. Interval valued intuitionistic fuzzy sets. *Fuzzy Sets Syst.* **1989**, *31*, 343–349.
13. Atanassov, K. Intuitionistic fuzzy sets. *Fuzzy Sets Syst.* **1986**, *20*, 87–96.
14. Huang, Y. *BCI-Algebra*; Science Press: Beijing, China, 2006.
15. Meng, J.; Jun, Y.B. *BCK-Algebras*; Kyungmoon Sa Co.: Seoul, Korea, 1994.
16. Iséki, K.; Tanaka, S. An introduction to the theory of BCK-algebras. *Math. Jpn.* **1978**, *23*, 1–26.

axioms

MDPI

Article

Fuzzy Analogues of Sets and Functions Can Be Uniquely Determined from the Corresponding Ordered Category: A Theorem

Christian Servin [1,†], Gerardo D. Muela [2,†] and Vladik Kreinovich [2,*,†]

1 Department of Computer Science and Information Technology, El Paso Community College, El Paso, TX 79915, USA; cservin@gmail.com
2 Department of Computer Science, University of Texas at El Paso, El Paso, TX 79968, USA; gdmuela@miners.utep.edu
* Correspondence: vladik@utep.edu; Tel.: +1-915-747-6951
† All the authors contributed equally to this work.

Received: 9 August 2017; Accepted: 9 January 2018; Published: 23 January 2018

Abstract: In modern mathematics, many concepts and ideas are described in terms of category theory. From this viewpoint, it is desirable to analyze what can be determined if, instead of the basic category of sets, we consider a similar category of fuzzy sets. In this paper, we describe a natural fuzzy analog of the category of sets and functions, and we show that, in this category, fuzzy relations (a natural fuzzy analogue of functions) can be determined in category terms—of course, modulo 1-1 mapping of the corresponding universe of discourse and 1-1 re-scaling of fuzzy degrees.

Keywords: fuzzy set; ordered category; category of fuzzy sets

1. Introduction

Category theory is one of the main tools of modern mathematics.

Many mathematical theories can be naturally described in terms of a directed graph, where vertices are objects studied in this theory (e.g., sets in set theory, topological spaces in topology, linear spaces in linear algebra), and edges relate different objects: e.g., functions map one set into another, continuous mappings map one topological space into another, linear mapping maps one linear space into another, etc. The corresponding graph is known as a *category*; see, e.g., [1].

In precise terms, a *category* is a tuple $(\mathrm{Ob}, \mathrm{Mor}, :, \mathrm{id}, \circ)$, where:

- Ob is the set whose elements are called *objects*,
- Mor is a set whose elements are called *morphisms*,
- $: \mathrm{Mor} \to \mathrm{Ob} \times \mathrm{Ob}$ is a mapping that assigns, to each morphism $f \in \mathrm{Mor}$ a pair of objects $(a, b) \in \mathrm{Ob} \times \mathrm{Ob}$; this is denoted by $f : a \to b$; the object a is called f's *domain*, and b is called f's *range*;
- id is a mapping that assigns, to each object $a \in \mathrm{Ob}$, a morphism $\mathrm{id}_a : a \to a$; and
- $\circ$ is a mapping that assigns, to each pair of morphisms $f : a \to b$ and $g : b \to c$ for which the range of f is equal to the domain of g, a new morphism $g \circ f : a \to c$ so that for every $f : a \to b$, we have $\mathrm{id}_b \circ f = f \circ \mathrm{id}_a = f$.

Because of its universal character, category theory plays an important role in modern mathematics [1]. Many new mathematical concepts are defined in category terms, and many original concepts are re-formulated in category terms—such a reformulation in very general terms often enables mathematicians to generalize their ideas and results to a more general context.

As we have mentioned, different areas of mathematics can be described in terms of different categories:

- Set theory is naturally described in terms of a category Set in which objects are sets and morphisms are functions.
- Topology is described in terms of a category Top in which objects are topological spaces and morphisms are continuous mappings.
- Linear algebra is naturally described in terms of a category Lin, in which objects are linear spaces, and morphisms are linear mappings, etc.

Many mathematical concepts can be reformulated in terms of an appropriate category; see, e.g., [2–11] and references therein.

What happens in the fuzzy case?

If we allow fuzzy sets (see, e.g., [12–16]), what is a natural analog of the category Set? In the category Set, morphisms from a to b are functions. In the crisp case, for each function $f : a \to b$ and for each element $x \in a$, we have a unique value of $y = f(x) \in b$.

Fuzzy means that for each $x \in a$, instead of a single value $y = f(x) \in b$, we may have different possible values $y \in b$, with different degrees of confidence. In general, we can have all possible values $y \in b$. For each $x \in a$ and for each $y \in b$, we have a degree $R_f(x, y) \in [0, 1]$ to which y is a possible value of $f(x)$. Thus, a natural fuzzy analog of a function is a fuzzy relation.

Composition $g \circ f$ of fuzzy relations $f : a \to b$ and $g : b \to c$ can be defined in the usual way. Namely, we want to know, for each pair of elements $x \in a$ and $c \in z$, to what extent there exists a $y \in b$ for which f brings us from a to b and g brings us from y to z. If we interpret "and" as min and there exists (an infinite "or") as max, then the above description translates into the following formula:

$$R_{g \circ f}(x, z) = \max_y \min(R_f(x, y), R_g(y, z)). \tag{1}$$

Since we have fuzzy relations, there is no need to explicitly describe the domain of each morphism: if for some $x \notin a$, the value $f(x)$ is not defined, this simply means that for this x, we have $R_f(x, y) = 0$ for all $y \in b$. Similarly, there is no need to describe the range.

Thus, without losing generality, we can assume that we have only one object—the universal set U—and that the relation $R_f(x, y)$ is defined for all $x \in U$ and $y \in U$. Morphisms are then fuzzy relations, with the usual composition relation (1).

Need for an ordered category.

In the crisp case, every property is either true or false.

As we gain more information, we may become more confident in our knowledge. For example, we may start with the situation in which, for a given x, several different values $f(x)$ are possible, but after acquiring new information, we become more and more confident that there is only one possible value y_0 of $f(x)$. This means that for the remaining value y_0, the degree of possibility $R_f(x, y_0)$ remains the same, but for all $y \neq y_0$, the corresponding degree $R_f(x, y)$ decreases. To capture this phenomenon, it is reasonable to supplement the category structure with the corresponding component-wise ordering between fuzzy relations (morphisms): $f \leq f'$ if and only if $R_f(x, y) \leq R_{f'}(x, y)$ for all x and y.

Formulation of the problem.

What can be defined based on this category-theory formulation? Can we uniquely determine the elements of the Universe of discourse U and the corresponding relations based on the categorical information?

What we do in this paper.

In this paper, as an answer to the above questions, we present an axiomatic description of fuzzy sets in the language of categories, with a proof of the soundness of this description.

2. Results

Towards a precise formulation of the problem. It is easy to see that if we have a 1-1 mapping $\pi : U \to U$ of the Universe of discourse U onto itself (i.e., a bijection), then the corresponding transformation $R(x, y) \to R(\pi(x), \pi(y))$ is an *automorphism* of the corresponding category in the sense that it preserves the identity, composition, and order.

Similarly, if we have a 1-1 monotonically increasing mapping $H : [0, 1] \to [0, 1]$, then the transformation $R(x, y) \to H(R(x, y))$ is also such an automorphism. Indeed, since we only consider order between degrees, monotonic transformation of degrees should not change anything.

It turns out that modulo this simple equivalence, we can uniquely determine all the elements $x \in U$ and all the relations $R(x, y)$ from the ordered category, i.e., in precise terms, that every automorphism is a composition of the automorphisms of the above two types. The proof of this result will be based on an explicit description of elements of U and relations $R_f(x, y)$ in category terms.

Let us describe the problem in precise terms.

Definition 1. *By an* ordered category, *we mean a category in which for every two objects a and b, there is a partial order $\leq$ on the set* $\mathrm{Mor}(a, b)$ *of all morphisms from a to b.*

Definition 2. *Let U be a set; we will call it the Universe of discourse. By a U-fuzzy ordered category, we mean an ordered category in which*

- *the only object is the set U,*
- *morphisms are fuzzy relations, i.e., mappings* $R : U \times U \to [0, 1]$,
- *the morphism* id *is defined as the mapping for which* $\mathrm{id}(x, x) = 1$ *and* $\mathrm{id}(x, y) = 0$ *for* $x \neq y$,
- *the composition of morphisms is defined by the formula*

$$(g \circ f)(x, z) = \max_y \min(f(x, y), g(y, z)),$$

and

- *the order between the morphisms is the component-wise order:* $f \leq g$ *means that* $f(x, y) \leq g(x, y)$ *for all x and y.*

The U-fuzzy ordered category will be denoted by F_U.

Comment. One can easily see that this is indeed a category, i.e., that the composition of morphisms is associative, and the composition of any morphism f with the identity morphism id is equal to f: $f \circ \mathrm{id} = \mathrm{id} \circ f = f$.

Definition 3. *An* automorphism *of an ordered category is a pair consisting of bijections* $F : \mathrm{Ob} \to \mathrm{Ob}$ *and* $G : \mathrm{Mor} \to \mathrm{Mor}$ *for which:*

- *for all f, a, and b, we have* $f : a \to b$ *if and only if* $G(f) : F(a) \to F(b)$;
- *for all f and g, we have* $G(f \circ g) = G(f) \circ G(g)$,
- *for all a, we have* $G(\mathrm{id}_a) = \mathrm{id}_{F(a)}$, *and*
- *for all f and g, we have* $f \leq g$ *if and only if* $G(f) \leq G(g)$.

Comment. This definition is a natural generalization of the standard definition of automorphism of categories (see, e.g., [17–19]) to ordered categories.

Proposition 1. *Let* $\pi : U \to U$ *be a bijection of U, and let* $H : [0, 1] \to [0, 1]$ *be an increasing bijection of the interval* $[0, 1]$. *Then, the mapping* $G_{\pi,H}$ *that maps each morphism* $f(x, y)$ *into a morphism* $(G_{\pi,H}(f))(x, y) = H(f(\pi(x), \pi(y)))$ *is an automorphism of the category* F_U.

Our main result is that these are the only automorphisms of the category F_U.

Theorem 1. *For every set U, every automorphism of the ordered category F_U has the form $G_{\pi,H}$ for some bijection $\pi : U \to U$ and for some monotonic bijection $H : [0,1] \to [0,1]$.*

Comment. This may not be very clear from the formulation of the result, but the proof will show that we can determine elements of the set U and values of the mappings $f(x,y)$ in category terms, i.e., we can indeed define fuzzy relations—a natural fuzzy analogue of functions—in category terms.

3. Proofs

3.1. Proof of the Proposition

This proposition is easy to prove: a permutation π does not change anything, and the increasing bijection does not change the order.

3.2. Proof of the Theorem

1°. First, we can describe the morphism f_0 for which $f_0(x,y) = 0$ for all x and y in ordered-category terms, as the only morphism f for which $f \le g$ for all morphisms g.

Indeed, clearly $f_0 \le g$ for all g. Vice versa, if $f \le g$ for all g, then, in particular, $f \le f_0$, i.e., $f(x,y) \le f_0(x,y) = 0$ for all x and y, and since $f(x,y) \in [0,1]$, this means that indeed $f(x,y) = 0$ for all x and y.

2°. Let us first characterize all the morphisms $f \ne f_0$ for which the set $\{g : g \le f\}$ is linearly ordered. Since an automorphism preserves order, every automorphism maps such morphisms into morphisms with the same property.

Specifically, we will prove that a morphism has this property if and only if we have $f(x,y) > 0$ only for one pair (x,y), and we have $f(x',y') = 0$ for all other pairs (x',y').

Indeed, one can easily check that for such morphisms f, the only morphisms $g \le f$ are the morphisms which also have $g(x',y') = 0$ for all pairs $(x',y') \ne (x,y)$. Such morphisms g are uniquely described by the corresponding value $g(x,y)$. For every two such morphisms g and g', depending on whether $g(x,y) \le g'(x,y)$ or $g'(x,y) \le g(x,y)$, we have $g \le g'$ or $g' \le g$, i.e., the set $\{g : g \le f\}$ is indeed linearly ordered.

Vice versa, let us prove that if a morphism has this property, then it has $f(x,y) > 0$ only for one pair (x,y). Indeed, if we have $f(x,y) > 0$ and $f(x',y') > 0$ for two different pairs $(x,y) \ne (x',y')$, then we would be able to construct two different morphisms $g \le f$ and $g' \le f$ for which $g \not\le g'$ and $g' \not\le g$. Namely, we take:

- $g(x,y) = f(x,y) > 0$ and $g(x'',y'') = 0$ for all pairs $(x'',y'') \ne (x,y)$, and
- $g'(x,y) = f(x',y') > 0$ and $g(x'',y'') = 0$ for all pairs $(x'',y'') \ne (x',y')$.

This contradicts our assumption that the set $\{g : g \le f\}$ is linearly ordered.

3°. Let us now describe, in ordered-category terms, morphisms f for which $f(x,x) > 0$ for some $x \in U$ and $f(x',y')$ for all other pairs $(x',y') \ne (x,x)$.

Indeed, out of all morphisms described in Part 2 of this proof, such morphisms can be determined by the additional condition that $f \circ f = f$. This condition is clearly satisfied for such morphisms, while for morphisms for which $f(x,y) > 0$ for some $y \ne x$, the composition $f \circ f$ is, as one can see, identically 0 and thus, different from f.

4°. One can see that two morphisms f and f' of the type described in Part 3 are connected by the relation $\le$ (i.e., $f \le f'$ or $f' \le f$) if and only if they correspond to the same element $a \in U$.

Thus, we can describe elements of the set U in ordered-category terms: as equivalent classes of morphisms of the type described in Part 3 with respect to the relation $(f \le f') \vee (f' \le f)$.

Hence, if we have an automorphism, elements are mapped into elements in a 1-1 way, i.e., indeed, we have a bijection of the Universe of discourse.

5°. Let us now show that the degrees from the interval $[0,1]$ can also be described—modulo increasing bijections of this interval—in ordered-category terms.

5.1°. Indeed, for each element $x \in U$, different degrees $v \in [0,1]$ can be associated with different morphisms f described in Part 3 of this proof, i.e., morphisms for which:

- $f(x,x) > 0$ for this element x and
- $f(x',y')$ for all pairs $(x',y') \neq (x,x)$.

Different degrees are then simply associated with different values $v = f(x,x)$.

This construction provides us with degrees at each element $x \in U$. To obtain a general description of degrees, we need to relate the values corresponding to different elements $x, x' \in U$.

5.2°. Let us denote, by $f_{x,v}$, the morphism for which:

- $f_{x,v}(x,x) = v$ and
- $f_{x,v}(x',y') = 0$ for all pairs $(x',y') \neq (x,x)$.

We want, for every $x \neq y$, to connect the values v and w corresponding to functions $f_{x,v}$ and $f_{y,w}$. This connection comes from the following auxiliary result:

$$w \le v \Leftrightarrow \exists f_{x\to y}\, \exists f_{y\to x}\, (f_{x\to y} \circ f_{x,v} \circ f_{y\to x} = f_{y,w}).$$

Indeed, by definition of a composition, the values of the composition $g \circ f$ cannot exceed the largest value of each of the composed relations g and f. Thus, if $f_{x\to y} \circ f_{x,v} \circ f_{y\to x} = f_{y,w}$, then the value $f_{y,w}(b,b) = w$ cannot exceed the maximum value v of the function $f_{x,v}$; thus, $w \le v$.

Vice versa, if $w \le v$, then we can take the following morphisms $f_{x\to y}$ and $f_{y\to x}$:

- $f_{x\to y}(x,y) = w$ and $f_{x\to y}(x',y') = 0$ for all other pairs $(x',y') \neq (x,y)$, and, similarly,
- $f_{y\to x}(y,x) = w$ and $f_{y\to x}(x',y') = 0$ for all other pairs $(x',y') \neq (y,x)$.

In this case, as one can easily check, we have $f_{x\to y} \circ f_{x,v} \circ f_{y\to x} = f_{y,w}$.

5.3°. Now that we know how to describe the relation $w \le v$ for functions $f_{x,v}$ and $f_{y,w}$ in ordered-category form, we can describe equality $v = w$ between the degrees v and w corresponding to morphisms $f_{x,v}$ and $f_{y,w}$ as $(v \le w)\ \&\ (w \le v)$, i.e., in view of Part 5.2, as:

$$(\exists f_{x\to y}\, \exists f_{y\to x}\, (f_{x\to y} \circ f_{x,v} \circ f_{y\to x} = f_{y,w}))\ \&\ (\exists g_{y\to x}\, \exists g_{x\to y}\, (g_{y\to x} \circ f_{y,w} \circ g_{x\to y} = f_{x,v})).$$

This enables us to identify degrees $v \in [0,1]$ in ordered-category terms—by identifying them with the functions $f_{x,v}$ and taking into account the above possibility to compare degrees at different elements x.

Hence, if we have an automorphism, degrees are mapped into degrees in a 1-1 and order-preserving way, i.e., indeed, we have a monotonic bijection $H : [0,1] \to [0,1]$.

6°. To complete the proof, we need to show how, for each morphism f and for every two elements a and b, we can describe the value $f(x,y)$ in ordered-category terms. This will complete the proof that the given automorphism has the form $G_{\pi,H}$ for the mappings π and H as identified in Sections 4 and 5 of this proof.

6.1°. Let us first prove the following auxiliary result:

$$\exists f_{y\to x}\, (f_{y\to x} \circ f_{y,1} \circ f \circ f_{x,1} = f_{x,v}) \Leftrightarrow v \le f(x,y).$$

Indeed, by definition of a composition, the composition $c \stackrel{\text{def}}{=} f \circ f_{x,1}$ has the following form:

- $c(x,y') = f(x,y')$ for all y' and

- $c(x', y') = 0$ for all y' and for all $x' \neq a$.

Similarly, the composition $c' \stackrel{\text{def}}{=} f_{y,1} \circ f \circ f_{x,1} = f_{y,1} \circ c$ has the following form:

- $c'(x, y) = f(x, y)$, and
- $c'(x', y') = 0$ for all other pairs $(x', y') \neq (x, y)$.

As we have argued in Part 5 of this proof, the value of a composition function cannot exceed the maximum value of each of the composed morphisms. Thus, for the composition $f_{y \to x} \circ f_{y,1} \circ f \circ f_{x,1} = f_{y \to x} \circ c'$, the maximum value cannot exceed the maximum value $f(x, y)$ of the morphism c'. Thus, if $f_{y \to x} \circ c' = f_{x,v}$, the maximum value v of the morphism $f_{x,v}$ cannot exceed $f(x, y)$: $v \leq f(x, y)$.

Vice versa, for every $v \leq f(x, y)$, we can construct a morphism $f_{y \to x}$ for which $f_{y \to x} \circ c' = f_{x,v}$: namely, we can take:

- $f_{y \to x}(y, x) = v$, and
- $f_{y \to x}(x', y') = 0$ for all pairs $(x', y') \neq (y, x)$.

One can easily check that in this case indeed $f_{y \to x} \circ c' = f_{x,v}$.

6.2°. For each morphism f and for every two elements x and y, we can identify the degree $f(x, y)$ as the largest degree v for which the inequality $v \leq f(x, y)$ holds.

Since, according to Part 6.1 of this proof, the inequality $v \leq f(x, y)$ can be described in ordered-category terms, we can thus conclude that the degree $f(x, y)$ can also be described in ordered-category terms.

The proposition is proven.

4. Conclusions

Many concepts of modern mathematics, starting from the basic notions of sets and functions, are described in terms of category theory. Many other mathematical concepts can be reformulated in category terms. Due to the general nature of category theory, such a reformulation often helps to extend notions and results from one area to different areas of mathematics.

Because of this potential advantage, it is reasonable to ask whether similar *fuzzy* notions can also be described in category terms. In this paper, we show that fuzzy relations—i.e., fuzzy analogues of functions—can indeed be described in category terms. Specifically, we show that, in the corresponding fuzzy category, we can describe both:

- elements of the original universe of discourse (modulo a 1-1 permutation), and
- fuzzy degrees (modulo a 1-1 monotonic mapping from the interval $[0, 1]$ onto itself).

This result shows the soundness of our axiomatic description of fuzzy sets in the language of categories.

At this moment, what we have is a very theoretical paper. However, we hope that, similarly to how the reformulation of crisp notions in category terms can help generalize the corresponding results, our reformulation will help extend fuzzy results to more general situations—and thus, will facilitate future applications.

Acknowledgments: This work was supported, in part, by the National Science Foundation grant HRD-1242122 (Cyber-ShARE Center of Excellence). The authors are greatly thankful to the editors and to the anonymous reviewers for their helpful suggestions.

Author Contributions: All the authors contributed equally to this work.

Conflicts of Interest: The authors declare no conflict of interest.

References

1. Awodey, S. *Category Theory*; Oxford University Press: Oxford, UK, 2010.
2. Dolgopol'sky, V.P.; Kreinovich, V.Y. Definability in Categories. 2. Semigroups and Ordinals. *Abstr. Papers Present. Am. Math. Soc.* **1980**, *1*, 618.
3. Dolgopol'sky, V.P.; Kreinovich, V.Y. Definability in Categories. 4. Separable Topological Spaces. *Abstr. Papers Present. Am. Math. Soc.* **1981**, *2*, 234.
4. Feferman, S.; Sieg, W. (Eds.) *Proofs, Categories and Computations*; College Publications: London, UK, 2010.
5. Kosheleva, O.M. Definability in Categories. 5. Categories. *Abstr. Papers Present. Am. Math. Soc.* **1981**, *2*, 234.
6. Kosheleva, O.M. Definability in Categories. 7. Ultimateness and Applications. *Abstr. Papers Present. Am. Math. Soc.* **1981**, *2*, 294.
7. Kosheleva, O.M.; Kreinovich, V.Y. Definability in Categories. 1. Groups. *Abstr. Papers Present. Am. Math. Soc.* **1980**, *1*, 472.
8. Kosheleva, O.M.; Kreinovich, V.Y. Definability in Categories. 6. Strong Definability. *Abstr. Papers Present. Am. Math. Soc.* **1980**, *1*, 555.
9. Kosheleva, O.M.; Kreinovich, V.Y. Definability in Categories. 3. Topological Spaces. *Abstr. Papers Present. Am. Math. Soc.* **1980**, *1*, 618.
10. Kosheleva, O.M.; Kreinovich, V. On definability in categories. In Proceedings of the Summaries of the First East European Category Seminar, Predela, Bulgaria, 13–18 March 1989; Velinov, Y., Lozanov, R., Eds.; pp. 27–28.
11. Kreinovich, V.; Ceberio, M.; Brefort, Q. In category of sets and relations, it is possible to describe functions in purely category terms. *Eur. Math. J.* **2015**, *6*, 90–94.
12. Belohlavek, R.; Dauben, J.W.; Klir, G.J. *Fuzzy Logic and Mathematics: A Historical Perspective*; Oxford University Press: New York, NY, USA, 2017.
13. Mendel, J.M. *Uncertain Rule-Based Fuzzy Systems: Introduction and New Directions*; Springer: Cham, Switzerland, 2017.
14. Nguyen, H.T.; Walker, E.A. *A First Course in Fuzzy Logic*; Chapman and Hall/CRC: Boca Raton, FL, USA, 2006.
15. Zadeh, L.A. Fuzzy sets. *Inf. Control* **1965**, *8*, 338–353.
16. Klir, G.; Yuan, B. *Fuzzy Sets and Fuzzy Logic*; Prentice Hall: Upper Saddle River, NJ, USA, 1995.
17. Hazewinkel, M. Equivalence of categories. In *Encyclopedia of Mathematics*; Springer Science and Business Media: Berin/Heidelberg, Germany, 2001.
18. Mac Lane, S. *Categories for the Working Mathematician*; Springer: New York, NY, USA, 1998.
19. Schröder, L. Categories: A free tour. In *Categorical Perspectives*; Kozlowski, J., Melton, A., Eds.; Springer Science and Business Media: Berin/Heidelberg, Germany, 2001; pp. 1–27.

axioms

MDPI

Article

Revision of the Kosiński's Theory of Ordered Fuzzy Numbers

Krzysztof Piasecki

Department of Investment and Real Estate, Poznan University of Economics and Business, al. Niepodleglosci 10, 61-875 Poznań, Poland; krzysztof.piasecki@ue.poznan.pl; Tel.: +48-721-771-572

Received: 5 January 2018; Accepted: 26 February 2018; Published: 2 March 2018

Abstract: Ordered fuzzy numbers are defined by Kosiński. In this way, he was going to supplement a fuzzy number by orientation. A significant drawback of Kosiński's theory is that there exist such ordered fuzzy numbers which, in fact, are not fuzzy numbers. For this reason, a fully formalized correct definition of ordered fuzzy numbers is proposed here. Also, the arithmetic proposed by Kosiński has a significant disadvantage. The space of ordered fuzzy numbers is not closed under Kosiński's addition. On the other hand, many mathematical applications require the considered space be closed under used arithmetic operations. Therefore, the Kosinski's theory is modified in this way that the space of ordered fuzzy numbers is closed under revised arithmetic operations. In addition, it is shown that the multiple revised sum of finite sequence of ordered fuzzy numbers depends on its summands ordering.

Keywords: ordered fuzzy number; arithmetic operator; linear space

---

1. Introduction

The ordered fuzzy numbers (OFNs) were introduced by Kosiński and his associates in a series of papers [1–4] as an extension of the notion of fuzzy numbers (FNs). For this reason, ordered fuzzy numbers are increasingly called Kosinski's numbers [5,6]. The monograph [7] is a competent source of information about the contemporary state of knowledge on OFN.

OFNs have already begun to find their use in operations research applied in economics and finance. Examples of such applications can already be found in the works [8–12].

It is a common view that the theory of fuzzy sets is very extensive and very helpful in many applications. On the other hand, the previous Kosiński's theory of ordered fuzzy number is poor (see [7]). Thus, it is obvious that considerations with use the theory fuzzy sets may be much more fruitful. Kosiński [4] has shown that there exist improper OFNs which cannot be represented by a pair of FN and orientation. It means that we cannot apply any knowledge of fuzzy sets to solve practical problems described by improper OFN. Therefore, considerations with improper use of OFNs may not be fruitful. Thus, the main purpose of this article is to revise Kosiński's theory. The presented study limits the family of all OFNs to the family of all proper OFNs which are represented by a pair of FN and orientation. This goal will be achieved by modifying the OFNs' definition.

Kosiński [1–3] has determined OFNs' arithmetic as an extension of results obtained by Goetschel and Voxman [13] for fuzzy numbers. Moreover, Kosiński [4] has shown that there exist such proper OFNs that their Kosiński's sum is equal to an improper OFN. This means that the family of all proper OFNs is not closed under Kosiński's addition. On the other hand, most mathematical applications require the considered objects' family be closed under used arithmetic operations. For this reason, the revised arithmetic is introduced for OFNs.

The linear space of FNs is usually determined using dot product and sum [14] defined by means of the Zadeh's Extension Principle [15–17]. The last goal of this paper is to investigate the linear space of

OFNs determined by dot products and revised sums of OFNs. In this article, the discussion on linear space properties is limited due to exploration the perceived disadvantages associated with the revised sums. Many mathematical applications require the finite multiple sums to be independent on summands ordering. Any associative and commutative sum satisfies this property. Therefore, the associativity and commutativity of revised sum will also be considered. All considerations are based on three counterexamples with use of trapezoidal or triangle OFNs. This approach is sufficient to falsify theses formulated for the general case of OFNs. On the other hand, all previously known uses of ordered fuzzy number uses are limited to applications of trapezoidal OFNs.

2. Elements of Fuzzy Numbers Theory

Zadeh [18] introduced the fuzzy sets theory to deal with linguistic variables problems. A fuzzy set is an extension of a crisp set. The crisp sets use only the notions of a full membership or non-membership, whereas fuzzy sets allow partial membership.

Objects of any cognitive-application activities are elements of the predefined space $\mathbb{X}$. The basic tool for imprecise classification of these elements is the notion of fuzzy set $\mathcal{A}$ described by its membership function $\mu_A \in [0;1]^{\mathbb{X}}$, as the set of ordered pairs

$$\mathcal{A} = \{(x, \mu_A(x)); x \in \mathbb{X}\}. \tag{1}$$

In multi-valued logic terms, the value $\mu_A(x)$ of membership function is interpreted as the true value of the sentence "$x \in \mathcal{A}$". By $\mathcal{F}(\mathbb{X})$ we denote the family of all fuzzy subsets of the space $\mathbb{X}$. In this work, the following notions will be applied for analyzing any fuzzy subset:

- $\alpha-$cut $[\mathcal{A}]_\alpha$ of the fuzzy subset $\mathcal{A} \in \mathcal{F}(\mathbb{X})$ determined for each $\alpha \in [0;1]$ by dependence

$$[\mathcal{A}]_\alpha = \{x \in \mathbb{X} : \ \mu_A(x) \geq \alpha\}; \tag{2}$$

- support closure $[\mathcal{A}]_{0^+}$ of the fuzzy subset $\mathcal{A} \in \mathcal{F}(\mathbb{X})$ given in a following way

$$[\mathcal{A}]_{0^+} = \lim_{\alpha \to 0^+} [\mathcal{A}]_\alpha. \tag{3}$$

An imprecise number is a family of values in which each considered value belongs to it to a varying degree. A commonly accepted model of imprecise number is the fuzzy number, defined as a fuzzy subset of the real line $\mathbb{R}$. The most general definition of fuzzy number is given as follows:

Definition 1. *[14] The fuzzy number (FN) is such a fuzzy subset $\mathcal{S} \in \mathcal{F}(\mathbb{R})$ with bounded support closure $[\mathcal{S}]_{0^+}$ that it is represented by its upper semi-continuous membership function $\mu_S \in [0;1]^{\mathbb{R}}$ satisfying the conditions:*

$$\exists_{x \in \mathbb{R}} \ \mu_S(x) = 1, \tag{4}$$

$$\forall_{(x,y,z) \in \mathbb{R}^3} \ x \leq y \leq z \Rightarrow \mu_S(y) \geq \min\{\mu_S(x); \mu_S(z)\}. \tag{5}$$

The set of all FN we denote by the symbol $\mathbb{F}$.

Dubois and Prade [19] first introduced the arithmetic operations on FN. In this paper, the symbol $\circ$ means one of the usual arithmetic operations: addition $+$, subtraction $-$, multiplication $*$, and division $/$. We assume that $\mathcal{X}$, $\mathcal{Y}$, $\mathcal{Z} \in \mathbb{F}$ are represented respectively by their membership functions μ_X, μ_Y, $\mu_Z \in [0;1]^{\mathbb{R}}$. In [19] any arithmetic operation $\circ$ on $\mathbb{R}$ is extended to arithmetic operation $\circledcirc$ on $\mathbb{F}$ by means of the identity

$$\mathcal{Z} = \mathcal{X} \circledcirc \mathcal{Y}, \tag{6}$$

where we have

$$\mu_Z(z) = \sup\Big\{\min\{\mu_X(x), \mu_Y(y)\} : z = x \circ y,\ (x,y) \in \mathbb{R}^2\Big\}. \tag{7}$$

These arithmetic operations are coherent with the Zadeh's Extension Principle [15–17].

Among other things, Dubois and Prade [20] have distinguished a special type of representation of FN called LR-type FN which may be generalized in a following way.

Definition 2. *Let us assume that for any nondecreasing sequence* $\{a,b,c,d\} \subset \mathbb{R}$ *the left reference function* $L_S \in [0,1]^{[a,b]}$ *and the right reference function* $R_S \in [0,1]^{[c,d]}$ *are upper semi-continuous monotonic functions satisfying the condition*

$$L_S(b) = R_S(c) = 1. \tag{8}$$

Then the identity

$$\mu_S(\cdot|a,b,c,d,L_S,R_S) = \begin{cases} 0, & x \notin [a,d] = [d,a], \\ L_S(x), & x \in [a,b] = [b,a], \\ 1, & x \in [b,c] = [c,b], \\ R_S(x), & x \in [c,d] = [d,c] \end{cases} \tag{9}$$

defines the membership function $\mu_S(\cdot|a,b,c,d,L_S,R_S) \in [0,1]^{\mathbb{R}}$ *of the FN* $\mathcal{S}(a,b,c,d,L_S,R_S)$ *which is called LR-type FN (LR-FN).*

Let us note that this identity describes additionally extended notation of numerical intervals, which is used in this work.

Remark 1. *In original version of the Dubois–Prade definition [20] LR-FN was defined only for the case of any sequence* $\{a,b,c,d\} \subset \mathbb{R}$ *fulfilling the condition*

$$a < b \leq c < d \tag{10}$$

and reference functions given as continuous surjections.

Further generalizations have resulted from requirements imposed by the needs of individual applications.

Goetschel and Voxman [13] showed that the Definition 1 can be replaced by following an equivalent definition of FN:

Definition 3 [13]**.** *FN is such fuzzy subset* $\mathcal{S} \in \mathcal{F}(\mathbb{R})$ *with bounded support closure* $[\mathcal{S}]_{0^+}$ *that it fulfils the condition*

$$\forall_{\alpha\in]0;1]} \exists_{(l_S(\alpha), r_S(\alpha))\in\mathbb{R}^2 : l_S(\alpha)\leq r_S(\alpha)} : \ [\mathcal{S}]_\alpha = [l_S(\alpha), r_S(\alpha)]. \tag{11}$$

Then Goetschel and Voxman [4] prove the following theorems:

Theorem 1 [13]**.** *For any FN* $\mathcal{S} \in \mathcal{F}(\mathbb{R})$*:*

$$l_S \text{ is a nondecreasing and bounded function on } [0;\ 1]; \tag{12}$$

$$r_S \text{ is a nondecreasing and bounded function on } [0;\ 1]; \tag{13}$$

$$l_S(1) \leq r_S(1); \tag{14}$$

$$\text{function } l_S \text{ and } r_S \text{ are left} - \text{continuous on }]\,0;\ 1] \text{ and right} - \text{continuous at } \{0\}. \tag{15}$$

If a pair of functions $(l_S, r_S) \in \mathbb{R}^{[0;1]} \times \mathbb{R}^{[0;1]}$ fulfils the conditions (12)–(15) then it is called a pair of Goetschel–Voxman functions.

Theorem 2 [13]. *For any pair of Goetschel–Voxman functions (l_S, r_S) there exists exactly one such FN $\mathcal{S} \in \mathcal{F}(\mathbb{R})$ that the condition (11) is fulfilled.*

Therefore, the pair of Goetschel–Voxman functions representing FN $\mathcal{S} \in \mathcal{F}(\mathbb{R})$ will be denoted by the symbol (l_S, r_S).

Definition 4. *For any upper semi-continuous and nondecreasing function $H \in [0;1]^{[H(0);H(1)]}$ its lower pseudo-inverse function $H^{\star} \in [H(0);H(1)]^{[0;1]}$ is given by the identity*

$$H^{\star}(x) = \min\{x \in [H(0);H(1)] : H(x) \geq \alpha\}. \tag{16}$$

Definition 5. *For any upper semi-continuous and nonincreasing function $H \in [0;1]^{[H(1);H(0)]}$ its lower pseudo-inverse function $H^{\star} \in [H(1);H(0)]^{[0;1]}$ is given by the identity*

$$H^{\star}(x) = \max\{x \in [H(0);H(1)] : H(x) \geq \alpha\}. \tag{17}$$

In a special case, we can conclude that:

Theorem 3 [13]. *For any LR-FN $\mathcal{S} = \mathcal{S}(a, b, c, d, L_S, R_S)$ additionally:*

$$\textit{function } l_S \textit{ and } r_S \textit{ are continuous on } [0,1], \tag{18}$$

$$l_s = L_S^{\star}, \tag{19}$$

$$r_S = R_S^{\star}. \tag{20}$$

The next theorem is proved in [13] with the use of following notions:

Definition 6. *For any bounded continuous and nondecreasing function $h \in [h(0),\, h(1)]^{[0;1]}$ its upper pseudo-inverse function $h^{\lhd} \in [0;1]^{[h(0),\, h(1)]}$ is given by the identity*

$$h^{\lhd}(x) = \max\{\alpha \in [0;1] : h(\alpha) = x\}. \tag{21}$$

Definition 7. *For any bounded continuous and nonincreasing function $h \in [h(1),\, h(0)]^{[0;1]}$ its upper pseudo-inverse function $h^{\lhd} \in [0;1]^{[h(1),\, h(0)]}$ is given by the identity*

$$h^{\lhd}(x) = \min\{\alpha \in [0;1] : h(\alpha) = x\}. \tag{22}$$

Theorem 4 [13]. *For any pair of Goetschel–Voxman functions (l_S, r_S) additionally fulfilling condition (18) there exists exactly one such LR-FN $\mathcal{S} \in \mathcal{F}(\mathbb{R})$ that condition (11) is fulfilled and we have*

$$\mathcal{S} = \mathcal{S}(l_S(0), l_S(1), r_S(1), r_S(0), l_S^{\lhd}, r_S^{\lhd}). \tag{23}$$

We assume that $\mathcal{X}, \mathcal{Y}, \mathcal{Z} \in \mathbb{F}$ are represented respectively by their pairs of Goetschel–Voxman functions (l_X, r_X), (l_Y, r_Y), $(l_Z, r_Z) \in \mathbb{R}^{[0;1]} \times \mathbb{R}^{[0;1]}$. Goetschel and Voxman [13] have proved that if $\mathcal{C} \in \mathbb{F}$ is given by the identity (6) then we have and

$$\forall_{\alpha\in[0;1]} : \quad l_Z(\alpha) = l_X(\alpha) \circ l_Y(\alpha) \text{ and } r_Z(\alpha) = r_X(\alpha) \circ r_Y(\alpha). \tag{24}$$

Let us note that if the bounded continuous and monotonic function $h \in [h(1),\ h(0)]^{[0;1]}$ is bijection then we have

$$\forall_{x\in[h(0),h(1)]} : \quad h^{-1}(x) = h^{\lhd}(x). \tag{25}$$

3. Kosiński's Ordered Fuzzy Numbers

The concept of ordered fuzzy numbers (OFN) was introduced by Kosiński and his co-writers in the series of papers [1–4] as an extension of concept of FN. Thus, any OFN should be determined by (1) as a fuzzy subset in the real line $\mathbb{R}$. On the other hand, Kosiński has defined OFN as a ordered pair of functions from the interval $[0,1]$ into $\mathbb{R}$. This pair is not a fuzzy subset in $\mathbb{R}$. Thus, we cannot accept Kosiński's original terminology. Then again, the intuitive Kosiński's approach to the notion of OFN is very useful. For these reasons, a revised definition of OFN, which fully corresponds with the intuitive Kosiński's will be presented below. The OFN concept of number is closely linked to the following ordered pair.

Definition 8. *For any sequence $\{a,b,c,d\} \subset \mathbb{R}$, the Kosiński's pair $S(a,b,c,d)$ is defined as any ordered pair (f_S, g_S) of monotonic continuous surjections $f_S : [0,1] \to UP_S = [a,b]$ and $g_S : [0,1] \to DOWN_S = [c,d]$ fulfilling the condition*

$$f_s(0) = a \text{ and } f_s(1) = b \text{ and } g_s(1) = c \text{ and } g_s(0) = d. \tag{26}$$

Remark 2. *In the original final version of the Kosiński's definition [4], the OFN is defined as ordered pair (f_S, g_S) of continuous bijections $f_S : [0,1] \to UP_S = [a,b]$ and $g_S : [0,1] \to DOWN_S = [c,d]$. Because OFN defined in such a way, is not fuzzy set, in Definition 8 Kosiński's term of OFN was replaced by the term "Kosiński's pair". Moreover, Kosiński has assumed implicitly that the sequence $\{a,b,c,d\}$ fulfils the condition (10) or the condition*

$$a > b \geq c > d. \tag{27}$$

Kosiński marked the additional conditions of the above definition only on graphs. Some example of such Kosiński's graphs are presented in Figure 1.

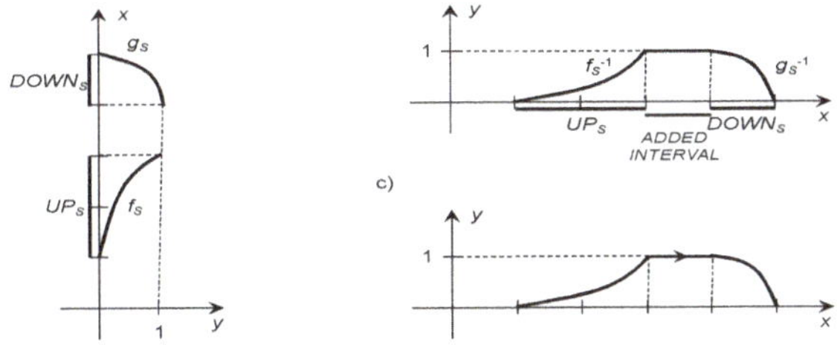

Figure 1. (**a**) Positively oriented Kosiński's pair; (**b**) Membership function of FN determined by positive oriented OFN; (**c**) Arrow denotes the positive orientation of OFN. Source: [4].

For any Kosiński pair (f_S, g_S) the function $f_S : [0,1] \rightarrow UP_S$ is called the up-function. Then the function $g_S : [0,1] \rightarrow DOWN_S$ is called the down-function. The up-function and the down-function are collectively referred to as Kosiński's maps.

Note that Kosiński developed his theory without using the results obtained by Goetschel and Voxman [13]. In this article, thanks to the use of these results, we can generalize Kosiński's theory to the case when Kosiński's maps are continuous surjections. This generalization is needed for individual applications.

By means of Goetschel–Voxman results, we can define OFN in such manner which fully corresponds to the intuitive OFN definition by Kosiński.

Definition 9. *For any sequence* $\{a,b,c,d\} \subset \mathbb{R}$ *the Kosiński's pair* $S(a,b,c,d) = (f_S, g_S)$ *determines Kosiński's ordered fuzzy number (K's-OFN)* $\overleftrightarrow{\mathcal{S}}(a,b,c,d,f_S^{\triangleleft},g_S^{\triangleleft})$ *defined explicitly by its membership relation* $\mu_S(\cdot|a,b,c,d,f_S^{\triangleleft},g_S^{\triangleleft}) \subset \mathbb{R} \times [0,1]$ *given by (9).*

The above definition is coherent to the intuitive Kosiński's approach of the OFN notion. Therefore, we agree with other scientists [5,6] that the OFN should be called the Kosiński's number. The space of all K's-OFN is denoted by the symbol $\breve{\mathbb{K}}$. For any OFN $\overleftrightarrow{S} \in \breve{\mathbb{K}}$, its up-function is denoted by the symbol f_S and its down-function is denoted by the symbol g_S.

Any sequence $\{a,b,c,d\} \subset \mathbb{R}$ satisfies exactly one of the following conditions

$$b < c \text{ or } (b = c \text{ and } a < d), \tag{28}$$

$$b > c \text{ or } (b = c \text{ and } a > d), \tag{29}$$

$$a = b = c = d. \tag{30}$$

If condition (28) is fulfilled then the K's-OFN $\overleftrightarrow{\mathcal{S}}(a,b,c,d,f_S^{\triangleleft},g_S^{\triangleleft})$ has a positive orientation. For this case, an example of graphs of Kosiński's maps is presented on the Figure 1a. Then the graph of membership function of FN $\mathcal{S}(a,b,c,d,f_S^{\triangleleft},g_S^{\triangleleft})$ is presented in Figure 1b. The graph of K's-OFN membership function with positive orientation is shown in Figure 1c. The membership function of K's-OFN has an extra arrow denoting the orientation, which provides supplementary information. Positively oriented K's-OFN is interpreted as such imprecise number, which may increase. The space of all positively oriented OFN is denoted by the symbol $\breve{\mathbb{K}}^+$. An example of positively oriented K's-OFN is presented in Figure 2a.

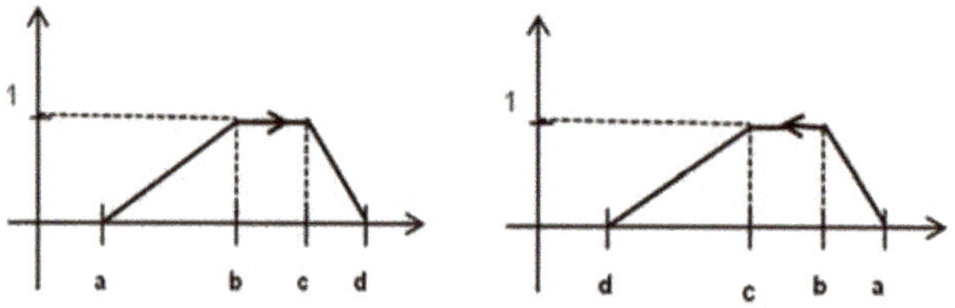

(a) (b)

Figure 2. The membership function of K's-OFN $\overleftrightarrow{\mathcal{S}}(a,b,c,d,f_S^{\triangleleft},g_S^{\triangleleft}) = \overleftrightarrow{T\nabla}(a,b,c,d)$ with: (**a**) positive orientation, (**b**) negative orientation. Source: Own elaboration

If the condition (29) is fulfilled, then the K's-OFN $\overleftrightarrow{\mathcal{S}}(a,b,c,d,f_S^{\triangleleft},g_S^{\triangleleft})$ has a negative orientation. Negatively oriented K's-OFN is interpreted as such imprecise number, which may decrease. The space of all negatively oriented OFN is denoted by the symbol $\breve{\mathbb{K}}^-$. Some example of negatively oriented

K's-OFN is presented in the Figure 2b. If condition (30) is fulfilled, then the K's-OFN $\overleftrightarrow{\mathcal{S}}(a, b, c, d, f_S^{\triangleleft}, g_S^{\triangleleft})$ describes $a \in \mathbb{R}$ which is not oriented.

If the sequence $\{a, b, c, d\} \subset \mathbb{R}$ is not monotonic, then the membership relation $\mu_S(\cdot|a, b, c, d, f_S^{\triangleleft}, g_S^{\triangleleft}) \subset \mathbb{R} \times [0, 1]$ is not a function. Then, this membership relation cannot be considered as a membership function of any fuzzy set. Therefore, for any non-monotonic sequence $\{a, b, c, d\} \subset \mathbb{R}$ the K's-OFN $\overleftrightarrow{\mathcal{S}}(a, b, c, d, f_S^{\triangleleft}, g_S^{\triangleleft})$ is called an improper K's-OFN [4].An example of negatively oriented improper K's-OFN is presented in Figure 3. The remaining K's-OFNs are called proper ones. Some examples of proper K's-OFNs are presented in Figure 2.

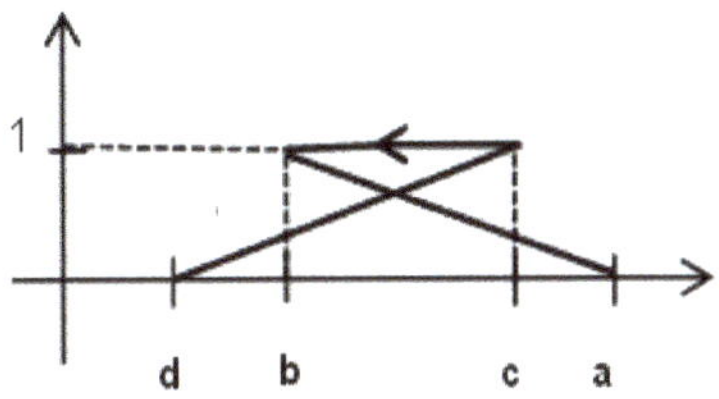

Figure 3. Membership relation of improper K's-OFN $\overleftrightarrow{\mathcal{S}}(a, b, c, d, f_S^{\triangleleft}, g_S^{\triangleleft}) = \overleftrightarrow{\mathcal{T}\nabla}(d, c, b, a)$. Source: Own elaboration.

The difference relation between K's-OFN $\overleftrightarrow{\mathcal{S}}(a, b, c, d, f_S^{\triangleleft}, g_S^{\triangleleft})$ and crisp zero is defined in the following way

$$\overleftrightarrow{\mathcal{S}}(a, b, c, d, f_S^{\triangleleft}, g_S^{\triangleleft}) \neq 0 \Leftrightarrow \min\{a,\ b,\ c,\ d\} \cdot \max\{a,\ b,\ c,\ d\ \} > 0. \tag{31}$$

We assume that $\overleftrightarrow{\mathcal{X}}, \overleftrightarrow{\mathcal{Y}}, \overleftrightarrow{\mathcal{Z}} \in \check{\mathbb{K}}$ are represented respectively by their Kosiński's pairs $(f_X, g_X), (f_Y, g_Y), (f_Z, g_Z) \in \mathbb{R}^{[0;1]}$. Any arithmetic operation $\circ$ on $\mathbb{R}$ is extended by Kosiński to arithmetic operation $\circledcirc$ on $\check{\mathbb{K}}$ by means of the identity

$$\overleftrightarrow{\mathcal{Z}} = \overleftrightarrow{\mathcal{X}} \circledcirc \overleftrightarrow{\mathcal{Y}}, \tag{32}$$

where we have

$$\forall_{\alpha \in [0;1]} \quad : \quad f_Z(\alpha) = f_X(\alpha) \circ f_Y(\alpha) \text{ and } g_Z(\alpha) = g_X(\alpha) \circ g_Y(\alpha). \tag{33}$$

The division $\oslash$ is defined only for K's-OFN $\overleftrightarrow{\mathcal{Y}} \neq 0$.

The above Kosinski's definition of arithmetic operations is coherent with the Goetschel and Voxman identity (24). On the other hand, due to the significant expansion of the arithmetic operations domain, Kosiński's definition is an important extension of the identity (24). Therefore, the above defined arithmetic operations will be called Kosiński's operations.

Due to (24), it is very easy to check that if proper K's-OFNs have identical orientation then results obtained by Kosiński's arithmetic are identical with the results obtained by means of arithmetic introduced by Dubois and Prade [19]. Moreover, Kosiński [4] has shown that if K's-OFNs have different orientation then results obtained by his arithmetic may be different form results obtained by arithmetic introduced by Dubois and Prade [19].

All our further considerations will be illustrated by examples with use of the following type of K's-OFNs.

Definition 10. *For any sequence* $\{a, b, c, d\} \subset \mathbb{R}$ *Kosiński's trapezoidal ordered fuzzy number (K's-TrOFN)* $\overleftrightarrow{T\nabla}(a, b, c, d)$ *is defined explicitly by its membership relation* $\mu_{Tr}(\cdot|a, b, c, d) \subset \mathbb{R} \times [0, 1]$ *given by the identity*

$$\mu_{T_r}(x|a,b,c,d) = \begin{cases} 0, & x \notin [a,d] = [d,a], \\ \frac{x-a}{b-a}, & x \in [a,b[=]b,a], \\ 1, & x \in [b,c] = [c,b], \\ \frac{x-d}{c-d}, & x \in]c,d] = [d,c[. \end{cases} \tag{34}$$

Examples of proper K's-TrOFNs are presented in Figure 2. An example of improper K's-TrOFN is given in Figure 3. For any K's-TrOFN $\overleftrightarrow{T\nabla}(a, b, c, d)$ its up-function f_{Tr} and down-function g_{Tr} are given by the identities

$$f_{Tr}(\alpha) = (b - a) \cdot \alpha + a, \tag{35}$$

$$g_{Tr}(\alpha) = (c - d) \cdot \alpha + d. \tag{36}$$

Among other things, it implies, that for any triple of K's-TrOFN we have

$$\overleftrightarrow{T\nabla}(a_1, b_1, c_1, d_1) \oplus \overleftrightarrow{T\nabla}(a_2, b_2, c_2, d_2) = \overleftrightarrow{T\nabla}(a_1 + a_2, b_1 + b_2, c_1 + c_2, d_1 + d_2). \tag{37}$$

4. Revision of the Kosiński's Theory

In general, if OFN is determined by such membership relation which is not a function then, it is called an improper OFN. On the other hand, if OFN is determined by such a membership relation that it is a function then it is called a proper OFN. Only in case of a proper OFN can we apply all knowledge of fuzzy sets to solve practical problems described by OFN. In this way, our practical considerations will be more fruitful. Therefore, in my opinion, limiting the OFNs' family to the family of proper OFNs is useful and absolutely necessary. Thus, the cognitive paradigm of limiting the OFNs' family to the family of proper OFNs is well justified. For this reason, the revised definition of OFN is proposed as follows.

Definition 11. *For any monotonic sequence* $\{a, b, c, d\} \subset \mathbb{R}$ *the Kosiński's pair* $S(a, b, c, d) = (f_S, g_S)$ *determines an ordered fuzzy number (OFN)* $\overleftrightarrow{S}(a, b, c, d, f_S^{\triangleleft}, g_S^{\triangleleft})$ *defined explicitly by its membership function* $\mu_S(\cdot|a, b, c, d, f_S^{\triangleleft}, g_S^{\triangleleft}) \in [0; 1]^{\mathbb{R}}$ *given by (9).*

It is obvious that any above defined OFN is proper. In agreement with the above definition, any proper K's-TrOFN is called a trapezoidal ordered fuzzy number (TrOFN). The space of all OFNs, the space of positively oriented OFNs, and the space of negatively oriented OFNs, are respectively denoted by the symbols $\mathbb{K}$, $\mathbb{K}^+$, $\mathbb{K}^-$. Thus we can now note the following relationships:

1. Any K's-OFN belonging to $\check{\mathbb{K}} \backslash \mathbb{K}$ is improper.
2. Any K's-OFN belonging to $\mathbb{K}$ is proper.
3. $\mathbb{K}^+ = \mathbb{K} \cap \check{\mathbb{K}}^+, \mathbb{K}^- = \mathbb{K} \cap \check{\mathbb{K}}^-$.

Let us look now at the following case.

Counterexample 1. *Let us calculate the Kosiński's sum of TrOFNs* $\overleftrightarrow{T\nabla}(1; 3; 7; 8)$ *and* $\overleftrightarrow{T\nabla}(5; 4; 4; 2)$. *Due to the identity (37), we have*

$$\overleftrightarrow{T\nabla}(1; 3; 7; 8) \oplus \overleftrightarrow{Tr}(5; 4; 4; 2) = \overleftrightarrow{Tr}(6; 7; 11; 10).$$

It means that the sum of TrOFNs may be equal to improper K's-TrOFN.

Similarly, the above conclusion was obtained by Kosiński [4]. We can generalize this conclusion as follows:

Conclusion: The space $\mathbb{K}$ of all OFNs is not closed under Kosiński's operations $\circledcirc$.

Therefore, we ought to modify arithmetic of OFN in such a way that the space $\mathbb{K}$ of all OFN will be closed under revised arithmetic operations.

Let us consider the OFNs $\overleftrightarrow{\mathcal{X}}, \overleftrightarrow{\mathcal{Y}}, \overleftrightarrow{\mathcal{W}} \in \mathbb{K}$ described as follows

$$\overleftrightarrow{\mathcal{X}} = \overleftrightarrow{\mathcal{X}}(a_X, b_X, c_X, d_X, f_X^{\triangleleft}, g_X^{\triangleleft}), \quad \overleftrightarrow{\mathcal{Y}} = \overleftrightarrow{\mathcal{X}}(a_Y, b_Y, c_Y, d_Y, f_Y^{\triangleleft}, g_Y^{\triangleleft}), \quad \overleftrightarrow{\mathcal{W}} = \overleftrightarrow{\mathcal{X}}(a_W, b_W, c_W, d_W, f_W^{\triangleleft}, g_W^{\triangleleft}). \tag{38}$$

Any arithmetic operations $\circ$ on $\mathbb{R}$ is extended now to revised arithmetic operation $\boxed{\circ}$ on $\mathbb{K}$ by the identity

$$\overleftrightarrow{\mathcal{W}} = \overleftrightarrow{\mathcal{X}} \boxed{\circ} \overleftrightarrow{\mathcal{Y}}, \tag{39}$$

where we have

$$\check{a}_W = a_X \circ a_Y, \tag{40}$$

$$b_W = b_X \circ b_Y, \tag{41}$$

$$c_W = c_X \circ c_Y, \tag{42}$$

$$\check{d}_W = d_X \circ d_Y, \tag{43}$$

$$a_W = \begin{cases} \min\{\check{a}_W, b_W\}, & (b_W < c_W) \vee (b_W = c_W \wedge \check{a}_W \leq \check{d}_W), \\ \max\{\check{a}_W, b_W\}, & (b_W > c_W) \vee (b_W = c_W \wedge \check{a}_W > \check{d}_W), \end{cases} \tag{44}$$

$$d_W = \begin{cases} \max\left\{\check{d}_W, c_W\right\}, & (b_W < c_W) \vee (b_W = c_W \wedge \check{a}_W \leq \check{d}_W), \\ \min\left\{\check{d}_W, c_W\right\}, & (b_W > c_W) \vee (b_W = c_W \wedge \check{a}_W > \check{d}_W), \end{cases} \tag{45}$$

$$\forall_{\alpha \in [0;1]} : \quad f_W(\alpha) = \begin{cases} f_X(\alpha) \circ f_Y(\alpha), & a_W \neq b_W, \\ b_W, & a_W = b_W, \end{cases} \tag{46}$$

$$\forall_{\alpha \in [0;1]} : \quad g_W(\alpha) = \begin{cases} g_X(\alpha) \circ g_Y(\alpha), & c_W \neq d_W, \\ c_W, & c_W = d_W. \end{cases} \tag{47}$$

Example 1. *Some cases of modified sum $\boxplus$ are presented below. Let us observe that in all these cases the Kosiński's sum $\oplus$ does not exist.*

$$\begin{aligned}
&\overleftrightarrow{Tr}(1;\ 2;\ 4;6) \boxplus \overleftrightarrow{Tr}(5;3;2;1) = \overleftrightarrow{Tr}(5;5;6;7), \ \overleftrightarrow{Tr}(6;\ 4;\ 2;1) \boxplus \overleftrightarrow{Tr}(1;2;3;5) = \overleftrightarrow{Tr}(7;6;5;5), \\
&\overleftrightarrow{Tr}(1;\ 2;\ 4;4) \boxplus \overleftrightarrow{Tr}(5;3;2;1) = \overleftrightarrow{Tr}(5;5;6;6), \ \overleftrightarrow{Tr}(4;\ 4;\ 2;1) \boxplus \overleftrightarrow{Tr}(1;2;3;5) = \overleftrightarrow{Tr}(6;6;5;5), \\
&\overleftrightarrow{Tr}(1;\ 2;\ 3;4) \boxplus \overleftrightarrow{Tr}(6;3;2;2) = \overleftrightarrow{Tr}(7;5;5;5), \ \overleftrightarrow{Tr}(1;\ 3;\ 7;8) \boxplus \overleftrightarrow{Tr}(5;4;4;2) = \overleftrightarrow{Tr}(6;7;11;11).
\end{aligned}$$

Theorem 5. *Any revised arithmetic operation $\boxed{\circ}$ satisfies the condition*

$$\forall_{\overleftrightarrow{\mathcal{X}}, \overleftrightarrow{\mathcal{Y}} \in \mathbb{K}^2} : \quad \overleftrightarrow{\mathcal{X}} \circledcirc \overleftrightarrow{\mathcal{Y}} \in \mathbb{K} \Rightarrow \overleftrightarrow{\mathcal{X}} \circledcirc \overleftrightarrow{\mathcal{Y}} = \overleftrightarrow{\mathcal{X}} \boxed{\circ} \overleftrightarrow{\mathcal{Y}}. \tag{48}$$

Proof. We take into account such OFNs $\overleftrightarrow{\mathcal{X}}, \overleftrightarrow{\mathcal{Y}}, \overleftrightarrow{\mathcal{W}} \in \mathbb{K}$ described by (38) that the condition (39) is fulfilled. Let K's-OFN $\overleftrightarrow{\mathcal{Z}} \in \check{\mathbb{K}}$ be given by the identity (32). Then, according to (33), the sum $\overleftrightarrow{\mathcal{Z}}$ is described by K's-OFN $\overleftrightarrow{\mathcal{Z}}(\check{a}_W, b_W, c_W, \check{d}_W, f_Z^{\triangleleft}, g_Z^{\triangleleft})$, where the sequence $\left\{\check{a}_W, b_W, c_W, \check{d}_W\right\}$ is determined by Equations (40)–(43) and the Kosiński's pair (f_Z, g_Z) is determined by (32). Additionally, let us note that

$$\check{a}_W = b_W \Rightarrow \forall_{\alpha\in[0;1]} \quad : \quad f_Z(\alpha) = b_W, \tag{49}$$

$$\check{d}_W = c_W \Rightarrow \forall_{\alpha\in[0;1]} \quad : \quad g_Z(\alpha) = c_W. \tag{50}$$

If $\overleftrightarrow{\mathcal{Z}} \in \mathbb{K}$ then the sequence $\left\{\check{a}_W, b_W, c_W, \check{d}_W\right\}$ is monotonic. For this reason, it is sufficient to consider the following three cases

$$\check{a}_W \le b_W \le c_W \le \check{d}_W, \tag{51}$$

$$\check{a}_W \ge b_W \ge c_W \ge \check{d}_W \quad \text{and} \quad \check{a}_W > \check{d}_W, \tag{52}$$

$$\check{a}_W \ge b_W \ge c_W \ge \check{d}_W \quad \text{and} \quad \check{a}_W = \check{d}_W\ . \tag{53}$$

For case (51), by (44) and (45), we obtain

$$a_W = \min\{\check{a}_W, b_W\} = \check{a}_W \quad \text{and} \quad d_W = \max\left\{\check{d}_W, c_W\right\} = \check{d}_W.$$

Thanks to this, from Equations (46), (47), (49) and (50) we obtain following conclusions

$$\forall_{\alpha\in[0;1]} \quad : \quad f_W(\alpha) = \left\{ \begin{array}{ll} f_X(\alpha) \circ f_Y(\alpha), & \check{a}_W < b_W \\ b_W, & \check{a}_W = b_W \end{array} \right\} = f_Z(\alpha),$$

$$\forall_{\alpha\in[0;1]} \quad : \quad g_W(\alpha) = \left\{ \begin{array}{ll} g_X(\alpha) \circ g_Y(\alpha), & c_W \ne \check{d}_W \\ c_W, & c_W = \check{d}_W \end{array} \right\} = g_Z(\alpha).$$

All this together implies that for the case (51) we have

$$\overleftrightarrow{\mathcal{Z}} = \overleftrightarrow{\mathcal{Z}}(a_W, b_W, c_W, d_W, f_W^{\triangleleft}, g_W^{\triangleleft}) = \overleftrightarrow{\mathcal{W}}. \tag{54}$$

For case (52), given by (44) and (45), we obtain

$$a_W = \max\{\check{a}_W, b_W\} = \check{a}_W \quad \text{and} \quad d_W = \min\left\{\check{d}_W, c_W\right\} = \check{d}_W.$$

Then using (46), (47), (49), and (50) we obtain

$$\forall_{\alpha\in[0;1]} \quad : \quad f_W(\alpha) = f_X(\alpha) \circ f_Y(\alpha) = f_Z(\alpha),$$

$$\forall_{\alpha\in[0;1]} \quad : \quad g_W(\alpha) = g_X(\alpha) \circ g_Y(\alpha) = g_Z(\alpha).$$

It implies that also for the case (52) we have (54). For the case (53) we have $\check{a}_Z = b_Z = c_Z = \check{d}_Z$. It means that the condition (53) implies the condition (51). The proof is completed. □

Along with the Kosiński's definition (32) and (33) and the Goetschel–Voxman combined together, the thesis the above theorem shows that the revised arithmetic operations are well defined for each corresponding operation on $\mathbb{R}$.

Theorem 6. *If for any pair* $(\overleftrightarrow{\mathcal{X}}, \overleftrightarrow{\mathcal{Y}}) \in \mathbb{K}^2$ *the K's-OFN* $\overleftrightarrow{\mathcal{Z}} = \overleftrightarrow{\mathcal{Z}}(a_Z, b_Z, c_Z, d_Z, f_Z^{\triangleleft}, g_Z^{\triangleleft})$ *is calculated by (32) then the OFN* $\overleftrightarrow{\mathcal{W}}$ *determined by (39) is given as follows*

$$\overleftrightarrow{\mathcal{W}} = \overleftrightarrow{\mathcal{W}}(a_W, b_Z, c_Z, d_W, f_W^{\triangleleft}, g_W^{\triangleleft}), \tag{55}$$

where we have

$$|a_W - a_Z| + |d_W - d_Z| = \gamma(\overleftrightarrow{\mathcal{Z}}) = \min\left\{|a - a_Z| + |d - d_Z| : \overleftrightarrow{\mathcal{S}}(a, b_Z, c_Z, d, f_S^{\triangleleft}, g_S^{\triangleleft}) \in \mathbb{K}\right\}, \quad (56)$$

$$\forall_{\alpha\in[0;1]} : \quad f_W(\alpha) = \begin{cases} f_Z(\alpha), & a_W \neq b_Z, \\ b_Z, & a_W = b_Z, \end{cases} \quad (57)$$

$$\forall_{\alpha\in[0;1]} : \quad g_W(\alpha) = \begin{cases} g_Z(\alpha), & c_Z \neq d_Z, \\ c_Z, & c_Z = d_Z. \end{cases} \quad (58)$$

Proof. The identity (57) immediately follows (46) together with (33) and (41). Similarly, taking into account (47) together with (33) and (42) we obtain the identity (58).

If $\overleftrightarrow{\mathcal{Z}} \in \mathbb{K}$ then any OFN $\overleftrightarrow{\mathcal{S}}(a_Z, b_Z, c_Z, d_Z, f_S^{\triangleleft}, g_S^{\triangleleft}) \in \mathbb{K}$. It implies that $a_W = a_Z$ and $d_W = d_Z$ because we have

$$\gamma(\overleftrightarrow{\mathcal{Z}}) = \min\{|a_Z - a_Z| + |d_Z - d_Z| : \overleftrightarrow{\mathcal{S}}(a_Z, b_Z, c_Z, d_Z, f_S^{\triangleleft}, g_S^{\triangleleft}) \in \mathbb{K}\} = |a_Z - a_Z| + |d_Z - d_Z|.$$

Taking into account all above simple conclusions, we obtain $\overleftrightarrow{\mathcal{W}} = \overleftrightarrow{\mathcal{W}}(a_Z, b_Z, c_Z, d_Z, f_Z^{\triangleleft}, g_Z^{\triangleleft})$ which is in accordance with the Theorem 5 If $\overleftrightarrow{\mathcal{Z}} \notin \mathbb{K}$, then we ought to consider the following 10 cases

$$b_Z > c_Z \geq d_Z \text{ and } a_Z < b_Z, \quad (59)$$

$$a_Z \geq b_Z > c_Z \text{ and } d_Z > c_Z, \quad (60)$$

$$b_Z > c_Z \text{ and } a_Z < b_Z \text{ and } d_Z > c_Z, \quad (61)$$

$$b_Z = c_Z \text{ and } a_Z \leq d_Z < b_Z, \quad (62)$$

$$b_Z = c_Z \text{ and } d_Z < a_Z < b_Z, \quad (63)$$

$$b_Z = c_Z \text{ and } d_Z \geq a_Z > b_Z, \quad (64)$$

$$b_Z = c_Z \text{ and } a_Z > d_Z > b_Z, \quad (65)$$

$$b_Z < c_Z \leq d_Z \text{ and } a_Z > b_Z, \quad (66)$$

$$a_Z \leq b_Z < c_Z \text{ and } d_Z < c_Z. \quad (67)$$

$$b_Z < c_Z \text{ and } a_Z > b_Z \text{ and } d_Z < c_Z. \quad (68)$$

For case (59), we see that if $a \geq b_Z$ then any OFN $\overleftrightarrow{\mathcal{S}}(a, b_Z, c_Z, d_Z, f_S^{\triangleleft}, g_S^{\triangleleft}) \in \mathbb{K}$. Thus, we obtain

$$\gamma(\overleftrightarrow{\mathcal{Z}}) = \min\{|a - a_Z| + |d_Z - d_Z| : \overleftrightarrow{\mathcal{S}}(a, b_Z, c_Z, d_Z, f_S^{\triangleleft}, g_S^{\triangleleft}) \in \mathbb{K},\ a \geq b_Z\} = |b_Z - a_Z| + |d_Z - d_Z|.$$

We see that $a_W = b_Z$ and $d_W = d_Z$. For case (60), we notice that if $d \leq c_Z$ then any OFN $\overleftrightarrow{\mathcal{S}}(a_Z, b_Z, c_Z, d, f_S^{\triangleleft}, g_S^{\triangleleft}) \in \mathbb{K}$. Thus we obtain

$$\gamma(\overleftrightarrow{\mathcal{Z}}) = \min\{|a_Z - a_Z| + |d - d_Z| : \overleftrightarrow{\mathcal{S}}(a_Z, b_Z, c_Z, d, f_S^{\triangleleft}, g_S^{\triangleleft}) \in \mathbb{K},\ d \leq c_Z\} = |a_Z - a_Z| + |c_Z - d_Z|.$$

We see that $a_W = a_Z$ and $d_W = c_Z$. For case (61), we see that if $a \geq b_Z$ and $d \leq c_Z$ then any OFN $\overleftrightarrow{\mathcal{S}}(a, b_Z, c_Z, d, f_S^{\triangleleft}, g_S^{\triangleleft}) \in \mathbb{K}$. Thus we obtain

$$\gamma(\overleftrightarrow{\mathcal{Z}}) = \min\{|a - a_Z| + |d - d_Z| : \overleftrightarrow{\mathcal{S}}(a, b_Z, c_Z, d, f_S^{\triangleleft}, g_S^{\triangleleft}) \in \mathbb{K},\ a \geq b_Z,\ d \leq c_Z,\} = |b_Z - a_Z| + |c_Z - d_Z|.$$

We see that $a_W = b_Z$ and $d_W = c_Z$. For case (62), we notice that if $a \geq b_Z$ then any OFN $\overset{\leftrightarrow}{S}(a, b_Z, b_Z, d_Z, f_S^{\lhd}, g_S^{\lhd}) \in \mathbb{K}$ and if $d \geq c_Z$ then any OFN $\overset{\leftrightarrow}{S}(a_Z, c_Z, c_Z, d, f_S^{\lhd}, g_S^{\lhd}) \in \mathbb{K}$ Thus we obtain

$$\begin{aligned}\gamma(\overset{\leftrightarrow}{Z}) &= \min\left\{\begin{array}{l}\min\left\{|a - a_Z| + |d_Z - d_Z| : \overset{\leftrightarrow}{S}(a, b_Z, b_Z, d_Z, f_S^{\lhd}, g_S^{\lhd}) \in \mathbb{K},\ a \geq b_Z\right\}, \\ \min\left\{|a_Z - a_Z| + |d - d_Z| : \overset{\leftrightarrow}{S}(a_Z, c_Z, c_Z, d, f_S^{\lhd}, g_S^{\lhd}) \in \mathbb{K},\ d \geq c_Z\right\}\end{array}\right\} \\ &= \min\{|b_Z - a_Z| + |d_Z - d_Z|,\ |a_Z - a_Z| + |c_Z - d_Z|\} = |a_Z - a_Z| + |c_Z - d_Z|.\end{aligned}$$

We see that $a_W = a_Z$ and $d_W = c_Z$. For case (63), we notice that if $a \geq b_Z$ then any OFN $\overset{\leftrightarrow}{S}(a, b_Z, b_Z, d_Z, f_S^{\lhd}, g_S^{\lhd}) \in \mathbb{K}$ and if $d \geq c_Z$ then any OFN $\overset{\leftrightarrow}{S}(a_Z, c_Z, c_Z, d, f_S^{\lhd}, g_S^{\lhd}) \in \mathbb{K}$ Thus we obtain

$$\begin{aligned}\gamma(\overset{\leftrightarrow}{Z}) &= \min\left\{\begin{array}{l}\min\{|a - a_Z| + |d_Z - d_Z| : \overset{\leftrightarrow}{S}(a, b_Z, b_Z, d_Z, f_S^{\lhd}, g_S^{\lhd}) \in \mathbb{K},\ a \geq b_Z\}, \\ \min\{|a_Z - a_Z| + |d - d_Z| : \overset{\leftrightarrow}{S}(a_Z, c_Z, c_Z, d, f_S^{\lhd}, g_S^{\lhd}) \in \mathbb{K},\ d \geq c_Z\}\end{array}\right\} \\ &= \min\{|b_Z - a_Z| + |d_Z - d_Z|,\ |a_Z - a_Z| + |c_Z - d_Z|\} = |a_Z - a_Z| + |c_Z - d_Z|.\end{aligned}$$

We see that $a_W = a_Z$ and $d_W = c_Z$. For case (64), we notice that if $a \leq b_Z$ then any OFN $\overset{\leftrightarrow}{S}(a, b_Z, b_Z, d_Z, f_S^{\lhd}, g_S^{\lhd}) \in \mathbb{K}$ and if $d \leq c_Z$ then any OFN $\overset{\leftrightarrow}{S}(a_Z, c_Z, c_Z, d, f_S^{\lhd}, g_S^{\lhd}) \in \mathbb{K}$ Thus we obtain

$$\begin{aligned}\gamma(\overset{\leftrightarrow}{Z}) &= \min\left\{\begin{array}{l}\min\{|a - a_Z| + |d_Z - d_Z| : \overset{\leftrightarrow}{S}(a, b_Z, b_Z, d_Z, f_S^{\lhd}, g_S^{\lhd}) \in \mathbb{K},\ a \leq b_Z\}, \\ \min\{|a_Z - a_Z| + |d - d_Z| : \overset{\leftrightarrow}{S}(a_Z, c_Z, c_Z, d, f_S^{\lhd}, g_S^{\lhd}) \in \mathbb{K},\ d \leq c_Z\}\end{array}\right\} \\ &= \min\{|b_Z - a_Z| + |d_Z - d_Z|,\ |a_Z - a_Z| + |c_Z - d_Z|\} = |a_Z - a_Z| + |c_Z - d_Z|.\end{aligned}$$

We see that $a_W = a_Z$ and $d_W = c_Z$. For case (65), we notice that if $a \leq b_Z$ then any OFN $\overset{\leftrightarrow}{S}(a, b_Z, b_Z, d_Z, f_S^{\lhd}, g_S^{\lhd}) \in \mathbb{K}$ and if $d \leq c_Z$ then any OFN $\overset{\leftrightarrow}{S}(a_Z, c_Z, c_Z, d, f_S^{\lhd}, g_S^{\lhd}) \in \mathbb{K}$ Thus we obtain

$$\begin{aligned}\gamma(\overset{\leftrightarrow}{Z}) &= \min\left\{\begin{array}{l}\min\left\{|a - a_Z| + |d_Z - d_Z| : \overset{\leftrightarrow}{S}(a, b_Z, b_Z, d_Z, f_S^{\lhd}, g_S^{\lhd}) \in \mathbb{K},\ a \leq b_Z\right\}, \\ \min\left\{|a_Z - a_Z| + |d - d_Z| : \overset{\leftrightarrow}{S}(a_Z, c_Z, c_Z, d, f_S^{\lhd}, g_S^{\lhd}) \in \mathbb{K},\ d \leq c_Z\right\}\end{array}\right\} \\ &= \min\{|b_Z - a_Z| + |d_Z - d_Z|,\ |a_Z - a_Z| + |c_Z - d_Z|\} = |a_Z - a_Z| + |c_Z - d_Z|.\end{aligned}$$

We see that $a_W = a_Z$ and $d_W = c_Z$. For case (66), we notice that if $a \leq b_Z$ then any OFN $\overset{\leftrightarrow}{S}(a, b_Z, c_Z, d_Z, f_S^{\lhd}, g_S^{\lhd}) \in \mathbb{K}$. Thus we obtain

$$\gamma(\overset{\leftrightarrow}{Z}) = \min\{|a - a_Z| + |d_Z - d_Z| : \overset{\leftrightarrow}{S}(a, b_Z, c_Z, d_Z, f_S^{\lhd}, g_S^{\lhd}) \in \mathbb{K},\ a \leq b_Z\} = |b_Z - a_Z| + |d_Z - d_Z|.$$

We see that $a_W = b_Z$ and $d_W = d_Z$. For case (67), we notice that if $d \geq c_Z$ then any OFN $\overset{\leftrightarrow}{S}(a_Z, b_Z, c_Z, d, f_S^{\lhd}, g_S^{\lhd}) \in \mathbb{K}$. Thus we obtain

$$\gamma(\overset{\leftrightarrow}{Z}) = \min\{|a_Z - a_Z| + |d - d_Z| : \overset{\leftrightarrow}{S}(a_Z, b_Z, c_Z, d, f_S^{\lhd}, g_S^{\lhd}) \in \mathbb{K},\ d \geq c_Z\} = |a_Z - a_Z| + |c_Z - d_Z|.$$

We see that $a_W = a_Z$ and $d_W = c_Z$. For case (68), we notice that if $a \leq b_Z$ and $d \geq c_Z$ then any OFN $\overset{\leftrightarrow}{S}(a, b_Z, c_Z, d, f_S^{\lhd}, g_S^{\lhd}) \in \mathbb{K}$. Thus we obtain

$$\gamma(\overset{\leftrightarrow}{Z}) = \min\{|a - a_Z| + |d - d_Z| : \overset{\leftrightarrow}{S}(a, b_Z, c_Z, d, f_S^{\lhd}, g_S^{\lhd}) \in \mathbb{K},\ a \leq b_Z,\ d \geq c_Z, \} = |b_Z - a_Z| + |c_Z - d_Z|.$$

We see that $a_W = b_Z$ and $d_W = c_Z$. For each above considered case, we have obtained results by means of the identities (44) and (45). The proof is completed. □

The space of all OFNs is closed under revised arithmetic operations. The last theorem shows that results obtained by using revised arithmetic operations are the best approximations of results obtained by means of Kosiński's arithmetic operations.

5. Linear Space of Ordered Fuzzy Numbers

Let us presume the OFNs $\overset{\leftrightarrow}{\mathcal{X}}$, $\overset{\leftrightarrow}{\mathcal{Y}}$, $\overset{\leftrightarrow}{\mathcal{W}} \in \mathbb{K}$ described by (38). For any pair, $(\beta, \overset{\leftrightarrow}{\mathcal{X}}) \in \mathbb{R} \times \mathbb{K}$ we define its dot product $\odot$ in a following way

$$\beta \odot \overset{\leftrightarrow}{\mathcal{X}} = \overset{\leftrightarrow}{\mathcal{S}}(\beta, \beta, \beta, \beta, f_S^{\lhd}, g_S^{\lhd}) \boxed{*} \overset{\leftrightarrow}{\mathcal{X}}. \tag{69}$$

Let us note that, equivalently, we can define the dot product $\odot$ as follows

$$\beta \odot \overset{\leftrightarrow}{\mathcal{X}} = \overset{\leftrightarrow}{\mathcal{S}}(\beta, \beta, \beta, \beta, f_S^{\lhd}, g_S^{\lhd}) \circledast \overset{\leftrightarrow}{\mathcal{X}} \tag{70}$$

because of the sequence $\{\beta \cdot a_X, \beta \cdot b_X, \beta \cdot c_X, \beta \cdot d_X\}$ is always a monotonic one.

The sum of OFNs is determined by revised addition $\boxed{+}$ defined by (39). The main link between the revised addition $\boxed{+}$ and the Kosiński's addition $\oplus$ is described in the Theorem 5. Here, we can indicate further additional relationships between the revised addition $\boxed{+}$ and the Kosiński's addition $\oplus$. Because for every OFNs $\overset{\leftrightarrow}{\mathcal{X}}$, $\overset{\leftrightarrow}{\mathcal{Y}} \in \mathbb{K}^+$ the sequence $\{a_X + a_Y, b_X + b_Y, c_X + c_Y, d_X + d_Y\}$ is always nondecreasing, we can say that

$$\forall_{\overset{\leftrightarrow}{\mathcal{X}}, \overset{\leftrightarrow}{\mathcal{Y}} \in \mathbb{K}^+} : \qquad \overset{\leftrightarrow}{\mathcal{X}} \oplus \overset{\leftrightarrow}{\mathcal{Y}} = \overset{\leftrightarrow}{\mathcal{X}} \boxed{+} \overset{\leftrightarrow}{\mathcal{Y}} \in \mathbb{K}^+. \tag{71}$$

Because for every OFNs $\overset{\leftrightarrow}{\mathcal{X}}$, $\overset{\leftrightarrow}{\mathcal{Y}} \in \mathbb{K}^-$ the sequence $\{a_X + a_Y, b_X + b_Y, c_X + c_Y, d_X + d_Y\}$ is always nonincreasing, we can say that

$$\forall_{\overset{\leftrightarrow}{\mathcal{X}}, \overset{\leftrightarrow}{\mathcal{Y}} \in \mathbb{K}^-} : \qquad \overset{\leftrightarrow}{\mathcal{X}} \oplus \overset{\leftrightarrow}{\mathcal{Y}} = \overset{\leftrightarrow}{\mathcal{X}} \boxed{+} \overset{\leftrightarrow}{\mathcal{Y}} \in \mathbb{K}^-. \tag{72}$$

If the sum is determined by revised addition $\boxed{+}$, then for any OFN $\overset{\leftrightarrow}{\mathcal{X}}$ its opposite element $-\overset{\leftrightarrow}{\mathcal{X}}$ is defined by the identity

$$-\overset{\leftrightarrow}{\mathcal{X}} = (-1) \odot \overset{\leftrightarrow}{\mathcal{X}}. \tag{73}$$

Then the neutral element $\overset{\leftrightarrow}{0}$ is described as follows

$$\overset{\leftrightarrow}{0} = \overset{\leftrightarrow}{0}(0; 0; 0; 0;\ f_0^{\lhd};\ g_0^{\lhd}). \tag{74}$$

For Kosiński's addition $\oplus$ its opposite elements and neutral elements are determined in the same way [4]. The identities (39)–(47) together with the commutativity of addition $+$ imply that the revised addition $\boxed{+}$ is commutative. The addition of FN and the Kosiński's sum $\oplus$ of OFN are associative and commutative [4,13,19]. The associativity of revised addition of $\boxed{+}$ will be investigated using the following special kind of TrOFN.

Definition 12. *For any monotonic sequence $\{a, b, c\} \subset \mathbb{R}$ a triangular OFN (TOFN) $\overset{\leftrightarrow}{\mathcal{T}}(a, b, c)$ is defined by the identity*

$$\overset{\leftrightarrow}{\mathcal{T}}(a, b, c) = \overset{\leftrightarrow}{\mathcal{T}\nabla}(a, b,\ b, c). \tag{75}$$

Let us take into account the following counterexample.

Counterexample 2. *Let the following four TOFN be given by*

$$\overset{\leftrightarrow}{\mathcal{A}} = \overset{\leftrightarrow}{\mathcal{T}}(10;40;70);\ \overset{\leftrightarrow}{\mathcal{B}} = \overset{\leftrightarrow}{\mathcal{T}}(110;100;60);\ \overset{\leftrightarrow}{\mathcal{C}} = \overset{\leftrightarrow}{\mathcal{T}}(50;65;105);\ \overset{\leftrightarrow}{\mathcal{D}} = \overset{\leftrightarrow}{\mathcal{T}}(120;90;67).$$

The number of different ways of associating three applications of the addition operator $\boxed{+}$ *is equal the Catalan number* $C_3 = 5$. *Therefore, we have the following five different associations of four summands* [21]

$$(\overset{\leftrightarrow}{\mathcal{A}}\boxed{+}\overset{\leftrightarrow}{\mathcal{B}})\boxed{+}(\overset{\leftrightarrow}{\mathcal{C}}\boxed{+}\overset{\leftrightarrow}{\mathcal{D}});\ ((\overset{\leftrightarrow}{\mathcal{A}}\boxed{+}\overset{\leftrightarrow}{\mathcal{B}})\boxed{+}\overset{\leftrightarrow}{\mathcal{C}})\boxed{+}\overset{\leftrightarrow}{\mathcal{D}};\ (\overset{\leftrightarrow}{\mathcal{A}}\boxed{+}(\overset{\leftrightarrow}{\mathcal{B}}\boxed{+}\overset{\leftrightarrow}{\mathcal{C}}))\boxed{+}\overset{\leftrightarrow}{\mathcal{D}};\ (\overset{\leftrightarrow}{\mathcal{A}}\boxed{+}(\overset{\leftrightarrow}{\mathcal{B}}\boxed{+}\overset{\leftrightarrow}{\mathcal{C}}))\boxed{+}\overset{\leftrightarrow}{\mathcal{D}};\ \overset{\leftrightarrow}{\mathcal{A}}\boxed{+}(\overset{\leftrightarrow}{\mathcal{B}}\boxed{+}(\overset{\leftrightarrow}{\mathcal{C}}\boxed{+}\overset{\leftrightarrow}{\mathcal{D}})).$$

In [22], it is shown that we have

$$(\overset{\leftrightarrow}{\mathcal{A}}\boxed{+}\overset{\leftrightarrow}{\mathcal{B}})\boxed{+}(\overset{\leftrightarrow}{\mathcal{C}}\boxed{+}\overset{\leftrightarrow}{\mathcal{D}}) = \overset{\leftrightarrow}{\mathcal{A}}\boxed{+}(\overset{\leftrightarrow}{\mathcal{B}}\boxed{+}(\overset{\leftrightarrow}{\mathcal{C}}\boxed{+}\overset{\leftrightarrow}{\mathcal{D}})) = \overset{\leftrightarrow}{\mathcal{T}}(275;295;302),$$

$$(\overset{\leftrightarrow}{\mathcal{A}}\boxed{+}(\overset{\leftrightarrow}{\mathcal{B}}\boxed{+}\overset{\leftrightarrow}{\mathcal{C}}))\boxed{+}\overset{\leftrightarrow}{\mathcal{D}} = \overset{\leftrightarrow}{\mathcal{A}}\boxed{+}((\overset{\leftrightarrow}{\mathcal{B}}\boxed{+}\overset{\leftrightarrow}{\mathcal{C}})\boxed{+}\overset{\leftrightarrow}{\mathcal{D}}) = \overset{\leftrightarrow}{\mathcal{T}}(290;295;302),$$

$$((\overset{\leftrightarrow}{\mathcal{A}}\boxed{+}\overset{\leftrightarrow}{\mathcal{B}})\boxed{+}\overset{\leftrightarrow}{\mathcal{C}})\boxed{+}\overset{\leftrightarrow}{\mathcal{D}} = \overset{\leftrightarrow}{\mathcal{T}}(290;295;312).$$

Many mathematical applications require a finite multiple sum to be independent of summands' ordering. Any associative and commutative sum satisfies this property. The results of above counterexample prove that revised addition $\boxed{+}$ of OFN is not associative. Therefore, we should check whether the multiple sum determined by $\boxed{+}$ depends on its summands' ordering.

Counterexample 3. *For all permutations of TOFN* $\overset{\leftrightarrow}{\mathcal{A}}, \overset{\leftrightarrow}{\mathcal{B}}, \overset{\leftrightarrow}{\mathcal{C}}, \overset{\leftrightarrow}{\mathcal{D}} \in \mathbb{K}$ *described in the Counterexample 2, we determine their multiple sum. In [22], it is shown that*

$$\begin{aligned}\overset{\leftrightarrow}{\mathcal{C}}\boxed{+}\overset{\leftrightarrow}{\mathcal{D}}\boxed{+}\overset{\leftrightarrow}{\mathcal{A}}\boxed{+}\overset{\leftrightarrow}{\mathcal{B}} &= \overset{\leftrightarrow}{\mathcal{C}}\boxed{+}\overset{\leftrightarrow}{\mathcal{D}}\boxed{+}\overset{\leftrightarrow}{\mathcal{B}}\boxed{+}\overset{\leftrightarrow}{\mathcal{A}} = \overset{\leftrightarrow}{\mathcal{D}}\boxed{+}\overset{\leftrightarrow}{\mathcal{A}}\boxed{+}\overset{\leftrightarrow}{\mathcal{B}}\boxed{+}\overset{\leftrightarrow}{\mathcal{C}} = \overset{\leftrightarrow}{\mathcal{D}}\boxed{+}\overset{\leftrightarrow}{\mathcal{C}}\boxed{+}\overset{\leftrightarrow}{\mathcal{A}}\boxed{+}\overset{\leftrightarrow}{\mathcal{B}} = \overset{\leftrightarrow}{\mathcal{D}}\boxed{+}\overset{\leftrightarrow}{\mathcal{C}}\boxed{+}\overset{\leftrightarrow}{\mathcal{B}}\boxed{+}\overset{\leftrightarrow}{\mathcal{A}}\\ &= \overset{\leftrightarrow}{\mathcal{T}}(275;295;302),\end{aligned}$$

$$\begin{aligned}\overset{\leftrightarrow}{\mathcal{A}}\boxed{+}\overset{\leftrightarrow}{\mathcal{C}}\boxed{+}\overset{\leftrightarrow}{\mathcal{B}}\boxed{+}\overset{\leftrightarrow}{\mathcal{D}} &= \overset{\leftrightarrow}{\mathcal{A}}\boxed{+}\overset{\leftrightarrow}{\mathcal{C}}\boxed{+}\overset{\leftrightarrow}{\mathcal{D}}\boxed{+}\overset{\leftrightarrow}{\mathcal{B}} = \overset{\leftrightarrow}{\mathcal{A}}\boxed{+}\overset{\leftrightarrow}{\mathcal{D}}\boxed{+}\overset{\leftrightarrow}{\mathcal{B}}\boxed{+}\overset{\leftrightarrow}{\mathcal{C}} = \overset{\leftrightarrow}{\mathcal{A}}\boxed{+}\overset{\leftrightarrow}{\mathcal{D}}\boxed{+}\overset{\leftrightarrow}{\mathcal{C}}\boxed{+}\overset{\leftrightarrow}{\mathcal{B}} = \overset{\leftrightarrow}{\mathcal{B}}\boxed{+}\overset{\leftrightarrow}{\mathcal{C}}\boxed{+}\overset{\leftrightarrow}{\mathcal{D}}\boxed{+}\overset{\leftrightarrow}{\mathcal{A}}\\ &= \overset{\leftrightarrow}{\mathcal{B}}\boxed{+}\overset{\leftrightarrow}{\mathcal{D}}\boxed{+}\overset{\leftrightarrow}{\mathcal{A}}\boxed{+}\overset{\leftrightarrow}{\mathcal{C}} = \overset{\leftrightarrow}{\mathcal{B}}\boxed{+}\overset{\leftrightarrow}{\mathcal{D}}\boxed{+}\overset{\leftrightarrow}{\mathcal{C}}\boxed{+}\overset{\leftrightarrow}{\mathcal{A}} = \overset{\leftrightarrow}{\mathcal{C}}\boxed{+}\overset{\leftrightarrow}{\mathcal{A}}\boxed{+}\overset{\leftrightarrow}{\mathcal{B}}\boxed{+}\overset{\leftrightarrow}{\mathcal{D}} = \overset{\leftrightarrow}{\mathcal{C}}\boxed{+}\overset{\leftrightarrow}{\mathcal{A}}\boxed{+}\overset{\leftrightarrow}{\mathcal{D}}\boxed{+}\overset{\leftrightarrow}{\mathcal{B}} = \overset{\leftrightarrow}{\mathcal{C}}\boxed{+}\overset{\leftrightarrow}{\mathcal{B}}\boxed{+}\overset{\leftrightarrow}{\mathcal{A}}\boxed{+}\overset{\leftrightarrow}{\mathcal{D}}\\ &= \overset{\leftrightarrow}{\mathcal{C}}\boxed{+}\overset{\leftrightarrow}{\mathcal{B}}\boxed{+}\overset{\leftrightarrow}{\mathcal{D}}\boxed{+}\overset{\leftrightarrow}{\mathcal{A}} = \overset{\leftrightarrow}{\mathcal{D}}\boxed{+}\overset{\leftrightarrow}{\mathcal{A}}\boxed{+}\overset{\leftrightarrow}{\mathcal{B}}\boxed{+}\overset{\leftrightarrow}{\mathcal{C}} = \overset{\leftrightarrow}{\mathcal{D}}\boxed{+}\overset{\leftrightarrow}{\mathcal{A}}\boxed{+}\overset{\leftrightarrow}{\mathcal{C}}\boxed{+}\overset{\leftrightarrow}{\mathcal{B}} = \overset{\leftrightarrow}{\mathcal{D}}\boxed{+}\overset{\leftrightarrow}{\mathcal{B}}\boxed{+}\overset{\leftrightarrow}{\mathcal{A}}\boxed{+}\overset{\leftrightarrow}{\mathcal{C}} = \overset{\leftrightarrow}{\mathcal{D}}\boxed{+}\overset{\leftrightarrow}{\mathcal{B}}\boxed{+}\overset{\leftrightarrow}{\mathcal{C}}\boxed{+}\overset{\leftrightarrow}{\mathcal{A}}\\ &= \overset{\leftrightarrow}{\mathcal{T}}(290;295;302),\end{aligned}$$

$$\begin{aligned}\overset{\leftrightarrow}{\mathcal{A}}\boxed{+}\overset{\leftrightarrow}{\mathcal{B}}\boxed{+}\overset{\leftrightarrow}{\mathcal{C}}\boxed{+}\overset{\leftrightarrow}{\mathcal{D}} &= \overset{\leftrightarrow}{\mathcal{A}}\boxed{+}\overset{\leftrightarrow}{\mathcal{B}}\boxed{+}\overset{\leftrightarrow}{\mathcal{D}}\boxed{+}\overset{\leftrightarrow}{\mathcal{C}} = \overset{\leftrightarrow}{\mathcal{B}}\boxed{+}\overset{\leftrightarrow}{\mathcal{A}}\boxed{+}\overset{\leftrightarrow}{\mathcal{C}}\boxed{+}\overset{\leftrightarrow}{\mathcal{D}} = \overset{\leftrightarrow}{\mathcal{B}}\boxed{+}\overset{\leftrightarrow}{\mathcal{A}}\boxed{+}\overset{\leftrightarrow}{\mathcal{D}}\boxed{+}\overset{\leftrightarrow}{\mathcal{C}} = \overset{\leftrightarrow}{\mathcal{B}}\boxed{+}\overset{\leftrightarrow}{\mathcal{C}}\boxed{+}\overset{\leftrightarrow}{\mathcal{A}}\boxed{+}\overset{\leftrightarrow}{\mathcal{D}}\\ &= \overset{\leftrightarrow}{\mathcal{T}}(290;295;312).\end{aligned}$$

The results of above counterexample prove that multiple sum calculated using addition $\boxed{+}$ depends on its summands' ordering.

6. Conclusions

The main goals of the article have been achieved as follows. The Definition 11 describes the OFN in a way that any OFN is proper. The arithmetic operations are revised using the identities (38)–(47). The assumption of conditions (44) and (45) guarantees that the family of all proper OFN is closed under revised arithmetic operations. An interpretation of the revised operations is described by Theorem 6, which shows that results obtained using revised arithmetic operations are the best approximation of results obtained by means of Kosiński's arithmetic operations. Thanks to this, we can apply all knowledge of fuzzy sets to solve practical problems described by proper OFN.

Unfortunately, the OFN linear space has one annoying drawback. Counterexample 3 shows that multiple sum determined by revised addition, depends on its summands' ordering. Therefore, the summands ordering should be clearly defined for each practical application of the multiple sum of finite sequence of OFN. This summands ordering must be sufficiently justified in the field of application. Some examples of such justification can be found in [23]. This proves that from a practical view point, the revised sum is useful in applications despite the fact that it is dependent on its summands' ordering.

Further research on the OFN theory should be devoted to the following problems:

- preorders on the family of all proper OFNs;
- equations determined by OFNs;
- imprecision measurement for OFN;
- reorientation of OFN.

Acknowledgments: The author is very grateful to the anonymous reviewers for their insightful and constructive comments and suggestions. Using these comments allowed me to improve this article.

Conflicts of Interest: The author declares no conflicts of interest.

References

1. Kosiński, W.; Słysz, P. Fuzzy numbers and their quotient space with algebraic operations. *Bull. Pol. Acad. Sci.* **1993**, *41*, 285–295.
2. Kosiński, W.; Prokopowicz, P.; Ślęzak, D. *Fuzzy Numbers with Algebraic Operations: Algorithmic Approach*; Klopotek, M., Wierzchoń, S.T., Michalewicz, M., Eds.; Physica Verlag: Heidelberg, Germany, 2002.
3. Kosiński, W.; Prokopowicz, P.; Ślęzak, D. Ordered fuzzy numbers. *Bull. Pol. Acad. Sci.* **2003**, *51*, 327–339.
4. Kosiński, W. On fuzzy number calculus. *Int. J. Appl. Math. Comput. Sci.* **2006**, *16*, 51–57.
5. Prokopowicz, P.; Pedrycz, W. The directed compatibility between ordered fuzzy numbers—A base tool for a direction sensitive fuzzy information processing. *Artif. Intell. Soft Comput.* **2015**, *119*, 249–259.
6. Prokopowicz, P. The directed inference for the Kosinski's fuzzy number model. In *Proceedings of the Second International Afro-European Conference for Industrial Advancement*; Springer: Berlin/Heidelberg, Germany, 2015.
7. Prokopowicz, P.; Czerniak, J.; Mikołajewski, D.; Apiecionek, Ł.; Slezak, D. *Theory and Applications of Ordered Fuzzy Number*; Springer: Berlin, Germany, 2017.
8. Kacprzak, D.; Kosiński, W.; Kosiński, W.K. Financial stock data and ordered fuzzy numbers. In *Proceedings of the International Conference on Artificial Intelligence and Soft Computing*; Springer: Berlin/Heidelberg, Germany, 2013.
9. Marszałek, A.; Burczyński, T. Ordered fuzzy candlesticks. In *Theory and Applications of Ordered Fuzzy Number*; Prokopowicz, P., Czerniak, J., Mikołajewski, D., Apiecionek, Ł., Slezak, D., Eds.; Springer: Berlin, Germany, 2017.
10. Roszkowska, E.; Kacprzak, D. The fuzzy SAW and fuzzy TOPSIS procedures based on ordered fuzzy numbers. *Inf. Sci.* **2016**, *369*, 564–584. [CrossRef]
11. Łyczkowska-Hanćkowiak, A. Behavioural present value determined by ordered fuzzy number. *SSRN Electr. J.* **2017**. [CrossRef]
12. Piasecki, K. Expected return rate determined as oriented fuzzy number. In *Proceedings of the 35th International Conference Mathematical Methods in Economics*; Prazak, P., Ed.; Gaudeamus: Hradec Kralove, Czech, 2017.
13. Goetschel, R.; Voxman, W. Elementary fuzzy calculus. *Fuzzy Set. Syst.* **1986**, *18*, 31–43. [CrossRef]
14. Dubois, D.; Prade, H. Fuzzy real algebra: Some results. *Fuzzy Set. Syst.* **1979**, *2*, 327–348. [CrossRef]
15. Zadeh, L.A. The concept of a linguistic variable and its application to approximate reasoning. Part I. Information linguistic variable. *Expert Syst. Appl.* **1975**, *36*, 3483–3488.
16. Zadeh, L.A. The concept of a linguistic variable and its application to approximate reasoning. Part II. *Inf. Sci.* **1975**, *8*, 301–357. [CrossRef]
17. Zadeh, L.A. The concept of a linguistic variable and its application to approximate reasoning. Part III. *Inf. Sci.* **1975**, *9*, 43–80. [CrossRef]
18. Zadeh, L.A. Fuzzy sets. *Inf. Control* **1965**, *8*, 338–353. [CrossRef]
19. Dubois, D.; Prade, H. Operations on fuzzy numbers. *Int. J. Syst. Sci.* **1978**, *9*, 613–629. [CrossRef]
20. Dubois, D.; Prade, H. *Fuzzy Sets and Systems: Theory and Applications*; Academic Press: New York, NY, USA, 1980.
21. Koshy, T. *Catalan Numbers with Applications*; Oxford University Press: London, UK, 2008.

22. Łyczkowska-Hanćkowiak, A.; Piasecki, K. On some difficulties in the addition of trapezoidal ordered fuzzy numbers. *arXiv* **2017**.
23. Łyczkowska-Hanćkowiak, A.; Piasecki, K. Present value of portfolio of assets with present values determined by trapezoidal ordered fuzzy numbers. *Oper. Res. Decis.*. (under review).

axioms

MDPI

Article

An Abstract Result on Projective Aggregation Functions

Juan C. Candeal

Departamento de Análisis Económico, Universidad de Zaragoza, Gran Vía 2, 50005 Zaragoza, Spain; candeal@unizar.es; Tel.: +34-976-761-822

Received: 1 March 2018; Accepted: 19 March 2018; Published: 20 March 2018

Abstract: A general characterization result of projective aggregation functions is shown, the proof of which makes use of the celebrated Arrow's theorem, thus providing a link between aggregation functions theory and social choice theory. The result can be viewed as a generalization of a theorem obtained by Kim (1990) for real-valued aggregation functions defined on the n-dimensional Euclidean space in the context of measurement theory. In addition, two applications of the core theorem of the article are shown. The first is a simple extension of the main result to the context of multi-valued aggregation functions. The second offers a new characterization of projective bijection aggregators, thus connecting aggregation operators theory with social choice.

Keywords: aggregation functions; aggregation operators; Arrow's theorem

MSC: 39B22; 91B14; 91E45

---

1. Introduction

This short note aims to further contribute to the study of the interplay existing among aggregation functions, aggregation operators and social choice. Specifically, this paper provides a general characterization result of projective aggregation functions defined on an abstract set by using a purely social choice approach; to wit, a suitable version of Arrow's theorem (see [1]).

An aggregation process basically consists of merging pieces of information (input data) to yield some output. The information is often symbolized by means of some numerical or abstract value, and the output obtained, via the data fusion, is made by means of the so-called aggregation functions. Classical aggregation functions include those of weighted means, t-norms or t-conorms, where all the input and output variables considered are real numbers. Generalizations of the concept of an aggregation function appear when considering, for example, aggregation with ordinal and nominal scales, aggregation on posets and lattices or aggregation of preferences (see, e.g., [1–5]).

All these developments have been closely linked to an increasing number of applications to many different topics such as image processing, decision making, pattern recognition, fuzzy control and social sciences, just to mention a few of them. In addition, aggregation operators theory has particular significance in all these fields of knowledge when considering, for instance, L-fuzzy sets or fuzzy sets of type 2 (for details, see [6–9]; see, also, [10–12] for certain applications of aggregation operators theory to social choice).

Aggregation functions play an important role in measurement theory, as well (see [13] for a detailed account of this discipline). In this context, a key feature has to do with the preservation of particular numerical scales used to measure both the input and the output variables, which are, in turn, real variables. The concept that reflects this property is the so-called comparison meaningfulness. Built on the seminal work of Luce (1959) (see [14]), it was shown by Kim (1990) in [15] that the only real-valued aggregation functions of n real variables that are continuous (continuity refers to the Euclidean topology), idempotent and comparison meaningful with respect to independent interval scales are the projections. This result was later generalized by other authors in several respects, noting that, in particular, continuity turned out to be an entirely redundant assumption in the previous statement. For details, see [16–21] and, more recently, [10].

In the current article, I provide a further generalization of the latter result in a much more abstract setting by studying conditions under which an aggregation function $W : X^n \to X$ turns out to be a projection, X being an arbitrary nonempty set. Whenever X has, at least, three elements, the two conditions found, that characterize such a property on W, are fullness and idempotency. This is the content of Theorem 2. Fullness tells us that, for a given n-tuple, which consists of nominal scales (a nominal scale is a bijection from X onto X) in X, the range of W over the subset of X^n obtained by the action of the given n-tuple over X, is the whole X. Thus, fullness turns out to be a property about how rich the codomain of W is under the action of a profile of nominal scales. Idempotency is a more familiar condition and tells us that whenever W is restricted to the diagonal of X^n, then the resulting function is the identity map. The main result of the paper, namely Theorem 2, appears in Section 3. It should be noted that its proof makes use of a particular version of Arrow's theorem. This version is included in Section 2 together with some basic definitions. The fact that the cardinality of X is, at least, three follows from the call to Arrow's result. In Section 4, certain consequences of Theorem 2 are shown. Firstly, a simple generalization of it is obtained in the framework of multi-valued aggregation functions, and secondly, a new characterization of projective bijection aggregators is presented. Although not developed in the article, these two consequences could be applied to the context of aggregation of preferences and aggregation of classifications, as seen, for example, in [22].

2. Preliminaries

Throughout the paper, the notations $n \in \mathbb{N}$ and $N := \{1, \ldots, n\}$ will be used. Typical elements of N will be denoted by j, k, i.

Let X be a nonempty set. Vectors of length n with values in X will be recurrently employed. The notation $a = (a_j)_{j \in N}$, or simply $a = (a_j)$, will stand for such a typical a vector.

The set of all bijections from X onto X will be denoted by $B(X)$, and the symbols ϕ, ψ, φ will represent typical elements of $B(X)$. Note that, if X is finite and $\phi \in B(X)$, then $\phi(X) := \{\phi(x) : x \in X\}$ can be viewed as a rearrangement of X. The notation $|X|$ stands for the cardinal number (for details on cardinals, see, e.g., [23]) of X.

An aggregation function is an X-valued function of n variables, each of them taking values in X; i.e., is a map $W : X^n \to X$. Certain properties on aggregation functions, that will be required in the main result of Section 3 are now presented.

Definition 1. *An aggregation function $W : X^n \to X$ is said to be:*

(1) *full if $\{W(\phi^1(x), \ldots, \phi^n(x)) : x \in X\} = X$, for every $\phi^1, \ldots, \phi^n \in B(X)$,*
(2) *idempotent if $W(x, \ldots, x) = x$, for every $x \in X$,*
(3) *projective if it is a projection, i.e., if there is $k \in N$ such that $W(a) = a_k$, for every $a = (a_j) \in X^n$.*

Another key and helpful concept that will be needed later is that of a bijection aggregator, or simply an aggregator. In order to define this concept, some previous notations will be useful. A profile is meant to be an element of $B(X)^n$. A typical profile will be denoted by $(\phi^1, \ldots, \phi^n)$. An aggregator is a map $F : B(X)^n \to B(X)$. I now present certain properties on bijection aggregators, some of which parallel those included in Definition 1 above for aggregation functions.

Definition 2. *An aggregator $F : B(X)^n \to B(X)$ is, or satisfies:*

(1) *binary independence if for every $(\phi^1, \ldots, \phi^n), (\psi^1, \ldots, \psi^n) \in B(X)^n$, $x, y \in X$ such that $\phi^j(x) = \psi^j(y)$, for all $j \in N$, it holds that $F(\phi^1, \ldots, \phi^n)(x) = F(\psi^1, \ldots, \psi^n)(y)$,*
(2) *unanimous if for every $(\phi^1, \ldots, \phi^n) \in B(X)^n$, $x, y \in X$ such that $\phi^j(x) = y$, for all $j \in N$, it holds that $F(\phi^1, \ldots, \phi^n)(x) = y$,*

(3) *projective if it is a projection, i.e., if there is* $k \in N$ *such that* $F(\phi^1, \ldots, \phi^n) = \phi^k$, *for every* $(\phi^1, \ldots, \phi^n) \in B(X)^n$.

I conclude this section by showing a version of Arrow's theorem that will be employed in a decisive manner in the proof of the main result of the paper. Arrow's theorem can be considered the cornerstone of social choice theory and has strongly influenced a large number of disciplines such as logic, mathematics, economics, psychology, sociology and politics, among others. In order to prepare such a version, certain concepts need to be introduced.

A transitive and complete (hence, reflexive) binary relation $\precsim$ defined on X is said to be a total preorder. A total order on X is an antisymmetric total preorder. Associated with $\precsim$, the asymmetric part, also called the strict part, denoted by $\prec$, is defined as the following binary relation on X: $x \prec y$ if and only if $x \precsim y$ and $\neg(y \precsim x)$ (or, equivalently, since $\precsim$ is a total preorder, $x \prec y$ if and only if $\neg(y \precsim x)$). Note that the asymmetric part of a total order defined on X turns out to be a linear order on X. Similarly, its symmetric part, denoted by $\sim$, is defined by: $x \sim y$ if and only if $x \precsim y$ and $y \precsim x$. The symmetric part of a total preorder turns out to be an equivalence relation on X. The binary relation defined on quotient set $X/\sim$ as: $[x] \precsim [y] \Leftrightarrow x \precsim y$, $(x, y \in X)$, is a total order on it (In the economics literature, a total preorder is also referred to as a preference and as a social welfare ordering in social choice theory. In addition, the asymmetric part and the symmetric part of a total preorder are called in economics the strict preference and the indifference relation, respectively. The set X is usually termed in these fields as the choice set, or the set of alternatives, or the set of social outcomes, and N stands for the individuals in the society.). A well ordering, or an ordinal, on X is a total order with the additional property that each nonempty subset of X has a first element.

Denote by $\mathcal{R}$, $\mathcal{T}$ and $\mathcal{W}$ the set of all total preorders, total orders and well orderings, respectively, defined on X. Obviously, $\mathcal{W} \subseteq \mathcal{T} \subseteq \mathcal{R}$.

Following Arrow (see [1]), a social welfare function turns out to be a map $G : \mathcal{R}^n \rightarrow \mathcal{R}$. Each n-tuple $(\precsim^1, \ldots, \precsim^n) \in \mathcal{R}^n$ is said to be a profile of individual preferences. The interpretation is that G maps profiles of individual preferences into social preferences. For each element of $G(\mathcal{R}^n)$, say $G(\precsim^1, \ldots, \precsim^n)$, the somewhat unpleasant notation $\precsim_{G(\precsim^1,\ldots,\precsim^n)}$ and $\prec_{G(\precsim^1,\ldots,\precsim^n)}$ for its asymmetric part, will be used.

Definition 3. *A social welfare function G satisfies:*

(i) *the condition of independence of irrelevant alternatives if for any two profiles of individuals preferences* $(\precsim_1^1, \ldots, \precsim_1^n), (\precsim_2^1, \ldots, \precsim_2^n) \in \mathcal{R}^n$ *and any two alternatives* $x, y \in X$ *such that* $x \precsim_1^j y \Leftrightarrow x \precsim_2^j y$, *and* $y \precsim_1^j x \Leftrightarrow y \precsim_2^j x$, *for all* $j \in N$, *it holds that* $x \precsim_{G(\precsim_1^1,\ldots,\precsim_1^n)} y \Leftrightarrow x \precsim_{G(\precsim_2^1,\ldots,\precsim_2^n)} y$ *and* $y \precsim_{G(\precsim_1^1,\ldots,\precsim_1^n)} x \Leftrightarrow y \precsim_{G(\precsim_2^1,\ldots,\precsim_2^n)} x$,

(ii) *the Pareto condition if for any profile* $(\precsim^1, \ldots, \precsim^n) \in \mathcal{R}^n$ *and any pair of alternatives* $x, y \in X$ *such that* $x \prec^j y$, *for all* $j \in N$, *it holds that* $x \prec_{G(\precsim^1,\ldots,\precsim^n)} y$.

Theorem 1. *Suppose that the cardinality of X is, at least, three. Let* $G : \mathcal{R}^n \rightarrow \mathcal{R}$ *be a social welfare function that satisfies the condition of independence of irrelevant alternatives and the Pareto condition. Then, there is* $k \in N$ *such that for any profile* $(\precsim^1, \ldots, \precsim^n) \in \mathcal{R}^n$ *and any pair of alternatives* $x, y \in X$, *it holds that* $x \prec^k y$ *entails* $x \prec_{G(\precsim^1,\ldots,\precsim^n)} y$ *[1].*

Arrow's statement can be reinforced provided that certain qualifications on the domain of the social welfare function G are done while maintaining the meaning of both conditions that appear in Definition 3 and the fact that $|X| \geq 3$. Let $\mathcal{S} \subseteq \mathcal{R}$. If the domain of G is restricted to $\mathcal{S}^n$, then the resulting map is called a partial social welfare function. A partial social welfare function $G : \mathcal{S}^n \rightarrow \mathcal{R}$ is said to be projective whenever there is $k \in N$ such that $G(\precsim^1, \ldots, \precsim^n) = \precsim^k$, for every $(\precsim^1, \ldots, \precsim^n) \in \mathcal{S}^n$. In the social choice literature, a projective partial social welfare function is usually referred to as a strong dictatorship, and individual k is called a strong dictator.

I now comment on two important cases where a partial social welfare function turns out to be projective. In both situations, the proof is a direct consequence of the original proof of Arrow's theorem (see [1] or [24] for an alternative view). The first one is $\mathcal{S} = \mathcal{T}$. The second one, which actually is the variant I will use later, involves ordinals. Let $\leq \in \mathcal{W}$ be a well ordering defined on X which does exist by Zorn's lemma. Let $\phi \in B(X)$. Consider the binary relation, denoted by $\phi_{\leq}$, on X defined by: $x \, \phi_{\leq} \, y \Leftrightarrow \phi(x) \leq \phi(y)$, $(x, y \in X)$. Clearly, $\phi_{\leq} \in \mathcal{W}$. Let $\mathcal{W}_{\leq}$ stand for the following subset of $\mathcal{W}$, $\mathcal{W}_{\leq} := \{\phi_{\leq} : \phi \in B(X)\}$. Let $\mathcal{S} = \mathcal{W}_{\leq}$. Then, a partial social welfare function $G : \mathcal{S}^n \to \mathcal{R}$ that satisfies the condition of independence of irrelevant alternatives and the Pareto condition is projective.

3. The Main Result

In this section, the statement of the major result of the paper is presented. As already said in the Introduction, it turns out to be a characterization of projective aggregation functions on an abstract set, which contains, at least, three elements. In addition, certain variants of this result are also included, which require slight modifications of the concept of being full.

Theorem 2. *Let $|X| \geq 3$. Then, for an aggregation function $W : X^n \to X$, the following conditions are equivalent:*

(i) W is projective.
(ii) W is idempotent and full.

Proof. That (i) entails (ii) is obvious, so I will focus on the other implication, which will be developed in four steps.

Step 1. For each $a = (a_j) \in X^n$, $W(a) \in \{a_1, \ldots, a_n\}$. Indeed, take the elements of $\{a_1, \ldots, a_n\}$ avoiding repetitions. Therefore, without loss of generality, it may be assumed that $a_i \neq a_j$, for all $i \neq j$. Let $A := \{a_1, \ldots, a_n\}$. Then, for each $j \in N$, consider the function $\phi^j : X \to X$ defined by: $\phi^j(a_i) = a_{i+j-1(\text{modulo } n)}$ and $\phi^j(x) = x$, for all $x \in X \setminus A$ (In this definition, I use the convention $a_0 = a_n$. Note that each ϕ^j maps the set $\{a_1, \ldots, a_n\}$ into itself in a bijective way starting from a_j following the usual order in N. In addition, observe that if there was any repetition in $\{a_1, \ldots, a_n\}$, say $a_k = a_l$, with $k \neq l$, then $\phi^k = \phi^l$.). Clearly, $\phi^j \in B(X)$, for every $j \in N$.

Now, because W is idempotent, it holds that $\{W(\phi^1(x), \ldots, \phi^n(x)) : x \in X \setminus A\} = X \setminus A$. Hence, since W is full, $\{W(\phi^1(x), \ldots, \phi^n(x)) : x \in A\} = A$. Thus, for $x = a_1$, it holds that $W(a) = W(a_1, \ldots, a_n) = W(\phi^1(a_1), \ldots, \phi^n(a_1)) \in \{a_1, \ldots, a_n\}$, which proves Step 1.

Step 2. Let there be given $a = (a_j), b = (b_j) \in X^n$ such that $a_j \neq b_j$, for all $j \in N$. Then, $W(a) \neq W(b)$. In order to see this, first take the elements of $\{a_1, \ldots, a_n\}$ and $\{b_1, \ldots, b_n\}$, avoiding repetitions. Therefore, as above, it may be assumed that $a_i \neq a_j$ and also that $b_i \neq b_j$, for all $i \neq j$. Let $A := \{a_1, \ldots, a_n\}$, $B := \{b_1, \ldots, b_n\}$ and $C := A \cup B$. Then, for each $j \in N$, consider a function $\phi^j : X \to X$ accomplishing the following requirements: ϕ^j is a bijection from C onto C such that $\phi^j(a_1) = a_j$, $\phi^j(b_1) = b_j$ and $\phi^j(x) = x$, for all $x \in X \setminus C$. Note that such a function can be easily built and also that $\phi^j \in B(X)$, for every $j \in N$.

Thus, since W is idempotent, it holds that $\{W(\phi^1(x), \ldots, \phi^n(x)) : x \in X \setminus C\} = X \setminus C$. Hence, since W is full, $\{W(\phi^1(x), \ldots, \phi^n(x)) : x \in C\} = C$. Note that, for $x = a_1 \in C$, $W(a) = W(a_1, \ldots, a_n) = W(\phi^1(a_1), \ldots, \phi^n(a_1))$ and for $x = b_1 \in C$, $W(b) = W(b_1, \ldots, b_n) = W(\phi^1(b_1), \ldots, \phi^n(b_1))$. Thus, necessarily, $W(a) \neq W(b)$.

Step 3. Let $\leq \in \mathcal{W}$ be a (fixed) well ordering on X. The inequalities "$<$" and "$>$", between elements of X, have the usual meanings. I now prove that W is strongly monotone in the following sense: for any $a = (a_j), b = (b_j) \in X^n$ such that $a_j < b_j$, for all $j \in N$, it holds that $W(a) < W(b)$. Suppose, by way of contradiction, that there are $a = (a_j), b = (b_j) \in X^n$ such that $a_j < b_j$, for all $j \in N$ and $W(a) > W(b)$ (note that the situation $W(a) = W(b)$ violates the condition of W being full as seen in Step 2 above). Let a_k, a_l and b_m be defined as follows: $a_k = \min_{j \in N} a_j$, $a_l = W(a)$

and $b_m = W(b)$. Note that the latter two numbers exist by Step 1, and also, it holds that $a_k < b_m < a_l$. Define $\alpha = (\alpha_j), \beta = (\beta_j) \in X^n$ in the following way:

$$\alpha_j = \begin{cases} a_l & \text{if } j \in A \\ a_k & \text{otherwise} \end{cases}, \qquad \beta_j = \begin{cases} a_l & \text{if } j \in B \\ b_m & \text{otherwise} \end{cases}$$

where $A := \{j \in N : a_j = a_l\}$ and $B := \{j \in N : b_j = b_m\}$. Note that $\alpha_j \neq \beta_j$, for all j. Thus, by Step 2, it holds that $W(\alpha) \neq W(\beta)$. Now, a contradiction will be reached by showing that $W(\alpha) = W(\beta)$. In order to see this, define $\gamma = (\gamma_j) \in X^n$ as follows: $\gamma_j = \begin{cases} a_k & \text{if } j \in A \\ a_l & \text{otherwise} \end{cases}$. Then, because $W(a) = a_l$, $W(\gamma) \in \{a_k, a_l\}$ and $\gamma_j \neq a_j$, for all j, it holds that $W(\gamma) = a_k$. Similarly, $W(\alpha) \in \{a_l, a_k\}$ and $\gamma_j \neq \alpha_j$, for all j. Hence, $W(\alpha) = a_l$. With a similar argument, by using now $W(b) = b_m$, $W(\beta) \in \{a_l, b_m\}$ and $\beta_j \neq b_j$, for all j, it holds that $W(\beta) = a_l$, which gives the desired contradiction.

Step 4. To finish the proof, consider the aggregator $F : B(X)^n \to B(X)$ defined as follows $F(\phi^1, \ldots, \phi^n)(x) = W(\phi^1(x), \ldots, \phi^n(x))$, for every $(\phi^1, \ldots, \phi^n) \in B(X)^n, x \in X$. By Step 2, joint to the fact that W is full, it follows that $F(B(X)^n) \subseteq B(X)$. I now prove that F satisfies binary independence. In order to see this, let $(\varphi^1, \ldots, \varphi^n), (\psi^1, \ldots, \psi^n) \in B(X)^n$ such that there are $x, y \in X$ for which $\varphi^j(x) = \psi^j(y)$, for all $j \in N$. Then, $F(\varphi^1, \ldots, \varphi^n)(x) = W(\varphi^1(x), \ldots, \varphi^n(x)) = W(\psi^1(y), \ldots, \psi^n(y)) = F(\psi^1, \ldots, \psi^n)(y)$, and so, F satisfies binary independence. Moreover, since by Step 3 above, W is strongly monotone, F satisfies the following monotonicity property: for every $(\phi^1, \ldots, \phi^n) \in B(X)^n$, $x, y \in X$ such that $\phi^j(x) < \phi^j(y)$, for all $j \in N$, it holds $F(\phi^1, \ldots, \phi^n)(x) < F(\phi^1, \ldots, \phi^n)(y)$.

Now, consider the set $\mathcal{S} = \mathcal{W}_{\leq}$ introduced in the last paragraph of Section 2. Define the partial social welfare function $G : \mathcal{S}^n \to \mathcal{R}$ as follows: $G(\phi^1_{\leq}, \ldots, \phi^n_{\leq}) := F(\phi^1, \ldots, \phi^n)_{\leq}$, for every $(\phi^1_{\leq}, \ldots, \phi^n_{\leq}) \in \mathcal{S}^n$. The two properties of F proven in the above paragraph clearly entail that G satisfies both the Pareto condition and the condition of independence of irrelevant alternatives. Therefore, by the second variant of Arrow's theorem mentioned at the end of Section 2, there is $k \in N$ such that $G(\phi^1_{\leq}, \ldots, \phi^n_{\leq}) = \phi^k_{\leq}$, for every $(\phi^1_{\leq}, \ldots, \phi^n_{\leq}) \in \mathcal{S}^n$. However, then, it clearly holds that $F(\phi^1, \ldots, \phi^n) = \phi^k$, for every $(\phi^1, \ldots, \phi^n) \in B(X)^n$. Hence, $W(a) = a_k$, for every $a = (a_j) \in X^n$, and the proof is finished. □

The next two examples show that fullness and idempotency are independent conditions.

Example 1. *(i) Let $X = \mathbb{R}$, and define $W : \mathbb{R}^2 \to \mathbb{R}$ as follows: $W(x, y) = \max\{x, y\}$. Obviously, W is a non-projective and idempotent aggregation function. However, it is not full since by taking the two bijections $\phi(x) = x$ and $\psi(x) = -x$, it holds that $\{W(\phi(x), \psi(x)) : x \in \mathbb{R}\} = [0, \infty) \subsetneq \mathbb{R}$.*
(ii) Let $X = \mathbb{R}$, and define $W : \mathbb{R}^2 \to \mathbb{R}$ as follows: $W(x, y) = x^3$. Obviously, W is a non-projective and full aggregation function, which fails to be idempotent.

Remark 1.
(i) *The assumption $|X| \geq 3$ cannot be ruled out from the statement of Theorem 2. Indeed, take $X = \{x, y\}$. Define $W : X^3 \to X$ as follows: $W(x, x, x) = x$, $W(y, y, y) = y$, $W(x, x, y) = x$, $W(x, y, x) = x$, $W(y, x, x) = x$, $W(y, y, x) = y$, $W(y, x, y) = y$, $W(x, y, y) = y$. Clearly, W so-defined is idempotent and full. However, it is not projective.*
(ii) *Alternative characterizations of projective functions can be shown provided that the concept of fullness is suitably modified. For instance, suppose that Z is a nonempty set such that $|Z| = |X|$, and denote by $B(Z, X)$ the set of all bijections from Z onto X. An aggregation function $W : X^n \to X$ is said to be full with respect to bijections if $\{W(b^1(z), \ldots, b^n(z)) : z \in Z\} = X$, for every $b^1, \ldots, b^n \in B(Z, X)$. Then, following the proof of Theorem 2 with the obvious modifications, the next result is in order: assume $|X| \geq 3$. Then, W is projective iff it is idempotent and full with respect to bijections.*
(iii) *In relation to Remark 1(ii) above, suppose now that $|Z| \geq |X|$. Let $S(Z, X)$ stand for the set of all surjections from Z onto X. An aggregation function $W : X^n \to X$ is said to be full with respect to surjections if $\{W(s^1(z), \ldots, s^n(z)) : z \in Z\} = X$, for every $s^1, \ldots, s^n \in S(Z, X)$. Translating into*

this context the arguments used in the proof of Theorem 2, the following result can be proven: assume $|X| \geq 3$. Then, W is projective iff it is idempotent and full with respect to surjections (This result requires consideration of a well ordering, say $\leq$, on X. Then, for every $s \in S(Z,X)$, define the binary relation, denoted by $s_{\leq}$, on Z as follows: $z\ s_{\leq}\ t \Leftrightarrow s(z) \leq s(t)$, $(z,t \in Z)$. Consider the following subset of binary relations on Z, $\mathcal{W}_{\leq} := \{s_{\leq} : s \in S(Z,X)\}$. Note that, for each $s \in S(Z,X)$, $s_{\leq}$ is a total preorder on Z that induces a well ordering on the quotient set $Z/\sim_{s_{\leq}}$. Then, the variant of Arrow's theorem used in this situation states that a partial social welfare function $G : (\mathcal{W}_{\leq})^n \to \mathcal{R}$ that satisfies the condition of independence of irrelevant alternatives and the Pareto condition is projective.). A similar result can be given by adapting the fullness condition to the context of injections from Z into X provided that $|Z| \leq |X|$.

4. Some Consequences of the Main Result

I conclude the paper with two applications of Theorem 2. The first one is a simple generalization of this result to the context of multi-valued aggregation functions. Let X be a nonempty set and $n \in \mathbb{N}$. Denote by $\mathcal{P}(X) = 2^X \setminus \emptyset$. Then, a multi-valued aggregation function is a map $G : X^n \to \mathcal{P}(X)$. A multi-valued aggregation function G is projective if there is $k \in \{1,\ldots,n\}$ such that $a_k \in G(a)$, for every $a = (a_j) \in X^n$. An aggregation function $V : X^n \to X$ is said to be a selection of G if $V(a) \in G(a)$, for every $a = (a_j) \in X^n$.

The next result, which is an obvious consequence of Theorem 2, is now in order.

Corollary 1. *Let $|X| \geq 3$. Then, for a multi-valued aggregation function $G : X^n \to \mathcal{P}(X)$, the following conditions are equivalent:*

(i) *G is projective.*
(ii) *G admits a selection that is idempotent and full.*

As the second application of Theorem 2, the following characterization of a (bijection) aggregator is shown.

Corollary 2. *Let $|X| \geq 3$. Then, for an aggregator $F : B(X)^n \to B(X)$, the following conditions are equivalent:*

(i) *F is projective.*
(ii) *F is unanimous and satisfies binary independence.*

Proof. That (i) implies (ii) is obvious. In order to see the converse, assume that $F : B(X)^n \to B(X)$ is a unanimous aggregator that satisfies binary independence. Define the aggregation function $W : X^n \to X$ as follows: for each $a = (a_j) \in X^n$, $W(a) = F(\phi^1,\ldots,\phi^n)(x)$, where $(\phi^1,\ldots,\phi^n) \in B(X)^n$, $x \in X$ are such that $\phi^j(x) = a_j$, for all $j \in N$. Since F satisfies binary independence, W is well defined (i.e., the value $F(\phi^1,\ldots,\phi^n)(x)$ does not depend on the particular $(\phi^1,\ldots,\phi^n) \in B(X)^n$ and $x \in X$ chosen). In addition, the unanimity of F entails that W is idempotent. Moreover, since the image of F lies on $B(X)$, it clearly follows that W is full. Hence, by Theorem 2, W is projective, and so, F is. □

Remark 2. *(i)* *It should be noted that, according to Remarks 1(ii) and 1(iii), alternative versions of Corollary 2 can be established by replacing $B(X)$ with $B(Z,X)$ and $S(Z,X)$, respectively.*

(ii) *Corollary 2 can be given an interpretation in the fuzzy framework. Assume U is a finite universe, with states $\{u_1,\ldots,u_m\}$, and denote by X a set of fuzzy numbers defined on U such that $|X| \geq 3$ (Note that each number in X can be identified with a vector in the m-dimensional unit cube where, for every $l \in \{1,\ldots,m\}$, the l-component might be interpreted as the uncertainty, or vagueness, that state u_l occurs. Then, X could be thought of as the set of possible uncertain situations to be managed.). Then, $B(X)$ represents the set of all possible rearrangements of the referenced uncertain situations. Further, suppose that there are n individuals, $n \in \mathbb{N}$, each of them presenting a particular arrangement of the uncertain situations, and a final arrangement needs to be reached as an aggregation procedure*

of the individual proposals. Then, Corollary 2 provides necessary and sufficient conditions for such an aggregation procedure to coincide with the proposal of some individual.

5. Conclusions

In this paper, a very general characterization result concerning projective aggregation functions is presented. This result is motivated by a theorem obtained by Kim (1990), for real-valued aggregation functions defined on the n-dimensional Euclidean space, in the measurement theory context. A key feature of the approach followed in the paper is that the proof of the core result is based on a particular version of Arrow's theorem, thus providing a link between aggregation functions theory and social choice theory. Some extensions and consequences are presented, as well; in particular, a generalization of the main result to the framework of multi-valued aggregation functions and a characterization of projective bijection aggregators. Although not developed in the article, it is pointed out that these consequences could have certain applications to the environment of the aggregation of preferences and aggregation of classifications.

Acknowledgments: This work has been partially supported by the Research Project ECO2012-34828 (Spain). I am grateful to Esteban Induráin, who suggested that I illustrate Corollary 2 with an example coming from the "fuzzy world". I also thank two referees for providing me with valuable comments that have led to an improved version of the manuscript. Technical support of the Assistant Editor L. Shen is also acknowledged.

Conflicts of Interest: The author declares no conflict of interest.

References

1. Arrow, K.J. *Social Choice and Individual Values*, 2nd ed.; Wiley: New York, NY, USA, 1963.
2. Zadeh, L.A. Fuzzy sets. *Inf. Control* **1965**, *8*, 338–353.
3. Zadeh, L.A. Quantitative fuzzy semantics. *Inf. Sci.* **1971**, *3*, 159–176.
4. Dubois, D.; Prade, H. *Fuzzy Sets and Systems: Theory and Applications*; Academic Press: New York, NY, USA, 1980.
5. Goguen, J.A. L-fuzzy sets. *J. Math. Anal. Appl.* **1967**, *18*, 145–174.
6. Aczél, J. *A Short Course on Functional Equations*; Kluwer: Alphen aan den Rijn, The Netherlands, 1987.
7. Calvo, T.; Mayor, G.; Mesiar, R. (Eds.) *Aggregation Operators: New Trends and Applications*; Springer: Berlin, Germany, 2002.
8. Bustince, H.; Fernández, J.; Mesiar, R.; Calvo, T. (Eds.) *Aggregation Functions in Theory and in Practise*; Springer: Berlin, Germany, 2002.
9. Tizhoosh, H. Image thresholding using type II fuzzy sets. *Pattern Recognit.* **2005**, *38*, 2363–2372.
10. Candeal, J.C. Aggregation operators, comparison meaningfulness and social choice. *J. Math. Psychol.* **2016**, *75*, 19–25.
11. De Miguel, L.; Campión, M.J.; Candeal, J.C.; Induráin, E.; Paternain, D. Pointwise aggregation of maps: Its structural functional equation and some applications to social choice. *Fuzzy Sets Syst.* **2017**, *325*, 137–151.
12. Candeal, J.C.; Induráin, E. Point-sensitive aggregation operators: Functional equations and applications to social choice. *Int. J. Uncertain. Fuzziness* **2017**, *25*, 973–986.
13. Krantz, D.; Luce, R.D.; Suppes, P.; Tversky, A. *Foundations of Measurement. Vol. I: Additive and Polynomial Representations*; Academic Press: New York, NY, USA, 1971.
14. Luce, R.D. On the possible psychophysical laws. *Psychol. Rev.* **1959**, *66*, 81–95.
15. Kim, S.-R. On the possible scientific laws. *Math. Soc. Sci.* **1990**, *20*, 19–36.
16. Marichal, J.L.; Mathonet, P. On comparison meaningfulness of aggregation functions. *J. Math. Psychol.* **2001**, *45*, 213–223.
17. Marichal, J.L. On order invariant synthesizing functions. *J. Math. Psychol.* **2002**, *46*, 661–676.
18. Marichal, J.L.; Mesiar, R.; Rückschlossová, A. A complete description of comparison meaningful functions. *Aequ. Math.* **2005**, *69*, 309–320.
19. Grabisch, M.; Marichal, J.L.; Mesiar, R.; Pap, E. *Aggregation Functions*; Encyclopedia of Mathematics and Its Applications 127; Cambridge University Press: Cambridge, UK, 2009.

20. Marichal, J.L.; Mesiar, R. A complete description of comparison meaningful aggregation functions mapping ordinal scales into an ordinal scale: A state of the art. *Aequ. Math.* **2009**, *77*, 207–236.
21. Aczél, J.; Roberts, F.S. On the possible merging functions. *Math. Soc. Sci.* **1989**, *17*, 205–243.
22. Maniquet, F.; Mongin, P. A theorem on aggregating classifications. *Math. Soc. Sci.* **2016**, *79*, 6–10.
23. Dugundji, J. *Topology*; Allyn and Bacon: Boston, MA, USA, 1966.
24. Campbell, J.C.; Kelly, J.R. Impossibility theorems in the Arrovian framework. In *Handbook of Social Choice and Welfare*; Arrow, K.J., Sen, A., Suzumura, K., Eds.; Elsevier: Amsterdam, The Netherlands, 2002; Volume 1, pp. 35–94.

axioms

MDPI

Article

Graphs in an Intuitionistic Fuzzy Soft Environment

Sundas Shahzadi and Muhammad Akram *

Department of Mathematics, University of the Punjab, New Campus, Lahore 54590, Pakistan; sundas1011@gmail.com

* Correspondence: m.akram@pucit.edu.pk

Received: 17 February 2018; Accepted: 14 March 2018; Published: 23 March 2018

Abstract: In this research article, we present a novel framework for handling intuitionistic fuzzy soft information by combining the theory of intuitionistic fuzzy soft sets with graphs. We introduce the notion of certain types of intuitionistic fuzzy soft graphs including neighbourly edge regular intuitionistic fuzzy soft graphs and strongyl edge irregular intuitionistic fuzzy soft graphs. We illustrate these novel concepts by several examples, and investigate some of their related properties. We present an application of intuitionistic fuzzy soft graph in a decision-making problem and also present our methods as an algorithm that is used in this application.

Keywords: regular intuitionistic fuzzy soft graphs; irregular intuitionistic fuzzy soft graph; edge irregular intuitionistic fuzzy soft graphs; decision-making

1. Introduction

Atanassov [1,2] introduced the concept of intuitionistic fuzzy sets as an extension of Zadeh's fuzzy set [3]. The concept of an intuitionistic fuzzy set can be viewed as an alternative approach in such cases where available information is not sufficient to define the impreciseness by the conventional fuzzy set. In fuzzy sets, only the degree of acceptance is considered, but an intuitionistic fuzzy set is characterized by a membership function and a non-membership function. The only requirement is that the sum of both values is not more than one. Intuitionistic fuzzy sets are being studied and used in different fields of science [4–7]. Mathematical modeling, analysis and computing of problems with uncertainty is one of the hottest areas in interdisciplinary research, involving applied mathematics, computational intelligence and decision-making. It is worth noting that the uncertainty that arises from various domains has a very variable nature and cannot be captured within a single mathematical framework. Molodtsov's soft set theory provide us a new mathematical tool for dealing with uncertainty from the viewpoint of parameterizations. Molodtsov [8] introduced the concept of soft set theory. He gave us a new technique for dealing with uncertainty from the viewpoint of parameters. Soft set theory has been extended with potential applications in several fields [9–14]. Maji et al. [15] introduced fuzzy soft sets. By using this definition of fuzzy soft sets, many interesting applications of fuzzy soft set theory have been discussed by researchers [16–21]. Maji et al. [22,23] introduced the concept of intuitionistic fuzzy soft set and presented some operations on intuitionistic fuzzy soft sets. The intuitionistic fuzzy soft set theory has been extended with applications in [24–26].

Fuzzy graph theory involves finding an increasing number of applications in modeling real time systems where the level of information inherent in the system varies with different levels of precision. Fuzzy models are becoming useful because of their aim in reducing the differences between the traditional numerical models used in engineering and sciences and the symbolic models used in expert systems. Kaufmann's initial definition of a fuzzy graph [27]. Rosenfeld [28] described the structure of fuzzy graphs obtaining analogs of several graph theoretical concepts. Bhattacharya [29] gave some remarks on fuzzy graphs and several concepts were introduced by Mordeson and Nair in [30]. Santhimasheswari and Sekar [31] introduced the idea of strongly edge irregular fuzzy graph. Several concepts and applications of fuzzy graphs and their extensions have been discussed in [32–39]. In this research article, we introduce the notion of certain types of intuitionistic fuzzy soft graphs including neighbourly edge regular intuitionistic fuzzy soft graphs and strongly edge irregular intuitionistic fuzzy soft graphs. We illustrate these novel concepts by several examples, and investigate some of their related properties. We present an application of intuitionistic fuzzy soft graph in the decision-making process. We describe our methods as algorithms that are used in our application.

For other notations and applications that are not mention in this paper, readers are refereed to [40–45].

2. Graphs in Intuitionistic Fuzzy Soft Environment

We consider $\mathcal{P}(V)$ as the set of all intuitionistic fuzzy sets (IFSs) of V and $\mathcal{P}(E)$ denotes the set of all IFSs of E. The notion of intuitionistic fuzzy soft graphs was presented in [34–36].

Definition 1. *An intuitionistic fuzzy soft graph on a nonempty set V is an ordered 3-tuple $\mathcal{G} = (\Phi, \Psi, M)$ such that*

(i) *M is a non-empty set of parameters,*
(ii) *(Φ, M) is an intuitionistic fuzzy soft set over V,*
(iii) *(Ψ, M) is an intuitionistic fuzzy soft relation on V, i.e., $\Psi : M \to \mathcal{P}(V \times V)$ where $\mathcal{P}(V \times V)$ is intuitionistic fuzzy power set,*
(iv) *$(\Phi(e), \Psi(e))$ is an intuitionistic fuzzy graph, for all $e \in M$.*

That is,

$$\Psi_\mu(e)(uv) \leq \min(\Phi_\mu(e)(u), \Phi_\mu(e)(v)),$$

$$\Psi_\nu(e)(uv) \leq \max(\Phi_\nu(e)(u), \Phi_\nu(e)(v))$$

such that $\Psi_\mu(e)(uv) + \Psi_\nu(e)(uv) \leq 1,\ \forall\, e \in M,\ u, v \in V$.

Note that $\Psi_\mu(e)(uv) = \Psi_\nu(e)(uv) = 0,\ \forall uv \in V \times V - E, e \in M$.

(Φ, M) is called an intuitionistic fuzzy soft vertex and (Ψ, M) is called an intuitionistic fuzzy soft edge. Thus, $((\Phi, M), (\Psi, M))$ is called an intuitionistic fuzzy soft graph if

$$\Psi_\mu(e)(uv) \leq \min(\Phi_\mu(e)(u), \Phi_\mu(e)(v)),\ \Psi_\nu(e)(uv) \leq \max(\Phi_\nu(e)(u), \Phi_\nu(e)(v))$$

such that $\Psi_\mu(e)(uv) + \Psi_\nu(e)(uv) \leq 1,\ \forall\, e \in M,\ u, v \in V$. In other words, an intuitionistic fuzzy soft graph is a parameterized family of intuitionistic fuzzy graphs.

Definition 2. *Let $G = (\Phi, \Psi, M)$ be an intuitionistic fuzzy soft graph of G^*. G is said to be a neighbourly edge irregular intuitionistic fuzzy soft graph if $H(e)$ is a neighbourly edge irregular intuitionistic fuzzy graph for all $e \in A$.*

Equivalently, an intuitionistic fuzzy soft graph G is called a neighborly edge irregular intuitionistic fuzzy soft graph if every two adjacent edges have distinct degrees in $H(e)$ for all $e \in M$.

Definition 3. *Let $G = (\Phi, \Psi, M)$ be an intuitionistic fuzzy soft graph of G^*. G is said to be a neighbourly edge totally irregular intuitionistic fuzzy soft graph if $H(e)$ is a neighbourly edge totally irregular intuitionistic fuzzy graph for all $e \in A$.*

Equivalently, an intuitionistic fuzzy soft graph G is called a neighborly edge totally irregular intuitionistic fuzzy soft graph if every two adjacent edges have distinct total degrees in $H(e)$ for all $e \in M$.

Example 1. *Consider a crisp graph $G = (V, E)$ such that $V = \{u_1, u_2, u_3, u_4\}$ and $E = \{u_1u_2, u_2u_3, u_3u_1, u_1u_4, u_2u_4, u_3u_4\}$. Let $P = \{e_1, e_2, e_3, e_4\}$ be a set of all parameters and $M = \{e_1, e_2, e_3\} \subset P$. Let (Φ, M) be an intuitionistic fuzzy soft set over V with intuitionistic fuzzy approximation function $\Phi : M \to \mathcal{P}(V)$*

defined by
$\Phi(e_1) = \{(u_1, 0.5, 0.2), (u_2, 0.6, 0.4), (u_3, 0.3, 0.1), (u_4, 0.5, 0.2)\}$,
$\Phi(e_2) = \{(u_1, 0.6, 0.4), (u_2, 0.7, 0.2), (u_3, 0.7, 0.1), (u_4, 0.5, 0.4)\}$,
$\Phi(e_3) = \{(u_1, 0.4, 0.6), (u_2, 0.3, 0.1), (u_3, 0.3, 0.6), (u_4, 0.4, 0.1)\}$.

Let (Ψ, M) *be an intuitionistic fuzzy soft set over E with intuitionistic fuzzy approximation function* $\Psi : M \to \mathcal{P}(E)$ *defined by*
$\Psi(e_1) = \{(u_1u_2, 0.4, 0.4), (u_2u_3, 0.2, 0.2), (u_1u_3, 0.0, 0.0), (u_1u_4, 0.2, 0.2), (u_2u_4, 0.3, 0.3), (u_3u_4, 0.0, 0.0)\}$,
$\Psi(e_2) = \{(u_1u_2, 0.4, 0.4), (u_2u_3, 0.0, 0.0), (u_1u_3, 0.2, 0.1), (u_1u_4, 0.2, 0.3), (u_2u_4, 0.0, 0.0), (u_3u_4, 0.4, 0.4)\}$,
$\Psi(e_3) = \{(u_1u_2, 0.1, 0.1), (u_2u_3, 0.0, 0.0), (u_1u_3, 0.2, 0.4), (u_1u_4, 0.1, 0.1), (u_2u_4, 0.0, 0.0), (u_3u_4, 0.2, 0.6)\}$.

Thus intuitionistic fuzzy graphs $H(e_1) = (\Phi(e_1), \Psi(e_1))$, $H(e_2) = (\Phi(e_2), \Psi(e_2))$ *and* $H(e_3) = (\Phi(e_3), \Psi(e_3))$ *of G corresponding to the parameters* e_1, e_2 *and* e_3, *respectively are shown in Figure 1.*

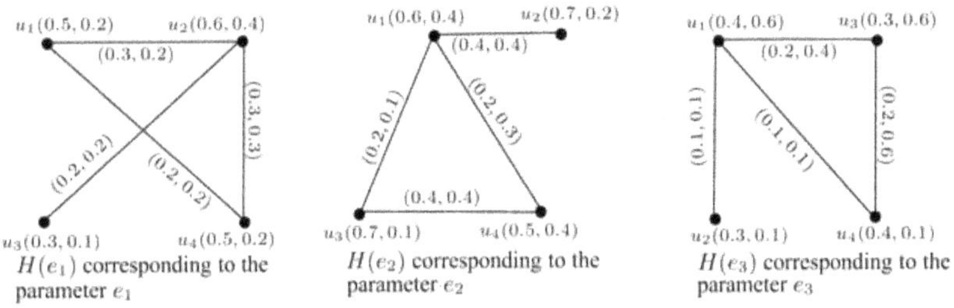

Figure 1. Intuitionistic fuzzy soft graph $G = \{H(e_1), H(e_2), H(e_3)\}$.

In intuitionistic fuzzy graph $H(e_i)$, *for* $i = 1, 2, 3$ *degree of edges are*
$deg_G(u_1u_2)(e_1) = (0.7, 0.7), deg_G(u_1u_4)(e_1) = deg_G(u_2u_3)(e_1) = (0.5, 0.6), deg_G(u_2u_4)(e_1) = (0.7, 0.6)$,
$deg_G(u_1u_2)(e_2) = deg_G(u_3u_4)(e_2) = (0.4, 0.4), deg_G(u_1u_3)(e_2) = (1.0, 1.1), deg_G(u_1u_4)(e_2) = (1.0, 0.9)$,
$deg_G(u_1u_2)(e_1) = (0.3, 0.5), deg_G(u_1u_4)(e_1) = (0.5, 1.1), deg_G(u_1u_3)(e_1) = (0.4, 0.8), deg_G(u_2u_4)(e_1) = (0.3, 0.5)$.

Clearly, every pair of adjacent edges in intuitionistic fuzzy graphs $H(e_i)$ *for* $i = 1, 2, 3$ *corresponding to the parameters* e_i *for* $i = 1, 2, 3$ *have distinct degrees. Hence G is the neighbourly edge irregular intuitionistic fuzzy soft graph.*

Theorem 1. *Let* $G = (\Phi, \Psi, M)$ *be a connected intuitionistic fuzzy soft graph on* $G^* = (V, E)$ *and* Ψ *is a constant function. If G is a neighbourly edge irregular(neighbourly edge totally irregular) intuitionistic fuzzy soft graph, then G is neighbourly edge totally irregular(neighbourly edge irregular) intuitionistic fuzzy soft graph.*

Proof. We assume that Ψ is a constant function, $\Psi_{e_i}(uv) = (c_i, \acute{c}_i)$, for all $uv \in E, e_i \in M$, where c_i and $\acute{c}_i$ are constants for $i = 1, 2, \ldots, k$. Let uv and vz be any pair of adjacent edges in E. Suppose G is a neighbourly edge irregular intuitionistic fuzzy soft graph, then $deg_G(uv)(e_i) \neq deg_G(vz)(e_i)$ for all $e_i \in M$, this implies
$(deg_\mu(uv)(e_i), deg_\nu(uv)(e_i)) \neq (deg_\mu(vz)(e_i), deg_\nu(vz)(e_i))$,
$(deg_\mu(uv)(e_i), deg_\nu(uv)(e_i)) + (c_i, \acute{c}_i) \neq (deg_\mu(vz)(e_i), deg_\nu(vz)(e_i)) + (c_i, \acute{c}_i)$,
$deg_G(uv)(e_i) + \Psi(uv)(e_i) \neq deg_G(vz)(e_i) + \Psi(vz)(e_i)$,
$tdeg_G(uv)(e_i) \neq tdeg_G(vz)(e_i)$, where uv and vz are adjacent edges in E. Hence, G is a neighbourly edge totally irregular intuitionistic fuzzy soft graph. □

Theorem 2. *Let $G = (\Phi, \Psi, M)$ be a connected intuitionistic fuzzy soft graph on $G^* = (V, E)$ and Ψ is a constant function. If G is a neighbourly edge totally irregular intuitionistic fuzzy soft graph, then G is neighbourly edge irregular intuitionistic fuzzy soft graph.*

Remark 1. *Let $G = (\Phi, \Psi, M)$ be a connected intuitionistic fuzzy soft graph on $G^* = (V, E)$ and Ψ is a constant function. If G is both neighbourly edge irregular intuitionistic fuzzy soft graph and neighbourly edge totally irregular intuitionistic fuzzy soft graph. Then Ψ need not be a constant function.*

Example 2. *Consider a simple graph $G^* = (V, E)$ such that $V = \{v_1, v_2, v_3, v_4\}$ and $E = \{v_1v_2, v_1v_3, v_1v_4, v_2v_3, v_2v_4, v_3v_4\}$. Let $M = \{e_1, e_2\}$ be set of parameters.*

Let $G = (H, M)$ be an intuitionistic fuzzy soft graph (see Figure 2), where intuitionistic fuzzy graphs $H(e_1)$ and $H(e_2)$ corresponding to the parameters e_1 and e_2, respectively are defined as follows:

$$H(e_1) = (\{(v_1, 0.2, 0.5), (v_2, 0.7, 0.3), (v_3, 0.5, 0.1), (v_4, 0.6, 0.3)\}, \{(v_1v_2, 0.1, 0.4), (v_1v_3, 0.2, 0.4), (v_1v_4, 0.1, 0.4), (v_2v_4, 0.5, 0.1)\}),$$
$$H(e_2) = (\{(v_1, 0.2, 0.4), (v_2, 0.2, 0.5), (v_3, 0.5, 0.1), (v_4, 0.5, 0.2)\}, \{(v_1v_2, 0.2, 0.3), (v_1v_4, 0.2, 0.4), (v_2v_3, 0.1, 0.4), (v_2v_4, 0.1, 0.3)\}).$$

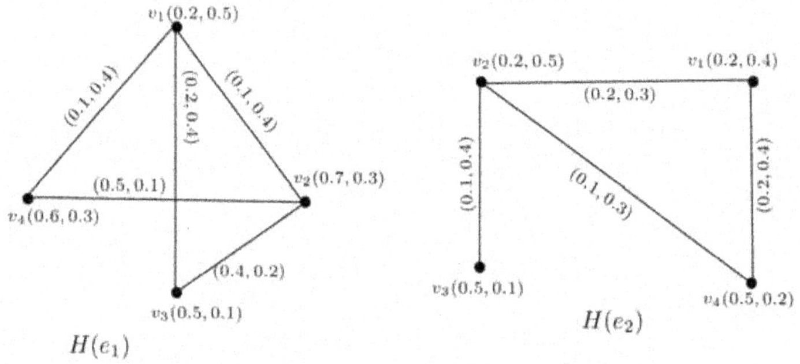

Figure 2. Intuitionistic fuzzy soft graph $G = \{H(e_1), H(e_2)\}$.

The degree of each edge in $H(e_1)$ corresponding to the parameter e_1 is $deg_G(v_1v_2)(e_1) = (1.2, 1.1)$, $deg_G(v_2v_4)(e_1) = (0.6, 1.0), deg_G(v_1v_3)(e_1) = (0.6, 1.0), deg_G(v_1v_4)(e_1) = (0.8, 0.9), deg_G(v_2v_3)(e_1) = (1.0, 0.7)$ and degree of each edge in $H(e_2)$ corresponding to the parameter e_2 is $deg_G(v_1v_2)(e_2) = (0.4, 1.2), deg_G(v_1v_4)(e_2) = (0.3, 0.7), deg_G(v_2v_3)(e_2) = (0.3, 0.6), deg_G(v_2v_4)(e_2) = (0.5, 1.1)$. It is easy to see that all the edges in $H(e_1)$ and $H(e_2)$ have distincts degrees. Therefore, the intuitionistic fuzzy soft graph $G = \{H(e_1), H(e_2)\}$ is a neighbourly edge irregular intuitionistic fuzzy soft graph.

The total degree of each edge in $H(e_1)$ corresponding to the parameter e_1 is $tdeg_G(v_1v_2)(e_1) = (1.3, 1.5)$, $deg_G(v_2v_4)(e_1) = (1.1, 1.1), deg_G(v_1v_3)(e_1) = (0.8, 1.4), deg_G(v_1v_4)(e_1) = (0.9, 1.3), deg_G(v_2v_3)(e_1) = (1.4, 0.9)$ and total degree of each edge in $H(e_2)$ corresponding to the parameter e_2 is $deg_G(v_1v_2)(e_2) = (0.6, 1.5), deg_G(v_1v_4)(e_2) = (0.5, 1.1), deg_G(v_2v_3)(e_2) = (0.4, 1.0), deg_G(v_2v_4)(e_2) = (0.6, 1.4)$. Clearly, all the edges in $H(e_1)$ and $H(e_2)$ have distinct total degrees. Therefore, intuitionistic fuzzy soft graph $G = \{H(e_1), H(e_2)\}$ is a neighbourly edge totally irregular intuitionistic fuzzy soft graph. Thus, G is both a neighbourly edge irregular and neighbourly edge totally irregular intuitionistic fuzzy soft graph but Ψ is not a constant function.

Theorem 3. *Let $G = (\Phi, \Psi, M)$ be a connected intuitionistic fuzzy soft graph on $G^* = (V, E)$ and Ψ is a constant function. If G is a neighbourly edge irregular intuitionistic fuzzy soft graph, then G is an irregular intuitionistic fuzzy soft graph.*

Proof. Let G be a connected intuitionistic fuzzy soft graph on $G^* = (V, E)$ and Ψ is a constant function, $\Psi(uv)(e_i) = (c_i, \acute{c}_i)$, where c_i and $\acute{c}_i$ are constants. Assume that G is a neighbourly edge irregular intuitionistic fuzzy soft graph. Consider uv and vz are two adjacent edges in G with distinct degrees,

$(deg_\mu(uv)(e_i), deg_\nu(uv)(e_i)) \neq (deg_\mu(vz)(e_i), deg_\nu(vz)(e_i))$,
$deg_\mu(uv)(e_i) \neq deg_\mu(vz)(e_i)$ or $deg_\nu(uv)(e_i) \neq deg_\nu(vz)(e_i)$,
$deg_\mu(u)(e_i) + deg_\mu(v)(e_i) - 2c_i \neq deg_\mu(v)(e_i) + deg_\mu(z)(e_i) - 2c_i$, or
$deg_\nu(u)(e_i) + deg_\nu(v)(e_i) - 2\acute{c}_i \neq deg_\nu(v)(e_i) + deg_\nu(z)(e_i) - 2\acute{c}_i$,
$deg_\mu(u)(e_i) \neq deg_\mu(z)(e_i)$, or $deg_\nu(u)(e_i) \neq deg_\nu(z)(e_i)$,
$(deg_\mu(u)(e_i), deg_\nu(u)(e_i),) \neq (deg_\mu(z)(e_i), deg_\nu(z)(e_i),)$,
$deg_G(u)(e_i) \neq deg_G(z)(e_i)$,

this implies there exist a vertex v which is adjacent to the vertices u and z have distinct degrees. Hence, G is an irregular intuitionistic fuzzy soft graph. □

Theorem 4. *Let $G = (\Phi, \Psi, M)$ be a connected intuitionistic fuzzy soft graph on $G^* = (V, E)$ and Ψ is a constant function. If G is a neighbourly edge totally irregular intuitionistic fuzzy soft graph, then G is an irregular intuitionistic fuzzy soft graph.*

Theorem 5. *Let $G = (\Phi, \Psi, M)$ be a connected intuitionistic fuzzy soft graph on $G^* = (V, E)$ and Ψ is a constant function. Then, G is a neighbourly edge irregular intuitionistic fuzzy soft graph if and only if G is highly irregular intuitionsistic fuzzy soft graph.*

Proof. Let $G = (\Phi, \Psi, M)$ be a connected intuitionistic fuzzy soft graph on G^* and Ψ is a constant function $\Psi(uv)(e_i) = (c_i, \acute{c}_i)$ for all $uv \in E$, where c_i and $\acute{c}_i$ are constants. Let u be a vertex adjacent with v, w and x. vu, uw and ux are adjacent edges in G. Suppose G is a neighbourly edge irregular intuitionistic fuzzy soft graph, this implies that every pair of adjacent edges in G have distinct degrees, then

$deg_G(vu)(e_i) \neq deg_G(uw)(e_i) \neq deg_G(ux)(e_i)$
$(deg_\mu(vu)(e_i), deg_\nu(vu)(e_i)) \neq (deg_\mu(uw)(e_i), deg_\nu(uw)(e_i)) \neq (deg_\mu(ux)(e_i), deg_\nu(ux)(e_i))$

Consider $(deg_\mu(vu)(e_i), deg_\nu(vu)(e_i)) \neq (deg_\mu(uw)(e_i), deg_\nu(uw)(e_i))$
$deg_\mu(vu)(e_i) \neq (deg_\mu(uw)(e_i)$ or $deg_\nu(vu)(e_i) \neq deg_\nu(uw)(e_i)$
$deg_\mu(u)(e_i) + deg_\mu(v)(e_i) - 2c_i \neq deg_\mu(u)(e_i) + deg_\mu(w)(e_i) - 2c_i$ or
$deg_\nu(u)(e_i) + deg_\nu(v)(e_i) - 2\acute{c}_i \neq deg_\nu(u)(e_i) + deg_\nu(w)(e_i) - 2\acute{c}_i$
$\Rightarrow deg_\mu(v)(e_i) \neq deg_\mu(w)(e_i)$ or $deg_\nu(v)(e_i) \neq deg_\nu(w)(e_i)$
$\Rightarrow (deg_\mu(v)(e_i), deg_\nu(v)(e_i)) \neq (deg_\mu(w)(e_i), \neq deg_\nu(w)(e_i))$
$\Rightarrow deg_G(v) \neq deg_G(w)$

Similarly, $deg_G(w) \neq deg_G(x) \Rightarrow deg_G(v) \neq deg_G(w) \neq deg_G(x)$
clealy, every vertex u is adjacent to the vertices v, w and x with distinct degrees. Hence G is highly irregular intuitionistic fuzzy soft graph.

Conversly, let vu and uw are any two adjacent edges in G. Suppose that G is highly irregular intuitionistic fuzzy soft graph, then every vertex adjacent to the vertices in $H(e_i)$ for all $e_i \in M$ having distinct degrees, such that

$deg_G(v) \neq deg_G(w)$
$(deg_\mu(v)(e_i), deg_\nu(v)(e_i)) \neq (deg_\mu(w)(e_i), deg_\nu(w)(e_i))$
$deg_\mu(v)(e_i) \neq deg_\mu(w)(e_i)$, or $deg_\nu(v)(e_i) \neq deg_\nu(w)(e_i)$
$(deg_\mu(v)(e_i) + deg_\mu(u)(e_i) - 2c_i) \neq (deg_\mu(w)(e_i) + deg_\mu(u)(e_i) - 2c_i)$ or
$(deg_\nu(v)(e_i) + deg_\nu(u)(e_i) - 2\acute{c}_i) \neq (deg_\nu(w)(e_i) + deg_\nu(u)(e_i) - 2\acute{c}_i)$
$deg_\mu(uv)(e_i) \neq deg_\mu(uw)(e_i)$ or $deg_\nu(uv)(e_i) \neq deg_\nu(uw)(e_i)$
$(deg_\mu(uv)(e_i), deg_\nu(uv)(e_i)) \neq (deg_\mu(uw)(e_i), deg_\nu(uv)(e_i))$

$\Rightarrow deg_G(uv) \neq deg_G(uw))$
$\Rightarrow$ every pair of adjacent edges have distinct degrees. Hence G is a neighbourly edge irregular intuitionistic fuzzy soft graph. □

Definition 4. *Let $G = (\Phi, \Psi, M)$ be an intuitionistic fuzzy soft graph on $G^* = (V, E)$. Then, G is said to be a strongly irregular intuitionistic fuzzy soft graph if every pair of vertices in G have distinct degrees.*

Theorem 6. *Let $G = (\Phi, \Psi, M)$ be a connected intuitionistic fuzzy soft graph on $G^* = (V, E)$ and Ψ is a constant function. If G is strongly irregular intuitionistic fuzzy soft graph, then G is a neighbourly edge irregular intuitionistic fuzzy soft graph.*

Proof. Let G be a connected intuitionistic fuzzy soft graph on $G^* = (V, E)$, and Ψ is a constant function $\Psi(xy)(e_i) = (c_i, \acute{c}_i)$ for all $e_i \in M, xy \in E$, where c_i and $\acute{c}_i$ are constants. Let xy and yz are any two adjacent edges in E. Suppose that G is strongly irregular intuitionistic fuzzy soft graph, this implies every pair of vertices in $H(e_i)$ for all $e_i \in M$ having distinct degrees, this implies

$deg_G(x) \neq deg_G(y) \neq deg_G(z)$
$deg_G(x) + deg_G(y) \neq deg_G(y) + deg_G(z)$
$(deg_\mu(x), deg_\nu(x)) + (deg_\mu(y), deg_\nu(y)) \neq (deg_\mu(y), deg_\nu(y)) + (deg_\mu(z), deg_\nu(z))$
$(deg_\mu(x) + deg_\mu(y), deg_\nu(x) + deg_\nu(y)) \neq (deg_\mu(y) + deg_\mu(z), deg_\nu(y) + deg_\nu(z))$
$deg_\mu(x) + deg_\mu(y) \neq deg_\mu(y) + deg_\mu(z)$ or $deg_\nu(x) + deg_\nu(y) \neq deg_\nu(y) + deg_\nu(z)$
$deg_\mu(x) + deg_\mu(y) - 2c_i \neq deg_\mu(y) + deg_\mu(z) - 2c_i$ or $deg_\nu(x) + deg_\nu(y) - 2\acute{c}_i \neq deg_\nu(y) + deg_\nu(z) - 2\acute{c}_i$
$deg_\mu(xy) \neq deg_\mu(yz)$ or $deg_\nu(xy) \neq deg_\nu(yz) \Rightarrow (deg_\mu(xy), deg_\nu(xy)) \neq (deg_\mu(yz), deg_\nu(yz))$
$\Rightarrow deg_G(xy) \neq deg_G(yz)$

Clearly, every pair of adjacent edges has distinct degrees. Hence G is a neighbourly edge irregular intuitionistic fuzzy soft graph. □

Theorem 7. *Let $G = (\Phi, \Psi, M)$ be a connected intuitionistic fuzzy soft graph on $G^* = (V, E)$ and Ψ is a constant function. If G is strongly irregular intuitionistic fuzzy soft graph, then G is a neighbourly edge totally irregular intuitionistic fuzzy soft graph.*

Remark 2. *Converse of the above theorems need not be true.*

We now study strongly edge irregular intuitionistic fuzzy soft graphs.

Definition 5. *Let $G = (\Phi, \Psi, M)$ be a connected intuitionistic fuzzy soft graph on $G^* = (V, E)$. Then G is said to be a strongly edge irregular intuitionistic fuzzy soft graph if every pair of edges has distinct degrees.*

Definition 6. *Let $G = (\Phi, \Psi, M)$ be a connected intuitionistic fuzzy soft graph on $G^* = (V, E)$. Then G is said to be a strongly edge totally irregular intuitionistic fuzzy graph if every pair of edges has distinct total degrees.*

Theorem 8. *Let $G = (\Phi, \Psi, M)$ be a connected intuitionistic fuzzy soft graph on $G^* = (V, E)$ and Ψ is constant function. If G is strongly edge irregular intuitionistic fuzzy soft graph, then G is strongly edge totally irregular intuitionistic fuzzy soft graph.*

Proof. We assume that Ψ is a constant function, let $\Psi(uv)(e_i) = (c_i, \acute{c}_i)$ for all $e_i \in M, uv \in E$, where c_i and $\acute{c}_i$ are constants. Let uv and wz be any pair of edges in E. Suppose that intuitionistic fuzzy soft graph $G = (\Phi, \Psi, M)$ is strongly edge irregular intuitionistic fuzzy soft graph. Then $deg_G(uv)(e_i) \neq deg_G(wz)(e_i)$, where uv and wz are any pair of edges in E,

$deg_G(uv)(e_i) + (c_i, \acute{c}_i) \neq deg_G(wz)(e_i) + (c_i, \acute{c}_i)$
$deg_G(uv)(e_i) + \Psi(uv)(e_i) \neq deg_G(wz)(e_i) + \Psi(wz)(e_i)$
$tdeg_G(uv)(e_i) \neq tdeg_G(wz)(e_i)$
for all $e_i \in M$, where uv and wz are any pair of edges in E. Hence G is strongly edge totally irregular intuitionistic fuzzy soft graph. □

Theorem 9. *Let $G = (\Phi, \Psi, M)$ be a connected intuitionistic fuzzy soft graph on $G^* = (V, E)$ and Ψ is constant function. If G is strongly edge totally irregular intuitionistic fuzzy soft graph, then G is strongly edge irregular intuitionistic fuzzy soft graph.*

Remark 3. *Let $G = (\Phi, \Psi, M)$ be a connected intuitionistic fuzzy soft graph on $G^* = (V, E)$. If G is both strongly edge irregular intuitionistic fuzzy soft graph and strongly edge totally irregular intuitionistic fuzzy soft graph. Then Ψ need not be a constant function.*

Theorem 10. *Let $G = (\Phi, \Psi, M)$ be a intuitionistic fuzzy soft graph on $G^* = (V, E)$. If G is strongly edge irregular intuitionistic fuzzy soft graph, then G is a neighbourly edge irregular intuitionistic fuzzy soft graph.*

Proof. Let G be an intuitionistic fuzzy soft graph on $G^* = (V, E)$. We assume that G is strongly edge irregular intuitionistic fuzzy soft graph, then every pair of edges in intuitionistic fuzzy soft graph G have distinct degrees. This implies every pair of adjacent edges have distinct degrees. Hence, G is a neighbourly edge irregular intuitionistic fuzzy soft graph. □

Theorem 11. *Let $G = (\Phi, \Psi, M)$ be an intuitionistic fuzzy soft graph on $G^* = (V, E)$. If G is strongly edge totally irregular intuitionistic fuzzy soft graph, then G is a neighbourly edge totally irregular intuitionistic fuzzy soft graph.*

Remark 4. *Converse of the above Theorems 10 and 11 need not be true.*

Theorem 12. *Let $G = (\Phi, \Psi, M)$ be a connected intuitionistic fuzzy soft graph on $G^* = (V, E)$ and Ψ is constant function. If G is strongly edge irregular intuitionistic fuzzy soft graph, then G is an irregular intuitionistic fuzzy soft graph.*

Proof. Let G be a connected intuitionistic fuzzy soft graph on $G^* = (V, E)$. We assume that Ψ is a constant function, let $\Psi(uv)(e_i) = (c_i, \acute{c}_i)$, for all $uv \in E, e_i \in M$, where c_i and $\acute{c}_i$ are constants. Consider G is strongly edge irregular intuitionistic fuzzy soft graph, then every pair of edges in G have distinct degrees. Let uv and vz are adjacent edges in G having distinct degrees, then for all $e_i \in M$

$deg_G(uv)(e_i) \neq deg_G(vz)(e_i)$
$deg_G(u)(e_i) + deg_G(v)(e_i) - 2\Psi(uv)(e_i) \neq deg_G(v)(e_i) + deg_G(z)(e_i) - 2\Psi(vz)(e_i)$
$deg_G(u)(e_i) + deg_G(v)(e_i) - 2(c_i, \acute{c}_i) \neq deg_G(v)(e_i) + deg_G(z)(e_i) - 2(c_i, \acute{c}_i)$
$deg_G(u)(e_i) + deg_G(v)(e_i) \neq deg_G(v)(e_i) + deg_G(z)(e_i)$
$deg_G(u)(e_i) + \neq deg_G(z)(e_i)$

this implies that G is an irregular intuitionistic fuzzy soft graph. □

Theorem 13. *Let $G = (\Phi, \Psi, M)$ be a connected intuitionistic fuzzy soft graph on $G^* = (V, E)$ and Ψ is constant function. If G is strongly edge totally irregular intuitionistic fuzzy soft graph, then G is an irregular intuitionistic fuzzy soft graph.*

Remark 5. *Converse of the above Theorems 12 and 13 need not be true.*

Theorem 14. *Let $G = (\Phi, \Psi, M)$ be a connected intuitionistic fuzzy soft graph on $G^* = (V, E)$ and Ψ is a constant function. If G is strongly edge irregular intuitionistic fuzzy soft graph, then G is highly irregular intuitionistic fuzzy soft graph.*

Proof. Let G be a connected intuitionistic fuzzy soft graph. We assume that Ψ is a constant function, let $\Psi(uv)(e_i) = (c_i, \acute{c}_i)$, for all $e_i \in M, uv \in E$, where c_i and $\acute{c}_i$ are constants. Let u be any vertex adjacent with vertices v, w and z. Then uv, uw, and ux are adjacent edges in G. Suppose that G is strongly edge irregular intuitionistic fuzzy soft graph, then every pair of edges in G has distinct degrees. This implies that every pair of adjacent edges in G has distinct degrees,

$deg_G(uv)(e_i) \neq deg_G(uw)(e_i) \neq deg_G(ux)(e_i)$
$deg_G(u)(e_i) + deg_G(v)(e_i) - 2\Psi(uv)(e_i) \neq deg_G(u)(e_i) + deg_G(w)(e_i) - 2\Psi(uw)(e_i) \neq deg_G(u)(e_i) + deg_G(x)(e_i) - 2\Psi(ux)(e_i)$
$deg_G(u)(e_i) + deg_G(v)(e_i) - 2(c_i, \acute{c}_i) \neq deg_G(u)(e_i) + deg_G(w)(e_i) - 2(c_i, \acute{c}_i) \neq deg_G(u)(e_i) + deg_G(x)(e_i) - 2(c_i, \acute{c}_i)$
$deg_G(u)(e_i) + deg_G(v)(e_i) \neq deg_G(u)(e_i) + deg_G(w)(e_i) \neq deg_G(u)(e_i) + deg_G(x)(e_i)$
$deg_G(v)(e_i) \neq deg_G(w)(e_i) \neq deg_G(x)(e_i)$

Clearly, there exists a vertex u which is adjacent to the vertices v, w and x that have distinct degrees. Hence, G is a highly irregular intuitionistic fuzzy soft graph. □

Theorem 15. *Let $G = (\Phi, \Psi, M)$ be a connected intuitionistic fuzzy soft graph on $G^* = (V, E)$ and Φ is a constant function. If G is strongly edge totally irregular intuitionistic fuzzy soft graph, then G is a highly irregular intuitionistic fuzzy soft graph.*

Remark 6. *The converse of the above Theorems 14 and 15 need not be true.*

Theorem 16. *Let $G = (\Phi, \Psi, M)$ be an intuitionistic fuzzy soft graph on $G^* = (V, E)$, a path on $2n(n > 1)$ vertices (see Figure 3). If the membership and non-membership values of the edges $s_1, s_2, s_3, \ldots, s_{2n-1}$ are, $(\Psi_{\mu_1}(e_i), \Psi_{\nu_1}(e_i)), (\Psi_{\mu_2}(e_i), \Psi_{\nu_2}(e_i)), \ldots, (\Psi_{\mu_{2n-1}}(e_i), \Psi_{\nu_{2n-1}}(e_i))$, respectively such that $\Psi_{\mu_1}(e_i) < \Psi_{\mu_2}(e_i) < \ldots, < \Psi_{\mu_{2n-1}}(e_i)$ and $\Psi_{\nu_1}(e_i) > \Psi_{\nu_2}(e_i) > \ldots > \Psi_{\nu_{2n-1}}(e_i)$ in intuitionistic fuzzy graph $H(e_i)$ corresponding to the parameter e_i, for all $e_i \in M$, then intuitionistic fuzzy soft graph G is strongly edge irregular and strongly edge totally irregular intuitionistic fuzzy soft graph.*

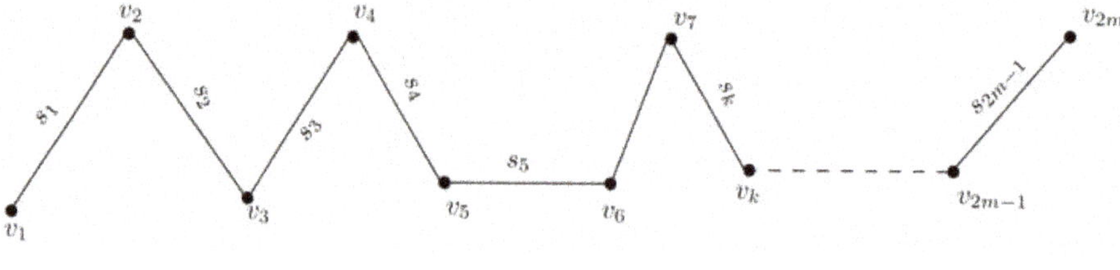

Figure 3. A path on $2n$ vertices.

Proof. Let $G = (\Phi, \Psi, M)$ be an intuitionistic fuzzy soft graph on $G^* = (V, E)$, a path on $2n(n > 1)$ vertices. Suppose that $\Psi(s_k)(e_i) = (\Psi_\mu(s_k)(e_i), \Psi_\nu(s_k)(e_i))$ be the membership and non-membership values of the edges s_k for $k = 1, 2, \ldots, 2n-1$, in intuitionistic fuzzy graph $H(e_i)$, for all $e_i \in M$. We assume that $\Psi_{\mu_1}(e_i) < \Psi_{\mu_2}(e_i) < \ldots, < \Psi_{\mu_{2n-1}}(e_i)$ and $\Psi_{\nu_1}(e_i) > \Psi_{\nu_2}(e_i) > \ldots > \Psi_{\nu_{2n-1}}(e_i)$ in intuitionistic fuzzy graph $H(e_i)$ corresponding to the parameter e_i, for all $e_i \in M$. The degree of each vertex in G is calculated as:

$deg_G(v_1)(e_i) = \Psi(s_1)(e_i)$
$deg_G(v_k)(e_i) = \Psi(s_{k-1})(e_i) + \Psi(s_k)(e_i)$ for $k = 2,3,\ldots,2n-1$
$deg_G(v_{2n})(e_i) = \Psi(s_{2n-1})(e_i)$ for all $e_i \in M$.

The degree of each edge in G is calculated as:

$deg_G(s_1)(e_i) = \Psi(s_2)(e_i)$
$deg_G(s_k)(e_i) = \Psi(s_{k-1})(e_i) + \Psi(s_{k+1})(e_i)$ for $k = 2,3,\ldots,2n-2$
$deg_G(s_{2n-1})(e_i) = \Psi(s_{2n-2})(e_i)$ for all $e_i \in M$.

Clearly, each edge in intuitionistic fuzzy graph $H(e_i)$, for all $e_i \in M$ has distinct degree; therefore, G is a strongly edge irregular intuitionistic fuzzy soft graph. The total degree of each edge in G is calculated as:

$tdeg_G(s_1)(e_i) = \Psi(s_2)(e_i) + \Psi(s_1)(e_i)$
$tdeg_G(s_k)(e_i) = \Psi(s_{k-1})(e_i) + \Psi(s_{k+1})(e_i) + \Psi(s_k)(e_i)$ for $k = 2,3,\ldots,2n-2$
$deg_G(s_{2n-1})(e_i) = \Psi(s_{2n-2})(e_i) + \Psi(s_{2n-1})(e_i)$ for all $e_i \in M$.

Since each edge in G has distinct total degree, therefore G is strongly edge totally irregular intuitionistic fuzzy soft graph. Hence, G is a strongly edge irregular and strongly edge totally irregular intuitionistic fuzzy soft graph. □

Theorem 17. *Let $G = (\Phi, \Psi, M)$ be an intuitionistic fuzzy soft graph on a simple graph $G^* = (V, E)$, be a star $K_{1,n}$ (Figure 4). If each edge in intuitionistic fuzzy graph $H(e_i)$, for all $e_i \in M$, has distinct membership and non-membership values, then G is strongly edge irregular and edge totally regular intuitionistic fuzzy soft graph.*

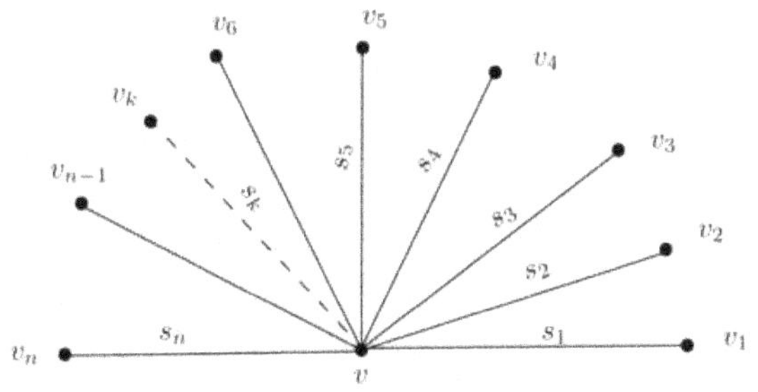

Figure 4. Star $K_{1,n}$.

Proof. Let $G = (\Phi, \Psi, M)$ be an intuitionistic fuzzy soft graph on G^*. Let $v, v_1, v_2, v_3, \ldots, v_n$ be vertices of G^*, $v_1, v_2, v_3, \ldots, v_n$ be the vertices adjacent to a vertex v. We assume that no two adjacent edges have the same membership and non-membership values in intuitionistic fuzzy graph $H(e_i)$, for all $e_i \in A$, then

$$\begin{aligned} deg_G(vv_k)(e_i) &= deg_G(v)(e_i) + deg_G(v_k)(e_i) - 2\Psi(vv_k)(e_i) \\ &= \Psi(vv_1)(e_i) + \Psi(vv_2)(e_i) + \ldots + \Psi(vv_n)(e_i) + \Psi(vv_k)(e_i) - 2\Psi(vv_k)(e_i) \\ deg_G(vv_k)(e_i) &= \Psi(vv_1)(e_i) + \Psi(vv_2)(e_i) + \ldots + \Psi(vv_n)(e_i) - \Psi(vv_k)(e_i). \end{aligned}$$

Clearly, all the edges in $H(e_i)$, for all $e_i \in A$ that have distinct degrees; therefore, G is a strongly edge irregular intuitionistic fuzzy soft graph. The total degree of each edge in $H(e_i)$, for all $e_i \in A$ is calculated as:

$$\begin{aligned} tdeg_G(vv_k)(e_i) &= deg_G(v)(e_i) + deg_G(v_k)(e_i) - \Psi(vv_k)(e_i) \\ &= \Psi(vv_1)(e_i) + \Psi(vv_2)(e_i) + \ldots + \Psi(vv_n)(e_i) + \Psi(vv_k)(e_i) - \Psi(vv_k)(e_i) \\ &= \Psi(vv_1)(e_i) + \Psi(vv_2)(e_i) + \ldots + \Psi(vv_n)(e_i). \end{aligned}$$

Clearly, all the edges in $H(e_i)$, for all $e_i \in A$ having same total degree, therefore G is edge totally regular intuitionistic fuzzy soft graph. □

Theorem 18. *Let $G = (\Phi, \Psi, M)$ be an intuitionistic fuzzy soft graph on $G^* = (V, E)$, a barbell graph $B_{m,n}$ (connecting two complete bipartite graphs $K_{1,m}$ and $K_{1,n}$ by a bridge Figure 5). If each edge in intuitionistic fuzzy graphs of G has distinct membership and non-membership values then G is a strongly edge irregular and edge totally regular intuitionistic fuzzy soft graph.*

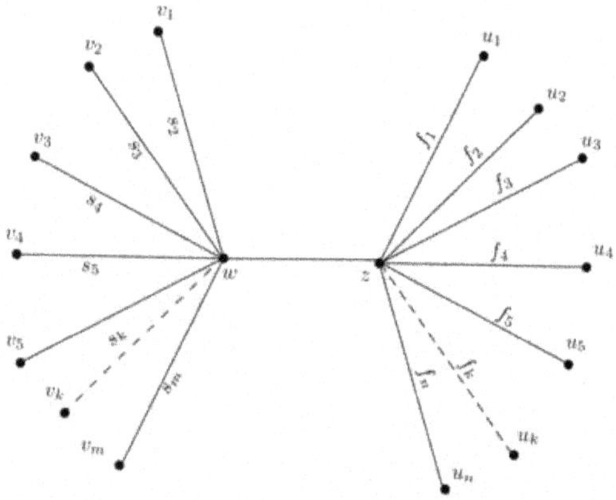

Figure 5. Barbell graph.

Proof. Let $G = (\Phi, \Psi, M)$ be an intuitionistic fuzzy soft graph on G^*. Suppose that G^* is a barbell graph $B_{(m,n)}$, then there exist a bridge, say wz, connecting m vertices $v_1, v_2, \ldots, v_m$ to a vertex w and n vertices $u_1, u_2, \ldots, u_n$ to a vertex z. Let $\Psi(wz)(e_i)$ be the value of bridge wz in intuitionistic fuzzy graph $H(e_i)$, for all $e_i \in M$. Let $\Psi(s_k)(e_i)$ be the value of the edges in $H(e_i)$, such that $\Psi(s_1)(e_i) < \Psi(s_2)(e_i) < \ldots < \Psi(s_m)(e_i)$ for all $e_i \in M, k = 1, 2, \ldots, m$. Let $\Psi(f_k)(e_i)$ be the value of the edges in $H(e_i)$, such that $\Psi(f_1)(e_i) < \Psi(f_2)(e_i) < \ldots < \Psi(f_n)(e_i)$ for all $e_i \in M, k = 1, 2, \ldots, n$. Suppose that $\Psi(s_1)(e_i) < \Psi(s_2)(e_i) < \ldots < \Psi(s_m)(e_i) < \Psi(f_1)(e_i) < \Psi(f_2)(e_i) < \ldots < \Psi(f_n)(e_i) < \Psi(wz)(e_i)$, for all $e_i \in M$. The degree of each edge in intuitionistic fuzzy soft graph G is defined as:

$$\begin{aligned}
deg_G(wz)(e_i) &= deg_G(w)(e_i) + deg_G(z)(e_i) - 2\Psi(wz)(e_i)\\
&= \Psi(wv_1)(e_i) + \Psi(wv_2)(e_i) + \ldots + \Psi(wv_m)(e_i) + \Psi(wz)(e_i) + \Psi(zu_1)(e_i) + \Psi(zu_2)(e_i) +\\
&\quad \ldots + \Psi(zu_n)(e_i) + \Psi(wz)(e_i) - 2\Psi(wz)(e_i)\\
&= \Psi(wv_1)(e_i) + \Psi(wv_2)(e_i) + \ldots + \Psi(wv_m)(e_i) + \Psi(zu_1)(e_i) + \Psi(zu_2)(e_i) + \ldots + \Psi(zu_n)(e_i),\\
deg_G(wv_k)(e_i) &= deg_G(w)(e_i) + deg_G(v_k)(e_i) - 2\Psi(wv_k)(e_i), \text{ for } k = 1, 2, \ldots, m\\
&= \Psi(wv_1)(e_i) + \Psi(wv_2)(e_i) + \ldots + \Psi(wv_m)(e_i) + \Psi(wz)(e_i) + \Psi(wv_k)(e_i) - 2\Psi(wv_k)(e_i)\\
&= \Psi(wv_1)(e_i) + \Psi(wv_2)(e_i) + \ldots + \Psi(wv_m)(e_i) + \Psi(wz)(e_i) - \Psi(wv_k)(e_i)\\
deg_G(zu_k)(e_i) &= deg_G(z)(e_i) + deg_G(u_k)(e_i) - 2\Psi(zu_k)(e_i) \text{ for } k = 1, 2, \ldots, n\\
&= \Psi(zu_1)(e_i) + \Psi(zu_2)(e_i) + \ldots + \Psi(zu_m)(e_i) + \Psi(wz)(e_i) + \Psi(zu_k)(e_i) - 2\Psi(zu_k)(e_i)\\
&= \Psi(zu_1)(e_i) + \Psi(zu_2)(e_i) + \ldots + \Psi(zu_m)(e_i) + \Psi(wz)(e_i) - \Psi(zu_k)(e_i).
\end{aligned}$$

Clearly, all the edges in G have distinct degrees therefore G is strongly edge irregular intuitionistic fuzzy soft graph. The total degree of each edge in intuitionistic fuzzy soft graph G is defined as:

$$\begin{aligned}
tdeg_G(w'1z)(e_i) &= deg_G(wz)(e_i) + \Psi(wz)(e_i)\\
&= \Psi(wv_1)(e_i) + \ldots + \Psi(wv_m)(e_i) + \Psi(zu_1)(e_i) + \ldots + \Psi(zu_n)(e_i) + \Psi(wz)(e_i)\\
tdeg_G(wv_k)(e_i) &= deg_G(wv_k)(e_i) + \Psi(wv_k)(e_i)\\
&= \Psi(wv_1)(e_i) + \Psi(wv_2)(e_i) + \ldots + \Psi(wv_m)(e_i) + \Psi(wz)(e_i) - \Psi(wv_k)(e_i) + \Psi(wv_k)(e_i)
\end{aligned}$$

$$= \Psi(wv_1)(e_i) + \Psi(wv_2)(e_i) + \ldots + \Psi(wv_m)(e_i) + \Psi(wz)(e_i) \text{ for } k = 1, 2, \ldots, m,$$

$$tdeg_G(zu_k)(e_i) = deg_G(zu_k)(e_i) + \Psi(zu_k)(e_i)$$
$$= \Psi(zu_1)(e_i) + \Psi(zu_2)(e_i) + \ldots + \Psi(zu_n)(e_i) + \Psi(wz)(e_i) - \Psi(zu_k)(e_i) + \Psi(zu_k)(e_i)$$
$$= \Psi(zu_1)(e_i) + \Psi(zu_2)(e_i) + \ldots + \Psi(zu_n)(e_i) + \Psi(wz)(e_i) \text{ for } k = 1, 2, \ldots, n.$$

Clearly, all the edges wv_k for $k = 1, 2, \ldots, m$ have the same total edge degree and all zu_k for $k = 1, 2, \ldots, n$ in G have the same total edge degree; therefore, G is not a strong edge totally irregular intuitionistic fuzzy soft graph. □

3. Application

In intuitionistic fuzzy soft set, the approximate functions are intuitionistic fuzzy subsets, the membership value and non-membership value are used to describe the parameterized family of intuitionistic fuzzy soft set for dealing with uncertainties. In many real applications, an intuitionistic fuzzy graph cannot be described comprehensively because of the uncertainty of knowledge. It is more reasonable to give simultaneously a membership degree and a non-membership degree for dealing with complicated and uncertain knowledge from the viewpoint of parameters. From such a point of view, intuitionistic fuzzy soft graphs are used in decision-making problems for better results. Intuitionistic fuzzy graphs are discussed in terms of the uncertainty in various real life phenomena, but intuitionistic fuzzy soft sets are provided for a more generalized and accurate approximate description of objects. We present an application of an intuitionistic fuzzy soft graph in a decision-making problem. The problem of object recognition has received paramount importance in recent times. The recognition problem may be viewed as a decision making-problem, where the final identification of the object is based upon the available set of information. We use the technique to calculate the score for the selected objects on the basis of k number of parameters $(e_1, e_2, \ldots, e_k)$ out of n number of objects $(C_1, C_2, \ldots, C_n)$. A software house wants to select the best software development project, a project with the greatest business impact, the least effort required and the highest probability of success. Let $V = \{$E-commerce development project, Custom development project, Applications development project, Game development project, Browser development project, Web development project, System integration development project$\}$ be the set of software development projects under consideration, $M = \{e_1 =$"ensure project success", $e_2 =$"technological feasibility", $e_3 =$"economical viability"$\}$ be a set of decision parameters correspond to the vertex set V, ssssss$E = \{$E-commerce development project Custom development project, E-commerce development project Applications development project, E-commerce development project Game development project, E-commerce development project Browser development project, Applications development project Web development project, Application development project Browser development project, Game development project Browser development project,Browser development project Web development project, Web development project System integration development project, Game development project System integration development project$\} \subseteq V \times V$ describes the relationship between two applicants corresponding to the given parameters e_1, e_2 and e_3. Our aim is to find out the best development project with the choice of parameters e_1 = "ensure project success", e_2 = "technological feasibility", e_3 = "economical viability". We consider an intuitionistic fuzzy soft graph $G = (\Phi, \Psi, M)$, where (Φ, M) is an intuitionistic fuzzy soft set over V which describes the membership and non-membership values of the projects based upon the given parameters. (Ψ, M) is an intuitionistic fuzzy soft set over $E \subseteq V \times V$ describes the membership and non-membership values of the relationship between two projects corresponding to the given parameters e_1, e_2 and e_3. An intuitionistic fuzzy soft graph $G = \{H(e_1), H(e_2), H(e_3)\}$ is given in Table 1.

The intuitionistic fuzzy graphs $H(e_1)$, $H(e_2)$ and $H(e_3)$ of intuitionistic fuzzy soft graph $G = \{H(e_1), H(e_2), H(e_3)\}$ corresponding to the parameters e_1 ="ensure project success", e_2 ="technological feasibility", and e_3 ="economical viability" are shown in Figure 6.

Table 1. Tabular representation of an intuitionistic fuzzy soft graph.

Φ	E-commerce	Custom	Browser	Applications	System Integration	Game	Web
e_1	(0.6, 0.2)	(0.5, 0.4)	(0.8, 0.1)	(0.3, 0.4)	(0.7, 0.2)	(0.4, 0.3)	(0.2, 0.3)
e_2	(0.3, 0.5)	(0.6, 0.3)	(0.6, 0.2)	(0.6, 0.2)	(0.4, 0.2)	(0.7, 0.3)	(0.8, 0.2)
e_3	(0.7, 0.2)	(0.4, 0.3)	(0.6, 0.2)	(0.6, 0.1)	(0.8, 0.1)	(0.5, 0.4)	(0.4, 0.3)
Ψ	**E-commerce Custom**	**E-commerce Browser**	**E-commerce Web**	**E-commerce System Integration**	**Custom Game**		
e_1	(0.4, 0.3)	(0.5, 0.2)	(0.2, 0.3)	(0.0, 0.0)	(0.4, 0.3)		
e_2	(0.3, 0.4)	(0.0, 0.0)	(0.2, 0.3)	(0.0, 0.0)	(0.5, 0.3)		
e_3	(0.0, 0.0)	(0.5, 0.2)	(0.4, 0.3)	(0.6, 0.2)	(0.3, 0.4)		
Ψ	**System Integration Custom**	**Applications Game**	**Applications System Integration**	**Applications Custom**			
e_1	(0.4, 0.3)	(0.2, 0.4)	(0.3, 0.4)	(0.0, 0.0)			
e_2	(0.4, 0.2)	(0.6, 0.3)	(0.0, 0.0)	(0.0, 0.0)			
e_3	(0.0, 0.0)	(0.5, 0.4)	(0.0, 0.0)	(0.2, 0.3)			
Ψ	**E-commerce Applications**	**System Integration Browser**	**Browser Applications**	**Game Web**	**Browser Custom**		
e_1	(0.3, 0.3)	(0.0, 0.0)	(0.0, 0.0)	(0.2, 0.3)	(0.4, 0.1)		
e_2	(0.3, 0.3)	(0.4, 0.2)	(0.6, 0.2)	(0.0, 0.0)	(0.6, 0.2)		
e_3	(0.6, 0.1)	(0.6, 0.2)	(0.0, 0.0)	(0.4, 0.3)	(0.4, 0.3)		
Ψ	**Browser Web**	**Game Browser**	**System Integration Web**	**Game System Integration**	**Applications Web**		
e_1	(0.0, 0.0)	(0.0, 0.0)	(0.2, 0.3)	(0.0, 0.0)	(0.0, 0.0)		
e_2	(0.5, 0.2)	(0.5, 0.2)	(0.0, 0.0)	(0.4, 0.3)	(0.0, 0.0)		
e_3	(0.0, 0.0)	(0.0, 0.0)	(0.4, 0.2)	(0.0, 0.0)	(0.5, 0.4)		

By taking the intersection of intuitionistic fuzzy graphs $H(e_1)$, $H(e_2)$ and $H(e_3)$, we obtain a resultant intuitionistic fuzzy graph $H(e), e = e_1 \wedge e_2 \wedge e_3$ (see Figure 7). The adjacency matrix of the resultant intuitionistic fuzzy graph is

$$H(e) = \begin{pmatrix} (0,0) & (0,0) & (0,0) & (0.3,0.3) & (0,0) & (0,0) & (0.2,0.3) \\ (0,0) & (0,0) & (0.4,0.3) & (0,0) & (0,0) & (0.3,0.4) & (0,0) \\ (0,0) & (0.4,0.3) & (0,0) & (0,0) & (0.4,0.4) & (0,0) & (0,0) \\ (0.3,0.3) & (0,0) & (0,0) & (0,0) & (0,0) & (0.2,0.4) & (0,0) \\ (0,0) & (0,0) & (0.4,0.4) & (0,0) & (0,0) & (0,0) & (0,0) \\ (0,0.0) & (0.3,0.4) & (0,0) & (0.2,0.4) & (0,0) & (0,0) & (0,0) \\ (0.2,0.3) & (0,0) & (0,0) & (0,0) & (0,0) & (0,0) & (0,0) \end{pmatrix}.$$

The score values of resultant intuitionistic fuzzy graph $H(e)$ is computed with the score function

$$S_{ij} = \frac{\sqrt{{\mu_j}^2 + (\nu_j - 1)^2 + {\pi_j}^2}}{\sqrt{{\mu_j}^2 + (\nu_j - 1)^2 + {\pi_j}^2} + \sqrt{(\mu_j - 1)^2 + (\nu_j)^2 + {\pi_j}^2}} \tag{1}$$

and $S_{ij} \in [0,1]$ the choice values are given in Table 2.

Table 2. Decision table with score values and choice values.

	E-commerce	Custom	Browser	Applications	System Integration	Game	Web	$\acute{m}_i$
E-commerce	0.5	0.5	0.5	0.5	0.5	0.5	0.471	3.471
Custom	0.5	0.5	0.54	0.5	0.5	0.46	0.5	3.5
Browser	0.5	0.54	0.5	0.5	0.5	0.5	0.5	3.54
Applications	0.5	0.5	0.5	0.5	0.5	0.433	0.5	3.433
System integration	0.5	0.5	0.5	0.5	0.5	0.5	0.5	3.5
Game	0.5	0.46	0.5	0.433	0.5	0.5	0.5	3.39
Web	0.47	0.5	0.5	0.5	0.5	0.5	0.5	3.47

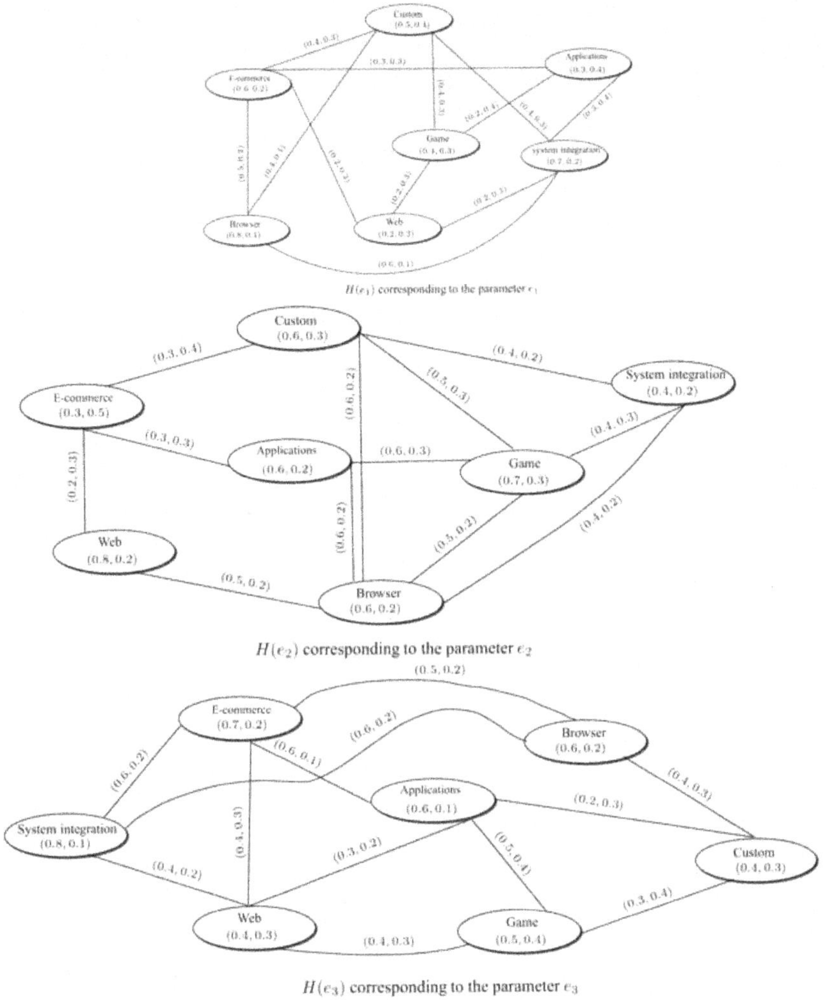

Figure 6. Intuitionistic fuzzy soft graph $G = \{H(e_1), H(e_2, H(e_3)\}$.

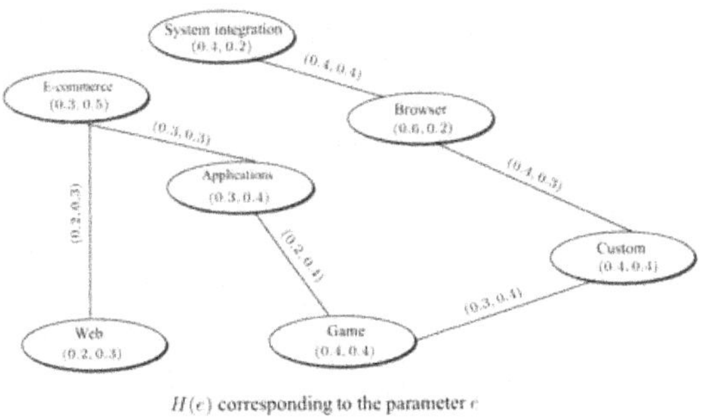

Figure 7. Intuitionistic fuzzy graph $H(e)$, where $e = e_1 \wedge e_2 \wedge e_3$.

From the Table 2, it follows that the maximum choice value is $\acute{m}_3 = 3.54$ and so the optimal decision is to select the browser development project after specifying weights for different parameters. We present our method as an algorithm (Algorithm 1) that is used in our application.

Algorithm 1 Find optimal decision for different parameters

Input: The choice parameters $e_1, e_2, \cdots, e_k$ for the selection of objects.
Input: The intuitionistic fuzzy soft set (Φ, M) over V and intuitionistic fuzzy soft relation (Ψ, M) on V.
Input: Adjacency matrices $H(e_1), H(e_2), \cdots, H(e_k)$ with respect to the parameters.

Compute the resultant adjacency matrix $H(e) = \bigcap_k H(e_k)$ for all $e_k \in M$.
Calculate the the score value of resultant adjacency matrix $H(e)$ by using score function S_{ij}.
Calculate the choice values $\acute{m}_i = \sum_j S_{ij}$ of each objects.
The decision is m_j if $\acute{m}_j = \max_i \acute{m}_i$.
If j has more than one value then any one may be chosen.

4. Conclusions and Future Work

Graph theory has various applications in different areas including mathematics, science and technology, biochemistry(genomics), electrical engineering (communication networks and coding theory), computer science (algorithms and computation) and operations research (scheduling). Soft set theory plays a significant role as a mathematical tool for mathematical modeling, system analysis and computing of decision making problems with uncertainty. An intuitionistic fuzzy soft model is a generalization of the fuzzy soft model which gives more precision, flexibility, and compatibility to a system when compared with the fuzzy soft model. We have applied the concept of intuitionistic fuzzy soft sets to graphs in this paper. We have presented certain types of intuitionistic fuzzy soft graphs. In the future we intend to extend our research of fuzzification to (1) Interval-valued fuzzy soft graphs; (2) Bipolar fuzzy soft hypergraphs, (3) Fuzzy rough soft graphs, and (4) Application of intuitionistic fuzzy soft graphs in decision support systems.

Author Contributions: Muhammad Akram and Sundas Shahzadi conceived and designed the experiments; Sundas Shahzadi wrote the paper.

Conflicts of Interest: The authors declare no conflict of interest.

References

1. Atanassov, K.T. *Intuitionistic Fuzzy Sets*; VII ITKR's Session; 1697/84; Deposed in Central for Science-Technical Library of Bulgarian Academy of Sciences: Sofia, Bulgaria, June 1983. (In Bulgarian)
2. Atanassov, K.T. *Intuitionistic Fuzzy Sets: Theory and Applications*; Studies in Fuzziness and Soft Computing; Physica: Heidelberg, Germany, 2012.
3. Zadeh, L.A. Fuzzy sets. *Inf. Control* **1965**, *8*, 338–353.
4. Marasini, D.; Quatto, P.; Ripamonti, E. Intuitionistic fuzzy sets in questionnaire analysis. *Qual. Quant.* **2016**, *50*, 767–790.
5. Bustince, H.; Barrenechea, E.; Pagola, M.; Fernandez, J.; Xu, Z.; Bedregal, B.; Montero, J.; Hagras, H.; Herrera, F.; De Baets, B. A historical account of types of fuzzy sets and their relationships. *IEEE Trans. Fuzzy Syst.* **2016**, *24*, 179–194.
6. Akram, M.; Ashraf, A.; Sarwar, M. Novel applications of intuitionistic fuzzy digraphs in decision support systems. *Sci. World J.* **2014**, *2014*, 904606.
7. Lei, Q.; Xu, Z.; Bustince, H.; Fernandez, J. Intuitionistic fuzzy integrals based on Archimedean t-conorms and t-norms. *Inf. Sci.* **2016**, *327*, 57–70.
8. Molodtsov, D.A. Soft set theory-first results. *Comput. Math. Appl.* **1999**, *37*, 19–31.
9. Akram, M.; Nawaz, S. Operations on soft graphs. *Fuzzy Inf. Eng.* **2015**, *7*, 423–449.

10. Alcantud, J.C.R.; Santos-Garca, G.; Hernndez-Galilea, E. Glaucoma diagnosis: A soft set based decision making procedure. In *Conference of the Spanish Association for Artificial Intelligence*; Springer: Cham, Switzerland, 2015; pp. 49–60.
11. Cağman, N.; Enginoglu, S. Soft matrix theory and its decision making. *Comput. Math. Appl.* **2010**, *59*, 3308–3314.
12. Feng, F.; Jun, Y.B.; Liu, X.Y.; Li, L.F. An adjustable approach to fuzzy soft set based decision making. *J. Comput. Appl. Math.* **2010**, *234*, 10–20.
13. Feng, F.; Akram, M.; Davvaz, B.; Fotea, V.L. A new approach to attribute analysis of information systems based on soft implications. *Knowl.-Based Syst.* **2014**, *70*, 281–292.
14. Som, T. On the theory of soft sets, soft relation and fuzzy soft relation. In Proceedings of the National Conference on Uncertainty: A Mathematical Approach, UAMA-06, Burdwan, India, 2006; pp. 1–9.
15. Maji, P.K.; Roy, A.R.; Biswas, R. Fuzzy soft sets. *J. Fuzzy Math.* **2001**, *9*, 589–602.
16. Roy, A.R.; Maji, P.K. A fuzzy soft set theoretic approach to decision making problems. *J. Comput. Appl. Math.* **2007**, *203*, 412–418.
17. Cağman, N.; Enginoglu, S.; Citak, F. Fuzzy soft set theory and its applications. *Iran. J. Fuzzy Syst.* **2011**, *8*, 137–147.
18. Akram, M.; Nawaz, S. On fuzzy soft graphs. *Ital. J. Pure Appl. Math.* **2015**, *34*, 497–514.
19. Akram, M.; Nawaz, S. Fuzzy soft graphs with applications. *J. Intell. Fuzzy Syst.* **2016**, *30*, 3619–3632.
20. Xiao, Z.; Gong, K.; Zou, Y. A combined forecasting approach based on fuzzy soft sets. *J. Comput. Appl. Math.* **2009**, *228*, 326–333.
21. Akram, M.; Ali, G.; Alshehri, N.O. A new multi-attribute decision-making method based on m-polar fuzzy soft rough sets. *Symmetry* **2017**, *9*, 271, doi:10.3390/sym9110271.
22. Maji, P.K.; Biswas, R.; Roy, A.R. Intuitionistic fuzzy soft sets. *J. Fuzzy Math.* **2001**, *9*, 677–691.
23. Maji, P.K.; Biswas, R.; Roy, A.R. On intuitionistic fuzzy soft sets. *J. Fuzzy Math.* **2004**, *12*, 669–683.
24. Cağman, N.; Karatas, S. Intuitionistic fuzzy soft set theory and its decision making. *J. Intell. Fuzzy Syst.* **2013**, *24*, 829–836.
25. Dinda, B.; Samanta, T.K.; Vidyapith, M.R. Relations on intuitionistic fuzzy soft sets. *Gen* **2010**, *1*, 74–83.
26. Jiang, Y.; Tang, Y.; Chen, Q. An adjustable approach to intuitionistic fuzzy soft sets based decision making. *Appl. Math. Model.* **2011**, *35*, 824–836.
27. Kauffman, A. *Introduction a la Theorie des Sous-Emsembles Flous*; Masson et Cie: Paris, France, 1973; Volume 1.
28. Rosenfeld, A. Fuzzy graphs. In *Fuzzy Sets and Their Applications*; Zadeh, L.A., Fu, K.S., Shimura, M., Eds.; Academic Press: New York, NY, USA, 1975; pp. 77–95.
29. Bhattacharya, P. Some remarks on fuzzy graphs. *Pattern Recognit. Lett.* **1987**, *6*, 297–302.
30. Mordeson, J.N.; Nair, P.S. *Fuzzy Graphs and Fuzzy Hypergraphs*; Second Edition 2001; Physica Verlag: Heidelberg, Germany, 1998.
31. Santhimaheswari, N.R.; Sekar, C. On strongly edge irregular fuzzy graphs. *Kragujev. J. Math.* **2016**, *40*, 125–135.
32. Radha, K.; Kumaravel, N. On edge regular fuzzy graphs. *Int. J. Math. Arch.* **2014**, *5*, 100–112.
33. Radha, K.; Kumaravel, N. Some properties of edge regular fuzzy graphs. *Jamal Acad. Res. J. Spec. Issue* **2014**, *121*, 121–127.
34. Shahzadi, S.; Akram, M. Edge regular intuitionistic fuzzy soft graphs. *J. Intell. Fuzzy Syst.* **2016**, *31*, 1881–1895.
35. Shahzadi, S.; Akram, M. Intuitionistic fuzzy soft graphs with applications. *J. Appl. Math. Comput.* **2016**, doi:10.1007/s12190-016-1041-8.
36. Akram, M.; Shahzadi, S. Novel intuitionistic fuzzy soft multiple-attribute decision-making methods. *Neural Comput. Appl.* **2016**, doi:10.1007/s00521-016-2543-x.
37. Akram, M.; Waseem, N.; Dudek, W.A. Certain types of edge m-polar fuzzy graphs. *Iran. J. Fuzzy Syst.* **2017**, *14*, 27–50.
38. Akram, M.; Siddique, S.; Davvaz, B. New concepts in neutrosophic graphs with application. *J. Appl. Math. Comput.* **2017**, doi:10.1007/s12190-017-1106-3.
39. Kóczy, L. Fuzzy graphs in the evaluation and optimization of networks. *Fuzzy Sets Syst.* **1992**, *46*, 307–319.
40. Medina, J.; Jeda-Aciego, O.M. Multi-adjoint t-concept lattices. *Inf. Sci.* **2010**, *180*, 712–725.
41. Nandhini, S.P.; Nandhini, E. Strongly irregular fuzzy graphs. *Int. J. Math. Arch.* **2014**, *5*, 110–114.

42. Pozna, C.; Minculete, N.; Precup, R.E.; Koczy, L.T.; Ballagi, A. Signatures: Definitions, operators and applications to fuzzy modelling. *Fuzzy Sets Syst.* **2012**, *201*, 86–104.
43. Basu, T.M.; Mahapatra, N.K.; Mondal, S.K. Different types of matrices in intuitionistic fuzzy soft set theory and their application in Predicting Terrorist Attack. *Int. J. Manag. IT Eng.* **2012**, *2*, 73–105.
44. Vrkalovic, S.; Teban, T.A.; Borlea, I.D. Stable Takagi-Sugeno fuzzy control designed by optimization. *Int. J. Artif. Intell.* **2017**, *15*, 17–29.
45. Garca, J.G.; Rodabaugh, S.E. Order-theoretic, topological, categorical redundancies of interval-valued sets, grey sets, vague sets, interval-valued intuitionistic sets, intuitionistic fuzzy sets and topologies. *Fuzzy Sets Syst.* **2005**, *56*, 445–484.

axioms

MDPI

Article

Quantiles in Abstract Convex Structures

Marta Cardin

Department of Economics, Ca'Foscari University of Venice, Sestiere Cannaregio 873, 30123 Venezia, Italy; mcardin@unive.it

Received: 14 May 2018; Accepted: 25 May 2018; Published: 28 May 2018

Abstract: In this short paper, we aim at a qualitative framework for modeling multivariate decision problems where each alternative is characterized by a set of properties. To this extent, we consider convex spaces as underlying universes and make use of lattice operations in convex spaces to formalize the notion of quantiles. We also put in evidence that many important models of decision problems can be viewed as convex spaces-based models. Several properties of aggregation operators are translated into this general setting, and independence and invariance are used to provide axiomatic characterizations of quantiles.

Keywords: convex space; aggregation operator; invariance; independence; quantile

1. Introduction

The aim of this paper is to propose a general unified framework for defining aggregation operators. Our framework is abstract and algebraic in nature and in this framework we generalize some results of the literature [1–4].

We consider convex structures where the notion considered here (see [5]) is not restricted to the context of vector spaces. The basic idea of our approach is to describe the space of alternatives in terms of a "topological" relation. We can prove that lattices, median spaces and interval spaces are convex spaces and also that to every property spaces (see [1,2,4]) is associated with a convex structure.

We then focus on aggregation operators $f\colon X^A \to X$ where X is convexity space and A is a nonempty set. We study aggregation operators that satisfy properties of monotonicity and independence and we consider aggregation operators that are based on *decisive* subsets of A. Moreover, we consider operators that are componentwise compatible with the structure of convexity space of X.

We propose also a particular version of Arrow's theorem thus considering a link between aggregation theory and social choice theory as in [6]. It appears that there are many connections between the work presented here with the results of [3,4,7–12]. Applications of these types of results can be found in in [4,13,14].

The structure of the paper is as follows. In Section 2 we introduce convex spaces and we provide the necessary definitions. Section 3 is devoted to describe some important examples. Finally in Sections 3 and 4 we study some classes of agggregation operators acting on abstract convex structures.

2. Abstract Convex Structures

The notion of convexity is a basic mathematical structure that is used to analyze many different problems and there are in the literature various kinds of generalized, topological, or axiomatically defined convexities. There are generalizations that are motivated by concrete problems and those that are stated from an axiomatic point of view, where the notion of abstract convexity is based on properties of a family of sets.

In this paper the general notion of abstract convexity structure that is studied in [5] is considered.

Definition 1. *A family $\mathcal{C}$ of subsets of a set X is a convexity on a set X if $\emptyset$ and X belong to $\mathcal{C}$ and $\mathcal{C}$ is closed under arbitrary intersections and closed under unions of chains.*

The elements of $\mathcal{C}$ are called convex sets of X and the pair $(X, \mathcal{C})$ is called a convex space.

Moreover, the convexity notion allows us to define the notion of the convex hull operator, which is similar to that of the closure operator in topology.

Definition 2. *If X is a set with a convexity $\mathcal{C}$ and A is a subset of X, then the convex hull of $A \subseteq X$ is the set*

$$convA = \bigcap\{C \in \mathcal{C} : A \subseteq C\}. \tag{1}$$

This operator enjoys certain properties that are identical to those of usual convexity: for instance $convA$ is the smallest convex set that contains set A. It is also clear that C is convex if and only if $convC = C$.

The convex hull of a set $\{x_1, \ldots, x_n\}$is called an n-polytope and is denoted by $[x_1, \ldots, x_n]$. A 2-polytope $[a, b]$ is called the segment joining a, b.

A convexity C is called N-ary ($N \in \mathbb{N}$) if $A \subseteq C$ whenever $convF \subseteq A$ for all $F \subseteq A$ where F has at most N elements. A 2-ary convexity is called an interval convexity.

We also consider biconvex spaces, i.e., triples of the form $(X, \mathcal{A}, \mathcal{B})$ where $\mathcal{A}, \mathcal{B}$ are two convexities on a set X, called the lower and the upper convexity. Obviously every convex space $(X, \mathcal{C})$ can be viewed as a biconvex space $(X, \mathcal{C}, \mathcal{C})$.

If X, Y are convex spaces with convexities $\mathcal{C}, \mathcal{D}$, respectively, we consider the following definition of a compatible map between two convex spaces.

Definition 3. *A map $\gamma \colon X \to Y$ is convex if $\gamma^{-1}(C) \in \mathcal{C}$ for every $\in \mathcal{C}$ and such that when $C_i \in \mathcal{C}$ for $i \in I$*

$$\gamma(\bigcap_{i \in I} C_i) = \bigcap_{i \in I} \gamma(C_i). \tag{2}$$

For a general theory of convexity we refer to [5].

3. Some Examples

We present some examples and classes of convex spaces. First of all we note that every real vector space together with the collection of all convex sets in the usual meaning, is a 2-arity convex space.

Ordered spaces The usual convexity on $\mathbb{R}$ can be defined in terms of ordering as follows: a set C is convex if and only if when $a, b \in C$ and $a \leq x \leq b$ implies $x \in C$. We can define in the same way a convexity on a partially ordered set (see [5], p. 6). Such a convexity is called the order convexity.

Lattices If $\langle L, \wedge, \vee \rangle$ is a lattice we denote by $\mathcal{L}$ and $\mathcal{U}$ the collections of all ideals and all filters respectively (the empty set and the whole lattice are treated as (non-proper) ideals and filters). Since the union of a chain of filters (ideals) is a filter (ideal), these are two convexities on L that will be called the lower and the upper lattice convexity respectively. Moreover there exists a convexity $\mathcal{C}$ generated by $\mathcal{L} \bigcup \mathcal{U}$ the least convexity containing all ideals and filters. This convexity will be called the lattice convexity on L.

Please note that if L is linearly ordered then G equals the order convexity. The convexity of the dual lattice is the same as the original one.

It is possible to consider lattices as convex spaces (with the lattice convexity) as well as bi-convex spaces (with the lower and upper lattice convexities). It is easy to check that a proper halfspace is either a prime filter or a prime ideal. It can be proved also that the lattice convexity is an interval convexity and that

$$[a, b] = \{x \in L : a \wedge b \leq x \leq a \vee b\}. \tag{3}$$

Median spaces A median space is a convexity space X with an interval convexity such that for each $a, b, c \in X$ there exists a unique point in $[a, b] \cap [a, c] \cap [b, c]$. We call it the median of a, b, c and denote by $m(a, b, c)$. This defines a map $m\colon X^3 \to X$, called the median operator on X. In any convexity space, every point in $[a, b] \cap [a, c] \cap [b, c]$ is called a median of a, b, c. There is a natural way to define the structure of a median space by means of the median operator (see [5]).

Property-based domains A property-based domain (as defined in [1]) is a pair $(X, \mathcal{H})$ where X is a non-empty set and $\mathcal{H}$ is a collection of non-empty subsets of X and if $x, y \in X$ and $x \neq y$ there exists $H \in \mathcal{H}$ such that $x \in H$ and $y \notin H$. The elements of $\mathcal{H}$ are referred to as properties and if $x \in H$ we say that x has property represented by the subset H. This definition is slightly more general than that of [3] and of [4], in fact it is not assumed that the set X is finite and we do not consider that the set H^c is a property if H is a property.

The "property space" model provides a very general framework for representing preferences and then aggregation of preferences. In every property-based domain we can define a convexity defined as follows. A subset $S \subseteq X$ is said to be convex if it is intersection of properties.

Arrowian framework The problem of preference aggregation can be viewed as a property-based domain and then as a convex space. We consider a set of alternatives A and a set $\mathcal{R}$ of binary relations in A. We can consider different requirements on the set $\mathcal{R}$ and so $\mathcal{R}$ can be the set of preorders or the set of linear orders in A.

If we define for each pair $a, b \in A$ the set

$$H_{a,b} = \{R \in \mathcal{R} : aRb\} \tag{4}$$

the family $\mathcal{H} = \{H_{a,b} : a, b \in A\}$ defines a property-based domain structure on the set $\mathcal{R}$. See [4] for more details on Arrowian framework.

4. Aggregation Functional over Convex Spaces

Aggregation operators are mathematical functions that are used to combine several inputs into a single representative outcome; see [15] for a comprehensive overview on aggregation theory. Aggregation operators play an important role in several fields such as decision sciences, computer and information sciences, economics and social sciences and there are a large number of different aggregation operators that differ on the assumptions on the inputs and about the information that we want to consider in the model.

Definition 4. *If N is an arbitrary nonempty set and X is a convex space, then an aggregation functional is a map* $F\colon X^A \to \mathcal{P}(X)$.

Our framework is very general, we do not assume that the sets X and A are finite or that the map $F\colon X^A \to \mathcal{P}(X)$ is surjective. Moreover, we consider the case in which there are more than one equivalent solutions and also the case in which there are no solutions. For each $c \in X$, we denote by **c** the constant c map in L^A.

The following properties of an aggregation functional are key to our analysis.

Monotonicity If $C \in \mathcal{C}$, $F(f) \subseteq C$ and $y \in C$ then $F(g) \subseteq C$ where $g(i) = y$ and $f(j) = g(j)$ if $j \neq i$.

Idempotence $c \in F(\mathbf{c})$ for every $c \in X$.

Independence If $C, D \in \mathcal{C}$ $F(f) \subseteq C$ and for all $i \in A$, $f(i) \in C$ if and only if $g(i) \in D$ we have that $F(g) \subseteq D$.

Invariance For every convex map $\gamma\colon X \to X$, $F(\gamma \circ f) = \gamma(F(f))$

5. Quantiles in Convex Spaces

We briefly consider aggregation functionals based on a complete lattices. As it is well known the quantile is a generalization of the concept of median and it plays an important role in statistical and economic literature. We study quantile in an ordinal framework and we consider an axiomatic representation of quantiles as in [1,7,8]. Here we provide a definition and characterization of quantiles for lattice-valued operators.

If A is a nonempty set and L a bounded lattice a non-additive measure on A with values in L is a function $m\colon 2^A \to L$ such that $m(\emptyset) = 0$, $m(A) = 1$ and $m(C) \leq m(D)$ whenever $C \subseteq D$.

Definition 5. *If α is an element of L, then the lattice-valued quantile of level α is the functional $Q_\alpha\colon L^A \to L$ defined by*

$$Q_\alpha(f,m) = \bigvee\{x : m(\{f \geq x\}) \geq \alpha\}.$$

It can be proved that this definition extends the well known definition of quantile for real-valued functions (see [8]).

We recall the definition of completely distributive lattice. A complete lattice L is said to be a completely distributive is the following distributive law holds

$$\bigwedge_{i\in I}\Big(\bigvee_{j\in J} x_{ij}\Big) = \bigvee_{f\in J^I}\Big(\bigwedge_{i\in I} x_{if(i)}\Big),$$

for every doubly indexed subset $\{x_{ij} : i \in I, j \in J\}$ of L. Please note that every complete chain (in particular, the extended real line and each product of complete chains) is completely distributive. Moreover, complete distributivity reduces to distributivity in the case of finite lattices.

A collection of sets $\mathcal{U} \subseteq 2^A$ is said to be an *upper set* in A if $X \in \mathcal{U}$ and $X \subset Y$ implies that $Y \in \mathcal{U}$. Then we can prove the following results.

Proposition 1. *Let L be a completely distributive lattice. An aggregation functional $F\colon L^A \to L$ is a lattice-valued quantile with respect to a non-additive measure $m\colon 2^A \to L$ if and only if there exists a upper set $\mathcal{U}$ such that*

$$F(f) = \bigvee\{x \in L : \textit{there exists } \; U \in \mathcal{U} \; \textit{ such that } \; f(i) \geq x \; \textit{ for every } \; i \in U\} \tag{5}$$

or if and only if there exists a upper set $\mathcal{U}$ such that

$$F(f) = \bigwedge\{x \in L : \textit{there exists } \; U \in \mathcal{U} \; \textit{ such that } \; f(i) \leq x \; \textit{ for every } \; i \in U\} \tag{6}$$

Proof of Proposition 1. By Proposition 1 in [7] if L is a completely distributive lattice an aggregation functional $F\colon L^A \to L$ is a lattice-valued quantile with respect to a non-additive measure $m\colon A \to L$ if and only if there exists a upper set $\mathcal{U}$ such that

$$F(f) = \bigvee_{U\in\mathcal{U}} \bigwedge_{i\in U} f(i)$$

or if and only if there exists a upper set $\mathcal{U}$ such that

$$F(f) = \bigwedge_{U\in\mathcal{U}} \bigvee_{i\in U} f(i).$$

Then we can prove that F is a lattice-valued quantile when $F(f) \geq x$ if and only if there exists $U \in \mathcal{U}$ such that $f(i) \geq x$ for every $i \in U$. Then we get

$$F(f) = \bigvee\{x \in L : \text{there exists } \; U \in \mathcal{U} \; \text{ such that } \; f(i) \geq x \; \text{ for every } \; i \in U\}.$$

The second statement follows similarly. □

Since we know that the lattice convexity is an interval convexity we can prove the following characterization of lattice-valued quantiles.

Proposition 2. *If L is a completely distributive lattice, then an aggregation functional $F\colon L^A \to L$ is a lattice-valued quantile with respect to a non-additive measure $m\colon \mathcal{A} \to L$ if and only if there exists an upper set $\mathcal{U}$ such that*

$$F(f) = \bigcap\{C \in \mathcal{C} : \{i : f(i) \in C\} \in \mathcal{U}\} \tag{7}$$

Then the elements in $F(f)$ belong to a convex set C if and only if $f(i)$ belongs to C for a "decisive " or a "large enough"set.

Let us define quantiles in an abstract convex structures.

Definition 6. *If N is an arbitrary nonempty set and X is a convex space, then a quantile is an aggregation functional $F\colon X^A \to \mathcal{P}(X)$ defined by*

$$F(f) = \bigcap\{C \in \mathcal{C} : \{i : f(i) \in C\} \in \mathcal{U}\} \tag{8}$$

where $\mathcal{U}$ is an upper set in A.

Furthermore, we can characterize from an axiomatic point of view quantiles in an abstract convex structure.

Proposition 3. *If N is an arbitrary nonempty set and X is a convex space, then a quantile is a monotone, idempotent and independent aggregation functional. Conversely an aggregation functional $F\colon X^A \to \mathcal{P}(X)$ that is monotone and independent is a quantile.*

Proof of Proposition 3. If C is a convex set in $\mathcal{C}$ and f is an element of X^A we define the set $N(f,C) = \{i \in N : f(i) \in C\}$. Let F be a quantile, $C \in \mathcal{C}$, f an element of X^A such that $F(f) \subseteq C$ and $y \in C$. If we define an element g of X^A by $g(i) = y$ and $f(j) = g(j)$ if $j \neq i$ then $N(f,C) \subseteq N(g,C)$ and so $F(g) \subseteq C$. Moreover, if $c \in C$ we have that $c \in F(\mathbf{c})$ if F is a quantile since $N(\mathbf{c},C) = A$.

By the definition of quantile if F is a quantile, $C, D \in \mathcal{C}$, $F(f) \subseteq C$ and for all $i \in A$, $f(i) \in C$ if and only if $g(i) \in D$ we can easily prove that $N(f,C) = N(g,D)$ and we get $F(g) \subseteq D$. So we have proved that quantiles are monotone , idempotent and independent functionals.

We note that functional F is monotone and independent if and only if $C, D \in \mathcal{C}$ $F(f) \subseteq C \in \mathcal{C}$ and for all $i \in N$, if $f(i) \in C$ then $g(i) \in D$ we have that $F(g) \in D$.

We say that a set $U \subseteq N$ is decisive with respect to an element $C \in \mathcal{C}$ if there exists $f \in X^A$ such that $N(f,C) = U$ and $F(x) \subseteq C$. Being F monotone and independent a set U is decisive with respect to C if and only if for every $f \in X^A$ such that $N(f,C) = U$, $F(f) \subseteq C$.

Since the functional F is monotone and independent then the set of decisive subset of N does not depend on the convex set C. If $\mathcal{U}$ is the family of decisive subsets of N for every $f \in X^A$, $F(f) \subseteq C$ if and only if $N(F,C) \in \mathcal{U}$. So we have proved that

$$F(f) = \bigcap\{C : N(f,C) \in \mathcal{U}\} = \bigcap\{C \in \mathcal{C} : \{i : f(i) \in C\} \in \mathcal{U}\}.$$

□

The following proposition presents another property of quantiles in convex spaces.

Proposition 4. *f N is an arbitrary nonempty set and X is a convex space, then a quantile is an invariant aggregation functional.*

Proof of Proposition 4. Let $F\colon X^A \to \mathcal{P}(X)$ be a quantile and $\gamma\colon X \to X$ a convex map.

Then $F(\gamma \circ f) = \bigcap\{C \in \mathcal{C} : \{i : (\gamma \circ f)(i) \in C\} \in \mathcal{U}\} = \bigcap\{\gamma(C), C \in \mathcal{C} : \{i : f(i) \in \gamma^{-1}C\} \in \mathcal{U}\} = \bigcap\{\gamma(D), D \in \mathcal{C} : \{i : f(i) \in D\} \in \mathcal{U}\} = \bigcap\{\gamma(C), C \in \mathcal{C} : \{i : f(i) \in C\} \in \mathcal{U}\}$.

If $\mathcal{D} = C \in \mathcal{C} : \{i : f(i) \in C\} \in \mathcal{U}\}$ being γ continuous

$$\bigcap_{C \in \mathcal{D}} \gamma(C) = \gamma(\bigcap_{C \in \mathcal{D}} C)$$

and we get that $F(\gamma \circ f) = \gamma(F(f))$. □

6. Concluding Remarks

We introduced a unified qualitative framework for studying aggregation operators. The approach presented in this paper has taken its inspiration from social choice theory and we generalize some results in social choice in certain respects. This setting has several appealing aspects, for it provides sufficiently rich structures studied in the literature , which allow the definition of quantiles from an ordinal point of view, and which do not depend on the usual arithmetical structure of the reals.

There are however many opportunities for much more detailed research in this area in particular from the point of view of aggregation theory. An obvious topic for future research is to analyze other aggregation functionals defined in convex spaces. There are several extensions avaiable within this framework, for instance, one could consider Sugeno type integral defined by a class of decisive sets.

Conflicts of Interest: The author declares no conflict of interest.

References

1. Cardin, M. Sugeno Integral on Property-Based Preference Domains. In *Advances in Fuzzy Logic and Technology 2017*; Advances in Intelligent Systems and Computing Series; Kacprzyk, J., Szmidt, E., Zadrożny, S., Atanassov, K., Krawczak, M., Eds.; Springer: Basel, Switzerland, 2017; Volume 641.
2. Cardin, M. Aggregation over Property-Based Preference Domains. In *Aggregation Functions in Theory and in Practice*; Advances in Intelligent Systems and Computing Series; Torra, V., Mesiar, R., Baets, B., Eds.; Springer: Basel, Switzerland, 2017; Volume 581.
3. Gordon, S. Unanimity in attribute-based preference domains. *Soc. Choice Welf.* **2015**, *44*, 13–29. [CrossRef]
4. Nehring, K.; Puppe, C. Abstract Arrowian aggregation. *J. Econ. Theory* **2010**, *145*, 467–494. [CrossRef]
5. Van de Vel, M.L.J. *Theory of Convex Structures*; North-Holland Mathematical Library Series; Elsevier: Amsterdam, The Netherlands, 1993; Volume 50.
6. Candeal, J.C. An Abstract Result on Projective Aggregation Functions. *Axioms* **2018**, *7*, 17. [CrossRef]
7. Cardin, M. A quantile approach to integration with respect to non-additive measures. In Proceedings of the International Conference on Modeling Decisions for Artificial Intelligence, Catalonia, Spain, 21–23 November 2012; Torra, V., Narukawa, Y., Lopez, B., Villaret, M., Eds.; Springer: Berlin/Heidelberg, Germany, 2012; pp. 139–148.
8. Cardin, M.; Couceiro, M. An ordinal approach to risk measurement. In *Mathematical and Statistical Methods for Actuarial Science and Finance*; Perna, C., Sibillo, M., Eds.; Springer: Milano, Italy, 2012; pp. 79–86.
9. Chambers, C. Ordinal Aggregation and Quantiles. *J. Econ. Theory* **2007**, *137*, 416–443. [CrossRef]
10. Halaš, R.; Mesiar, R.; Pócs, J. A new characterization of the discrete Sugeno integral. *Inf. Fusion* **2016**, *29*, 84–86. [CrossRef]
11. Halaš, R.; Mesiar, R.; Pócs, J. Congruences and the discrete Sugeno integrals on bounded distributive lattices. *Inf. Sci.* **2016**, *367*, 443–448. [CrossRef]
12. Leclerc, B.; Monjardet, B. Aggregation and Residuation. *Order* **2013**, *30*, 261–268. [CrossRef]
13. Daniëls, T.; Pacuit, E. A General Approach to Aggregation Problems. *J. Logic Comput.* **2009**, *19*, 517–536. [CrossRef]

14. Monjardet, B. Arrowian characterization of latticial federation consensus functions. *Math. Soc. Sci.* **1990**, *20*, 51–71. [CrossRef]
15. Grabisch, M.; Marichal, J.L.; Mesiar, R.; Pap, E. *Aggregation Functions Encyclopedia of Mathematics and its Applications*; Cambridge University Press: Cambridge, UK, 2009.

MDPI
St. Alban-Anlage 66
4052 Basel
Switzerland
Tel. +41 61 683 77 34
Fax +41 61 302 89 18
www.mdpi.com

Axioms Editorial Office
E-mail: axioms@mdpi.com
www.mdpi.com/journal/axioms

www.ingramcontent.com/pod-product-compliance
Lightning Source LLC
Chambersburg PA
CBHW051856210326
41597CB00033B/5921